MW00993430

Hazards Analysis

Reducing the Impact of Disasters

Second Edition

Hazards Analysis

Reducing the Impact of Disasters

Second Edition

John C. Pine

CRC Press is an imprint of the
Taylor & Francis Group, an **informa** business

First published 2015 by CRC Press

Published 2019 by CRC Press
Taylor & Francis Group
6000 Broken Sound Parkway NW, Suite 300
Boca Raton, FL 33487-2742

ISBN-13: 978-1-4822-2891-5 (hbk)

Library of Congress Cataloging-in-Publication Data

Pine, John C., 1946-
[Natural hazards analysis]
Hazards analysis : reducing the impact of disasters / John C. Pine. -- Second edition.
pages cm
Summary: "The impacts of natural and man-made disasters have been increasing at exponential rates over the past few decades. Our global interconnectedness and the scale of disaster events has grown to the point where catastrophic disasters have regional, national, and even global economic consequences. This volume presents a systematic process of hazards identification, vulnerability determination, and consequence assessment for the natural, built, and human environment. The book aims to examine strategies that may be taken at the individual, organization, community, and regional levels to reduce the adverse consequences of disasters and to foster sustainability. "-- Provided by publisher.
Includes bibliographical references and index.
ISBN 978-1-4822-2891-5 (hardback)
1. Disasters--Social aspects. 2. Emergency management. 3. Preparedness. 4. Community organization. 5. Disasters--Risk assessment. 6. Risk management. I. Title.

HV553.P528 2015
363.34'7--dc23 2014024300

Visit the Taylor & Francis Web site at
http://www.taylorandfrancis.com

and the CRC Press Web site at
http://www.crcpress.com

Contents

Preface

The social, economic, and environmental impacts from natural and human-caused disasters have increased during the recent decades. The impacts from these disasters are seen on a global scale and are driving efforts to reduce adverse consequences on vulnerable populations and communities. We have attempted to reduce the costs of these disasters but continue to carry out policies that encourage development in vulnerable geographic areas of the world. With threats imposed by global warming, we see that many communities are subject to increasing levels of risk and can achieve sustainable development through appropriately informed planning, protection, mitigation, and recovery strategies. Strategies to enhance community resilience to hazards must be based on research that integrates complex social, economic, and environmental systems.

Hazards Analysis: Reducing the Impacts of Disasters provides a structure and process for understanding the nature of natural and human-caused hazards and strategies for building sustainable communities. The key to sustainability is acknowledging the unique nature of the local community and how geography, social systems, the economy, and infrastructures influence a community's capacity to withstand and recover from a disaster. This book demonstrates how we use hazards analysis to identify and prioritize risks and develop approaches to community hazard mitigation. The role of hazard risk management is stressed for public, private and nonprofit organizations. Problem solving, decision making, and risk communications are stressed to ensure that we are in a position to identify key problems associated with hazards and the risks that they present. Throughout this text, we stress that hazards analysis is not an isolated process but one that engages the local community to ensure that all appreciate risks associated with hazards. We see that the hazards-analysis process is an ongoing one and must be adapted to reflect the changing nature of our communities.

We currently apply various models and tools to characterize hazards but continue to fully appreciate the social, economic, cultural, and environmental impacts of disasters. We acknowledge that the natural, social, and engineering sciences are interconnected and enable us to understand a very dynamic landscape. *Hazards Analysis: Reducing the Impacts of Disasters* provides a framework for understanding the nature and potential impacts of disasters while promoting community adaptation and resiliency policies.

The key to understanding the dynamics of risk and vulnerability is through a comprehensive hazards analysis. A multidisciplinary approach is advocated in this text to clarify how to deal with uncertainty in ever-changing social, economic, and natural landscapes. Our intent is to demonstrate how we can use modeling, spatial analysis tools in geographic information systems, remote sensing technology, and risk analysis to clarify the interdependencies between human, economic, and natural systems.

We believe that a better understanding of the nature of hazards and our vulnerabilities can facilitate community planning, hazard mitigation, and community resilience to disasters.

Primary Learning Objectives

- To explain the role and uses of hazards analysis in organizational decision making and the development of public policy
- To characterize the nature of hazards and their impacts on communities
- To be able to compare and contrast various hazard models and how they may be used to characterize natural and human-caused hazards
- To explain how we measure the direct and indirect social, economic and environmental impacts of disasters
- To describe the role and uses of spatial analysis and other geographic information system tools in a hazards analysis
- To clarify how individuals perceive risk and how our perceptions influence decision-making at the individual, business and community levels
- To explain how risk management and hazard mitigation strategies may be used to reduce or minimize losses from disasters
- To explain the role of the hazards analysis process in building sustainable community

Acknowledgements

This text is an outgrowth of my teaching and research over the past 25 years at both Louisiana State University (LSU) and Appalachian State University (ASU). I began teaching "Hazards Analysis" for both graduate and undergraduate students in 1996 at LSU and continue to offer this class at ASU. The hazards analysis process has been used with public, private, and nonprofit organizations. We appreciate the opportunity to work with organizations including the Blue Ridge Parkway, Grandfather Mountain Stewardship Foundation, Doe Mountain Recreational Authority, and Biltmore Estate. We have learned to appreciate the local nature of both hazards and their potential impacts and the need to develop sound risk management and hazard mitigation policies.

Many of the processes and tools that we present in this text were used in projects with local government entities as well as with federal agencies such as the Federal Emergency Management Agency (FEMA), the U.S. Environmental Protection Agency (EPA), the National Oceanographic Atmospheric Administration (NOAA), the National Science Foundation, and the Organization of American States.

I especially want to acknowledge and thank my colleagues: Gavin Smith at the University of North Carolina in Chapel Hill (UNC CH), for his contributions in Chapters 9 and 10. He has given us his insights on "Planning for Sustainable and Disaster-Resilient Communities," and "Creating Disaster-Resilient Communities: A New Natural Hazards Risk Management Framework." Greg Shaw at George Washington University (GWU) prepared the chapter on the "Hazards Risk Management Process." Kevin Shirley at Appalachian State University collaborated with me on "Risk Analysis: Assessing the Risks of Hazards." Steven Guillot at Vanderbilt University contributed to the chapter concerning "Risk Communication." Steven passed away in December, 2013, and I will miss his extensive consultation and friendship over many years.

I would like to also acknowledge colleagues in the engineering, computer, and social sciences for helping me to appreciate the interdependencies of our natural, social, and economic systems. I have learned in our collaborative efforts how hazards have impacted families, businesses, agencies, and communities. My appreciation to so many community partners for their willingness to share their insights on how we can view vulnerability and enhance community sustainability.

John C. Pine
Director
Research Institute for Environment, Energy and Economics
Appalachian State University
Boone, North Carolina

Editor

John C. Pine serves as the Director of the Research Institute for Environment, Energy & Economics (RIEEE), and professor, department of Geography and Planning, Appalachian State University, Boone, North Carolina. He joined the Appalachian faculty in 2009 after serving thirty years at Louisiana State University in Baton Rouge where he directed the graduate and undergraduate Disaster Science and Management Program. At Louisiana State University, he was a professor in the Department of Geography and Anthropology and the department of Environmental Sciences. His research on disasters and emergency management centers on emergency planning, risk assessment, and disaster recovery. He has worked for many years with public agencies at the federal, state and local level as well as non-profit and private entities to identify strategies to enhance preparedness and community sustainability. His publications focus on hazards and disasters including *Technology and Emergency Management from* John Wiley (2007) and *Tort Liability Today* from the Public Risk Management Association (2005). He currently serves on the boards of directors of the National Committee for the New River, the Learning Lodge at Grandfather Mountain and an advisory board for the American Meteorological Society. His publications have been included in *The Journal of Disaster Studies, Policy and Management, Disasters, Journal of Race and Society, International Journal of Mass Emergencies and Disasters, Oceanography, Journal of Emergency Management, Natural Disaster Review, Journal of Environmental Health, and the Journal of Hazardous Materials.* He received his doctorate in higher education administration and public administration from the University of Georgia, Athens, in 1979.

Office Address: 234 I.G. Greer, Appalachian State University, Boone, NC 28608
E-Mail: pinejc@appstate.edu Phone: (828) 262-2764 (Office)
Web Site: http://www.rieee.appstate.edu

Contributors

Stephen L. Guillot, Jr.
Vanderbilt University
Nasheville, Tennessee

John C. Pine
Appalachian State University
Boone, North Carolina

Greg Shaw
George Washington University
Washington, District of Columbia

Kevin L. Shirley
Appalachian State University
Boone, North Carolina

Gavin Smith
University of North Carolina
Chapel Hill, North Carolina

Chapter 1

Introduction to Hazards Analysis

John C. Pine

Objectives

The study of this chapter will enable you to:

1. Clarify why hazards analysis is critical in reducing losses from disasters.
2. Compare and contrast hazards terminology.
3. Examine extreme events as a primary driver of disasters and community losses.
4. Explain alternative hazard paradigms that include social, political, economic, and environmental systems.
5. Define the hazards analysis process and its links to hazards risk management and comprehensive emergency management.
6. Explain why communicating risk is so critical in a hazards analysis.

Key Terms

Community
Consequence assessment
Disaster
Hazards
Hazards analysis
Risk
Vulnerability
Vulnerability assessment

Issue

What factors influence how public officials and agencies and businesses understand the nature of hazards and their impacts?

Introduction

Disasters are natural- and human-caused events that have the potential to cause damage to a community, region, or a nation. Events associated with a disaster can overwhelm response resources and have damaging economic, social, or environmental impacts. The capacity of a community, region, or nation to deal with disaster impacts provides a basis for characterizing and classifying an event as a crisis that must be addressed by local resources or that requires outside assistance and support. The process of assessing the nature and impacts of hazards as well as strategies for mitigating or adapting to potential adverse impacts from a disaster is the foundation of hazards analysis. Hazards analysis provides a comprehensive fact base for the development of emergency preparedness, response and recovery plans as well as the establishment of comprehensive community goals and public policies. Unfortunately, few communities have established a comprehensive hazards analysis framework that lead to sustainable and resilient communities (Shoubridge 2012).

Over the past 25 years, we have seen escalating costs associated with the direct economic impacts of natural disasters. Although the number of injuries and causalities has been dropping in recent years, the property damage has increased dramatically (Abramovitz 2001; Mileti 1999). Mileti (1999) notes that disaster losses have been increasing and will likely in the future. He sees that damages will grow to an average of $50 billion annually—about $1 billion per week. Some experts believe that this is a relatively conservative estimate of losses since there is little inclusion of indirect losses (i.e., loss of jobs, market share, productivity, etc.). Mendes-Victor and Gonçalves (2012) note that "disasters are not natural; they are also consequences of decision, often seemingly unconnected to their ultimate consequences, of collectivities of people, and are caused by their inability or unwillingness to adopt sustainable patterns of living."

The rising cost of disasters has also paralleled the movements of our population to coastal regions, thus increasing their vulnerability to hazards. In addition, we have seen widespread adverse impacts of disasters in the form of massive displacement, economic losses, and suffering from all parts of our society. Hurricane Katrina in 2005, Irene in 2010, and Sandy in 2013 clearly demonstrated that many members of our community suffered from the flooding and storm surge. Post-storm after-action reports have consistently noted that governments at all levels were ill prepared to deal with such a massive disaster.

This book challenges us to first examine the nature of a community and the hazards that could impact our social, economic, and ecological systems. In addition, we identify an approach to the development of a broad-based hazards risk management strategy to reduce risk and mitigate losses. This book provides a framework for identifying and understanding hazards and vulnerabilities, as well as the need for risk management and mitigation strategies for building sustainable and resilient communities.

Terminology of Hazards

The concepts of hazards and risks include multiple definitions of key terms, such as "hazards," "disaster," "risk assessment," and "hazards analysis." These terms are complex and may require clarification as drivers of natural- and human-caused disasters evolve. Many experts who study hazards, disasters, and risks acknowledge that our use of many terms has changed. For example, Kaplan (1997) describes two theorems of communication, which explain the confusion resulting from different and conflicting definitions of terms used in risk analysis and assessment. The theorems state the following: Theorem (1) 50 percent of the problems in the world result from people using the same words with different meanings; Theorem (2) the other 50 percent comes from people using different words with the same meaning. This confusion has lead to organizations such as the Federal Emergency Management Agency (FEMA) and the International Association of Emergency Managers, the United States Environmental Protection Agency (1986), and other federal agencies to increase the professionalism in the field by recognizing the need for a common set of definitions.

A hazard refers to a potential harm that threatens our social, economic, and natural capital on a community, region, or country scale. Hazards may refer to many types of natural events (flood, hurricane, earthquake, wildfires, etc.), technological (hazardous materials spills, nuclear accident, power outage, etc.), or are human induced (biochemical, bombing, weapons, mass destruction, or terrorism). Compounded hazards are those that result from a combination of the above hazard types such as urban fires resulting from earthquakes, failures of dams or levees that result from flooding, or landslides that result from wildfires and heavy rains.

FEMA (1997) describes hazards as "events or physical conditions that have the potential to cause fatalities, injuries, property damage, infrastructure damage, agricultural losses, damage to the environment, interruption of business, or other types of harm or loss." A hazard may be measured by its physical characteristics, likelihood, or consequences. Water from heavy rains, levee breach, or dam break would be the source of the hazard. The likelihood could be considered a low risk or not likely; it could be a medium risk or one that has a high likelihood of occurring. A hazard has the potential to cause fatalities, injuries, property damage, infrastructure or agricultural loss, damage to the environment, interruption of business, or other types of harm.

Cutter (2001) notes that hazards evolve from interactions between natural, human, and technological systems but are also characterized by the areas of their origin. For example, the hazard may arise from a hurricane but flooding magnified not only from excessive rainfall but also from long-term nonsustainable agricultural or forest practices. Since a disaster could evolve from the interactions between social, natural, and technological systems, the classification of a complex hazard could be difficult. As a further illustration of the difficulty in classifying disasters, a hurricane or flood occurring in a community might also lead to an accidental release of a hazardous chemical from a container in floodwaters. In this case, we have the potential of two disasters—one natural and the second human caused or technological in nature. This suggests that we view hazards within a broader social, political, historic, economic, and environmental context to fully appreciate how hazards can cause damage to community resources.

The United Nations defines a disaster as, "a serious disruption of the functioning of society, causing widespread human, material, or environmental losses which exceed the ability of affected society to cope using only its own resources (United Nations, Department of Humanitarian Affairs 1992)." All disasters, small or large, are the result of a hazard being realized. There is a caveat to this definition, however, in that the realized hazard must overwhelm the response capability of a community to be considered disastrous (FEMA 1997).

Pearce (2000) suggests that any definition of disaster must reflect a given locality's capacity to respond. He goes on to state that the hazard event must be unusual and that the social, economic, political, and ecological impacts must be significant. He defines disasters:

> A disaster is a non-routine event that exceeds the capacity of the affected area to respond to it in such a way as to save lives; to preserve property; and to maintain the social, ecological, economic, and political stability of the affected region (p. 87).

Disasters are measured in terms of lives lost, injuries sustained, or property damaged, and must be distinguished from routine emergency events that can result in property damage or fatalities. For instance, a house fire may require a response by a jurisdiction's fire department and result in loss of life or property. However, as fires are common emergency occurrences, they are managed by local response agencies and are normally not considered a disaster. For a fire to be considered a disaster, it must overwhelm the capacity of the local responders.

Common breakdown of hazards include atmospheric climatic hazards, such as rain, lightning, snow, wind and dust storms, hailstorms, snow avalanches, heat waves, hail, snowstorms, and fog (Bryant 2005; FEMA 1997; Hewitt 1983). They also include geologic and seismic hazards such as landslides, avalanches, land subsidence, erosion, earthquakes, tsunamis, and volcanic and shifting sands. Hydrologic hazards make up the third type of natural hazard and include events such as flooding, storm surges, coastal erosion, waves, sea ice, and sea level rise. Hewitt (1998) explains that compounded hazards include tropical cyclones, thunderstorms, whiteouts, tornadoes, rain and wind storms, blizzards, drought, freezing rains, and wildfires; each combines several natural hazards and are not just the result from a single hazard.

Not all hazards result in disasters for a hazard event could decrease potential damaging impacts so as to minimize losses (Gruntfest et al. 1978; Hewitt 1997; Lindell and Meier 1994). The rate or speed of onset of the event could give communities notice needed to minimize deaths and injuries by ordering an evacuation for a flooded area. Availability of perceptual cues (such as wind, rain, or ground movement) provides notice of a pending disaster. The intensity of a disaster could vary spatially so as to have damage impacts in areas with no social or economic impacts. Technology such as weather radar allows us to see where a heavy storm is moving so as to provide warning to the local area. The areal extent of the damage zone or its size (geographic area influenced) and its duration could influence any damaging impacts and the community's capacity to deal with the hazard event. Wind damage from a tornado could be limited to nonpopulated areas and not cause injuries or property damage. Finally, the predictability of the event or notice of occurrence is also critical in allowing those affected by the event to seek safety. Despite our efforts to reduce our vulnerability to disasters, we see that property losses, deaths, and injuries continue to increase. Numerous studies have documented that increased losses are growing (Abramovitz 2001; Mileti 1999).

1. Population growth in high-hazard areas.
2. Marginalized land is being developed making us more susceptible to hazard impacts.
3. Larger concentrated populations in urbanized areas increase the potential for human and property loss; people are less familiar with hazards in their surroundings; growth may not be ecologically sustainable; more buildings and infrastructure may be damaged if an event occurs.
4. Inequality: people are not impacted by hazards equally; economic disparities cause large numbers of impoverished people to be at risk.
5. Climate change: immense potential for loss as sea levels rise; weather and climate patterns will change.
6. Political change: political unrest can directly cause loss (e.g., civil war) and/or make a region more susceptible to hazard impact due to lack of preparedness and/or inability to cope.
7. Economic growth: directly related to technological hazards, producing increased levels of many pollutants; usually results in fewer deaths from hazards, but increased economic loss; more property is at risk to hazards, but preparedness and mitigation measures minimize loss of human life.

Today we see significant disruptions in social, economic, and natural systems that are associated with policies and practices that evolve over different time frames. Economic disruptions can result from short-term economic drivers or prospective losses that may be associated with evolving natural conditions associated with climate change.

The terms "risk" and "hazard" are often used interchangeably and inconsistently. Differences result as emergency managers, risk managers, urban and regional planners, insurance specialists, and lay people develop meaning of the terms independently. These definitions can even be in conflict with each other. For example, it is not uncommon for the word risk to be used informally in a way that means "venture" or "opportunity," whereas in the field of risk management, the connotation is always negative. However, even among risk managers, the exact definition of risk varies considerably (Kedar 1970).

The risk of disaster is typically described in terms of the probabilities of events occurring within a specified period of time; for example, 5, 10, or 20 years; a specific magnitude or intensity (or higher); or a range such as low, medium, or high risk. For example, the risk of floods is commonly described by FEMA in terms of 100- and 500-year floods, indicating the average frequency of major flooding over those periods of time and the maximum area that has been inundated each time. Risk has the common meaning of danger (involuntary exposure to harm), peril (voluntary exposure to harm), venture (a business enterprise), and opportunity (positive connotation—it is worth attempting something if there is potential for gain). In a business context, it refers to probability considerations but is primarily concerned with uncertainty.

Views of Extreme Natural Events as Primary Causes of Disasters

Tobin and Montz (1997) provide a very insightful perspective on how we might view natural hazards and disasters. They see that one way of viewing disasters is that all or almost all responsibilities for disasters and their impacts are attributed to the processes of the geophysical world. In this approach, the root cause of death and destruction is caused by extreme natural events rather than human interface with the environment. Under this view of disasters, those who suffer losses are seen as powerless victims who have limited control and simply react to the immediate physical forces and processes associated with disasters. The physical world is thus viewed as external force, separate from human actions. This perspective was noted by Burton and Kates (1978) who see natural hazards as elements of the physical environment harmful to man and caused by forces external to him.

This perspective of disasters from Tobin and Montz is significant for the outputs of a hazards analysis. If the view of individuals in a high-risk area is just limited to the physical world, then there is little that can be done to minimize destructive hazard impacts. Quarantelli (1998) notes his early views of disasters and their origins. "The earliest workers in the area, including myself, with little conscious thought and accepting common sense views, initially accepted as a prototype model the notion that disasters were an outside attack upon social systems that 'broke down' in the face of such an assault from outside" (p. 266).

Cook et al. (2009) comment on the linkages between natural and human systems in characterizing hazards as natural events. They examined the Dust Bowl drought as a human-induced land degradation event rather than a natural disaster.

Steinberg (2000) also commented on this view of nature and disasters as extreme events that are beyond our control.

> ... [T]hese events are understood by scientists, the media, and technocrats as primarily accidents – unexpected, unpredictable happenings that are the price of doing business on this planet. Seen as freak events cut off from people's everyday interactions with the environment, they are positioned outside the moral compass of our culture (p. xix).

This view of how people view hazards is clarified by the concept of "bounded rationality," inadequate information, and ability to make sound choices in the face of risk. Tobin and Montz (1997) clarify its application to disasters by explaining that "bounded rationality" refers to the fact that "behavior is generally rational or logical but is limited by perception and prior knowledge" (p. 5). Burton et al. (1978) note "... [I]t is rare indeed that individuals have access to full information in appraising either natural events or alternative courses of action. Even if they were to have such information, they would have trouble processing it and taking appropriate action to reduce losses. The bounds on rational choice is dealing with natural hazards, as with all human decisions, are numerous" (p. 52).

An integrated assessment of risks focuses on risks from salinization, typhoon and flood, sedimentation, coastal erosion, sand drift, sea level rise, earthquake, environmental contamination, or land cracking. This type of assessment integrates multihazard process for a community. Schmidt et al. (2011) notes that few studies have addressed multirisk assessments including alternative hazard types and their impacts. Their approach provides for an assessment of alternative hazards and their impacts but does not provide an integrated multirisk assessment process.

A Changing Hazards Paradigm

Our efforts to understand the nature of hazards, their impacts, the likelihood of their occurrence, and how we use this information in hazard mitigation or other public policy decisions has resulted in alternative approaches to hazards analysis.

FEMA and later the NT1 approach suggests a data-based quantitative emphasis on the characteristics of potential hazards, their likelihood of occurrence, and a prioritization of alternatives to address threats. A hazard in this context is viewed in single events with specific causal events.

A more quantitative approach to assessing risk was stressed by the National Research Council (1983). This approach has four elements including: risk identification, dose–response assessment, exposure assessment, and risk characterization. This model was used as the standard beginning in 1980 as part of the Superfund legislation and institutionalized as part of the evaluation of abandoned superfund sites. This emphasis on quantitative analysis is also reflected in United Nations vulnerability and risk assessment processes (Coburn 1994). Cutter (2003) notes that most of the risk assessments used probability estimators and other statistical techniques. The United States Environmental Protection Agency's (EPA) approach was broadened in 1987 to look beyond just exposure as its carcinogenic potential to look at noncancer human health risk, ecological risk, and welfare risk. This process was revised in 2001 and characterized as a "relative-risk" approach that moved away from pollution control and technology fixes to one of risk reduction and sustainable approaches to pollution management.

Cutter (2003) notes that these processes for risk assessment are fraught with methodology concerns that include uncertainty especially with variability in individuals and ecosystems, and limited environmental data. Risk assessments must link good science with communities. This broader view of risk that includes communication and interaction between the scientist and those impacted by the assessment of hazards is very constructive.

Adger (2006) examines disasters and their adverse impacts in two ways. Two alternative models are provided to explain the complex nature of hazards and their impacts. A pressure and release model (PAR) examines the relationship between processes and dynamics that bring about unsafe conditions and their interface with disaster events, such as earthquakes, floods, or tropical cyclones. The emphasis in the PAR is the driving social forces and conditions that bring about vulnerability of people in place. The second model emphasizes access to resources and takes into consideration the role of both political and economic conditions as the basic causes of unsafe conditions. Bull-Kamanga et al. (2003) stress the importance of local processes for risk identification and reduction. The need for a hazards analysis to include local players enriches an accurate characterization of both the community and local conditions that may influence disaster impacts. The emphasis on underlying social conditions and its role in hazards analysis is also stressed by Weichselgartner (2001). In his view, a disaster is a product of a cumulative set of human decisions over long periods and that these decisions either create greater risk or reduce risk. Mitigation must stress the underlying human conditions and not just adjust the physical environment. Vulnerability analysis thus must take a broad view of the conditions that are present prior to a disaster as well as the physical environment of a community and the characteristics of the hazard.

Critical Thinking: We build levees and flood walls, and establish building codes and base flood elevations in an effort to protect property from hazards. How can hazards analysis help to foster greater awareness of risks associated with hazards? How can a hazards analysis have a constructive influence on decision making at the individual, family, and community levels?

Cutter (1996) and Cutter et al. (1997, 2003, and 2010) also suggest that a broader perspective is needed to fully appreciate the complexities associated with hazards and their numerous impacts. This hazards-of-place model of vulnerability involves a comprehensive understanding of hazard potential along with an examination of the geographic context, social conditions, and both biophysical and social vulnerabilities. A place-based view of vulnerability is then determined from these elements. A set of indicators can be used to examine vulnerability and take into account population variables and infrastructure lifelines. The goal in this approach is to assess social vulnerability and community sustainability for response and recovery. Cutter (1996) stresses that to understand a community's hazard potential, one must consider that a disaster is influenced by socioeconomic indicators, individual characteristics, and the community's geographic context impacted by the hazard. Others have also developed criteria to examine community sustainability. Miles and Chang (2006) examine community recovery capacity by using social and infrastructure variables. They stress the need for modeling recovery processes in understanding community resiliency capacity.

An emphasis of modeling hazards and measuring the resilience of communities is also seen in Bruneau et al. (2003). The quantitative measures combined with characterization of a hazard result in information that may be used in guiding mitigation and preparedness efforts. Their measurement of the local community centers around four characteristics including: robustness of systems to withstand loss of function, redundancy of elements of the system to suffer loss, resourcefulness and the capacity to mobilize resources when the system is threatened, and rapidity or the speed to achieve goals in a timely manner.

Hazards Analysis

There are many perspectives on "hazard analysis" that vary from FEMA's approach of knowing what could happen, the likelihood of it, and having some idea of the magnitude of the problems that could arise (FEMA 1997). FEMA's introduction of the HAZUS modeling software in 1997 reflects their interest in physical processes at the community or regional level. This approach, unfortunately, is limited to modeling one hazard at a time and fails to address a multihazards environment (Cutter 1996). A process approach to hazards analysis addresses adverse impacts of hazards (Long and John 1993) and stresses the role of hazard identification, risk screening, and the development of mitigation measures to control losses. The Coastal Engineering Research Center and the University of Virginia along with the

U.S. Geological Survey (USGS) used a quantitative hazards analysis approach in examining risk and exposure to coastal hazards (Anders et al. 1989). They evaluated the U. S. coastline for risk and exposure to coastal hazards by examining the characteristics of the hazards, coastal geographic features, population demographics, and civic infrastructure.

Researchers from Oak Ridge National Laboratory identified a coastal vulnerability index that includes risks from sea level rise in coastal communities (Daniels et al. 1992; Gornitz and White 1992; Gornitz et al. 1994). This study weighed the characteristics of coastal hazards, local geographic conditions, and the likelihood of extreme weather impacting local areas. Their approach to hazards analysis focused on characteristics of the hazard event, local geographic conditions, and demographic factors. They emphasized the use of model outputs in mitigating disaster impacts. Multihazard impacts were examined by Preuss and Hebenstreit (1991) to understand the impacts of both earthquakes and associated tsunamic flood events. This risk-based urban planning approach was designed to allow assignment of risk factors for vulnerabilities on a community basis.

The National Response Team, composed of 14 Federal Agencies, adopted a common approach to community level hazards analysis and planning (1987). This approach uses a process format in providing communities with a broad understanding of hazards and risks. National Respons Team-1 defines hazard analysis as a three-step process: (1) hazard identification, (2) vulnerability analysis, and (3) risk analysis. This approach to hazards analysis stresses the need for broad-based information to support community decision making to reduce vulnerability and minimize risk to people and property.

Critical Thinking: How do you ensure that an analysis of hazards is comprehensive and the potential impacts are sensitive to human, cultural, economic, political, and natural systems?

Hazard Identification

Hazards identification as noted in Figure 1.1 provides specific information on the nature and characteristics of the hazardous event and the community. It further examines an event's potential for causing injury to life or damage to property and the environment. Hazard identification takes advantage of the use of environmental modeling to characterize hazards and disaster impacts. As part of the EPA hazards analysis process, community involvement is encouraged through a broad-based team represented by local response agencies, the media, community public health units, medical treatment organizations, schools, public safety, and businesses. The formation of local emergency planning committees provides the basis for broad input in preparedness efforts.

Hazards analysis		
Hazards identification	**Vulnerability analysis**	**Risk analysis**
Identify hazards	Vulnerability zone	Likelihood of incident
Location	Human populations	Severity
Quantity of chemical	Critical facilities	Consequences
Nature of hazards	Environment	

Figure 1.1 U. S. Environmental Protection Agency hazards analysis process.

Vulnerability Analysis

Vulnerability analysis in Figure 1.1 is a measure of a community's propensity to incur loss. Vulnerability analysis may focus on physical, political, economic, and social vulnerabilities. Vulnerability is, in other words, the susceptibility to hazard risks. Vulnerability can also be a measure of resilience. According to Emergency Management Australia (2000), vulnerability is "The degree of susceptibility and resilience of the community and environment to hazards". Vulnerability analysis identifies the geographic areas that may be affected, individuals who may be subject to injury or death, and what facilities, property, or environment may be susceptible to damage from the event.

1. The extent of the vulnerable zones (i.e., an estimation of the area that may be affected in a significant way).
2. The population, in terms of numbers, density, and types of individuals (e.g., employees; neighborhood residents; people in hospitals, schools, nursing homes, prisons, and day care centers) that could be within a vulnerable zone.

Vulnerability analysis as viewed by EPA examines who and what is vulnerable and why (1986).

Critical Thinking: What types of private and public property might be damaged in a natural- or human-caused disaster? What essential support systems (e.g., communication or public services) and facilities and corridors could be affected? What property is more likely to be affected in a disaster?

When assessing risks, experts must factor in vulnerability. The vulnerability assessment is a measure of the exposure or susceptibility and resilience of a community to hazards. We stress that understanding vulnerability by itself is insufficient to plan for disasters. It must be accompanied by understanding the nature and characteristics of hazards. Hazards identification and characterization is thus a component of a full hazards analysis. Crozier and Glade (2006) note that vulnerability analysis is different from consequence analysis. Where vulnerability examines the potential for loss, consequence analysis clarifies what will be the impact. The analysis for consequence assessment is far more detailed and models many more factors that affect outcomes.

Risk Analysis

EPA in Figure 1.1 describes risk analysis as an assessment of the likelihood (probability) of an accidental release of a hazardous material and the consequences that might occur, based on the estimated vulnerable zones. The risk analysis is a judgment of probability and severity of consequences based on the history of previous incidents, local experience, and the best available current technological information. It provides an estimation of the following:

1. The likelihood (probability) of a disaster based on the history of current conditions and consideration of any unusual environmental conditions (e.g., areas in flood plains) or the possibility of multiple incidents such as a hurricane with tornadoes (e.g., flooding or fire hazards)
2. Severity of consequences of human injury that may occur (acute, delayed, and/or chronic health effects), the number of possible injuries and deaths, and the associated high-risk groups
3. Severity of consequences on critical facilities (e.g., hospitals, fire stations, police departments, and communication centers)
4. Severity of consequences of damage to property (temporary, repairable, permanent)
5. Severity of consequences of damage to the environment (recoverable, permanent)

Risk in this view is the product of the likelihood of a hazards occurring and the adverse consequences from the event. Simply stated,

$$\text{Risk} = \text{Likelihood} \times \text{Consequence}$$

Critical Thinking: Increasing numbers of people are moving into vulnerable areas as illustrated by population growth in coastal regions, wildland–urban areas, and sensitive mountain environments. This creates stress on resources and land use. Larger numbers of people may move into more sensitive environments. The best land may be developed, leaving development in areas that are marginal and more susceptible to hazards.

Linking Hazards Analysis to Risk and Comprehensive Emergency Management

Alexander (2000) describes two approaches to dealing with risks, one a community hazards mitigation approach that is based on a comprehensive hazards analysis, large-scale planning, and decisions at local community level. Hazards mitigation or comprehensive risk management strategies are developed and implemented at the local or regional community level. He suggests an additional perspective that includes extensive risk communication with the community. In this approach, he suggests that we establish a greater understanding and appreciation of local hazards and risk by the public and support grassroots democratic involvement. He suggests that communities include citizens in the hazards analysis process and proceed from nonstructural to structural protection, not vice versa (p. 27).

Alexander suggests that decision making at the individual, household, neighborhood, organizational, or community level should be made by informed individuals. Rational choices may be based on information from a comprehensive hazards analysis. Risk communication is thus a critical part of the hazards analysis process and a positive contribution to decision-making processes. Through risk communication and public participation in the hazards analysis process, risk management and hazard mitigation strategies may be adopted by citizens and the community to reduce vulnerability.

Individual citizens and communities thus make decisions that either increase or reduce our vulnerability to hazards. The key is to acknowledge the interface between environmental hazards and human actions and that actions can be initiated to reduce vulnerability through hazard mitigation and hazards risk management. As a result, the hazards analysis process must include opportunities for public involvement and risk communication and that decision making is essential in adopting effective risk management and hazards mitigation strategies by individuals, organizations, and a community. Through this approach, we reject a perspective that adopts the causality of environmental determinism where we have no power to reduce our vulnerability to environmental hazards.

This comprehensive approach to hazards analysis views disasters and hazards beyond just their geophysical processes and examine how social, economic, and political processes impact hazardousness. The importance of a community assessment as a part of hazards identification, which includes a close look at sociocultural, economic, and political systems may be seen vividly in the impacts from Hurricane Katrina in New Orleans. The damages from Hurricane Katrina revealed significant vulnerabilities associated with poverty, education, housing, employment, and governance. Unfortunately, what was revealed in New Orleans is present in many urban coastal cities. In fact, Mileti (1999) stresses the need for a community assessment and including the delineation of hazard areas within the hazards analysis process.

> Rebuilding that generally keeps people and property out of harm's way is increasingly viewed as an essential element of any disaster recovery program. Rebuilding that fails to acknowledge the location of high-hazards areas is not sustainable, nor is housing that is not built to withstand predictable physical forces. Indeed, disasters should be viewed as providing unique opportunities for change – not only to building local capability for recovery – but for long-term sustainable development as well (Mileti 1999, p. 237–238).

Hazards analysis process builds on the EPA approach of hazard identification, vulnerability assessment, and risk analysis by stressing the need for the use of the results of a risk analysis in hazard adaptation adjustments (Figure 1.2). Including risk communication, citizen participation, problem solving, risk management, hazard mitigation, and ongoing assessment are all parts of comprehensive emergency management. This approach stresses an action orientation through the adoption and implementation of comprehensive hazards risk management and hazards mitigation strategies and monitoring the effectiveness of hazard adjustments that are adopted and implemented. The ultimate goal of these hazard adjustments is to build resilient and sustainable organizations and communities.

Pelling (2011) distinguishes short-term coping capacity or coping strategies with long-term adaptive capacity or adaptation. Short-term coping strategies focus on the design and implementation of risk management and preparedness plans that might mitigate immediate impacts from disasters. Long-term adaptive capacity strategies concentrate on changing those practices and underlying institutions that generate the root or proximate causes of risk. Engle (2011) stresses the need to assess and measure adaptive capacity as part of the ongoing change process.

Further, we stress that the hazards analysis process includes an intentional assessment or monitoring of the impact of our hazard mitigation strategies and hazard risk management strategies. We want to know what are the short- and long-term results of our actions that might include increasing minimum base flood elevation requirements, strengthening building code requirements, or enhancing building inspections. Second, we stress that the hazards analysis process is not static but an ongoing one. We see that the ongoing review of our hazard mitigation and risk management policies could lead to program changes to strengthen or enhance opportunities to build more sustainable and hazard-resilient communities and organizations.

Including adaptation and coping strategies in the hazards analysis process suggests an action-oriented element to our understanding of risk and the need to develop strategies to cope or manage our organizations and communities to reduce our vulnerability. Kalaugher et al. (2013) stress this emphasis on the development of strategies to deal with complex adaptive socioecological system interactions. They note that the scale of the systems examined may vary nationally, regionally, or locally but can result in specific adaptive strategies.

Critical Thinking: One of the best hazards analysis efforts conducted on a large-scale basis was completed in 2007 by the Louisiana Coastal Protection and Restoration Authority. The plan includes a comprehensive assessment of the hazards in a coastal environment, impact assessments (social, economic, and ecological), public participating, and recommendations on strategies to protect the social, economic, and environmental assets of the state. This is an outstanding example of a region-wide hazards analysis and included federal, state, and local government agency collaboration.

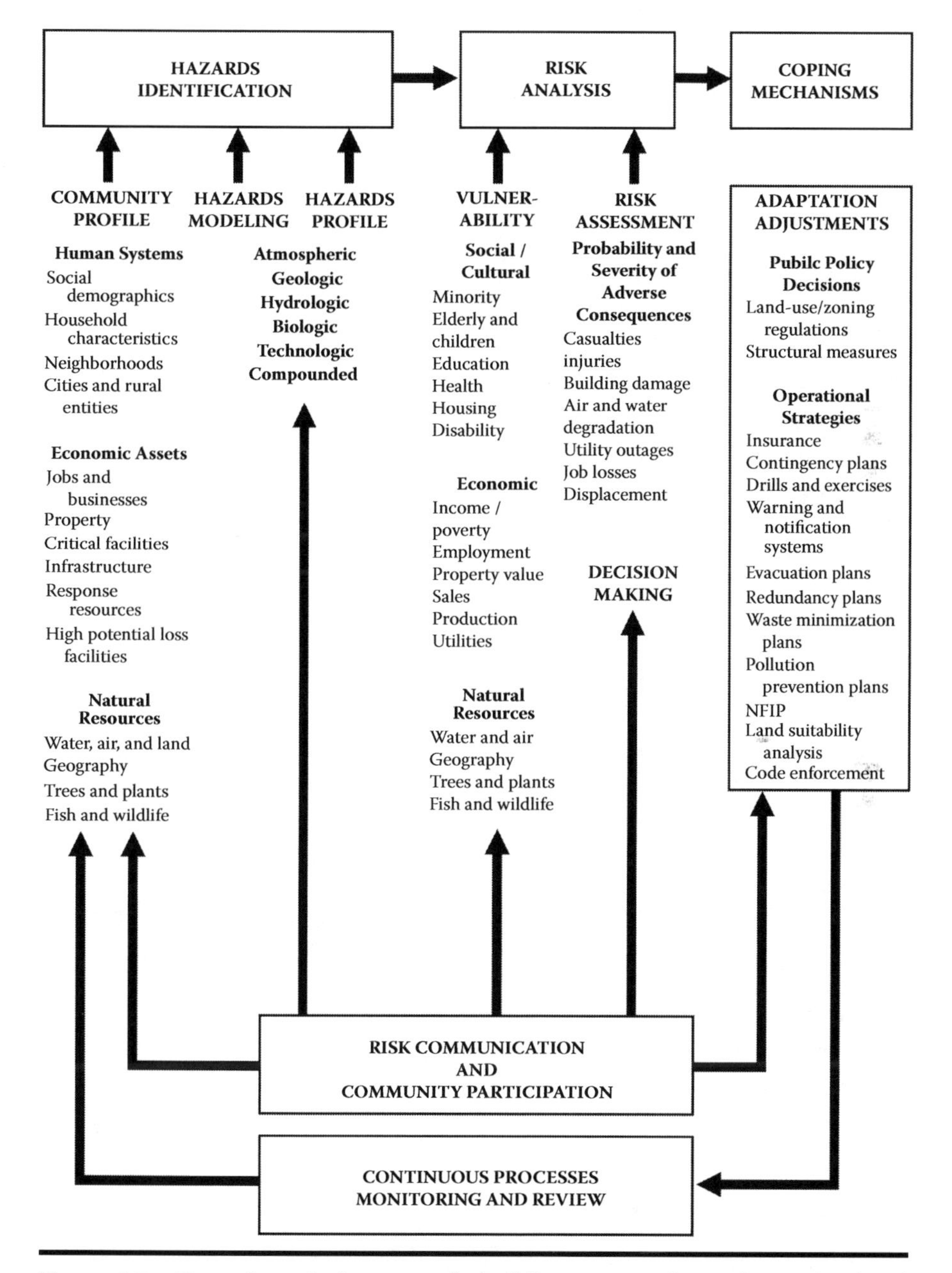

Figure 1.2 Hazards analysis process in building community and organizational resilience.

Communicating Risk from a Hazards Analysis

Through hazards adaptation and adjustments, we stress the role of risk communication and community and organizational participation in the hazards analysis process. Understanding of hazards will not be the sole result of just telling people of hazards, but allowing them to participate in the hazards analysis process at the neighborhood, community, and organizational levels. People support what they help build, and citizens as well as employees will advocate risk management and hazard mitigation strategies that they understand and help formulate. They will likely oppose what is imposed on them.

Engle (2011) adds to an examination of the need for enhancing community-adaptive capacity by stressing that adaptive capacity improves the opportunity of systems to manage varying ranges and magnitudes of climate impacts, while allowing for flexibility to rework approaches if deemed at a later date to be on an undesirable trajectory.

The results of a hazards analysis are not just for planning and mitigation of hazards but should be shared with the public as reflected in Figure 1.2. Any community has risks associated with natural- or human-caused hazards. Hazards that typically cause minimal damage are usually accepted as inevitable and little is done to reduce the risk. Such hazards may be viewed as nuisances, rather than real threats to live and property. Some communities are willing to accept more risk than others. Factors such as the political culture and the socioeconomic level of the community determine the levels and kinds of risk that may be accepted. For example, poor communities may be willing to accept more risk from environmental hazards because the economic base of the community will not directly support the allocation of resources for structural or nonstructural hazards mitigation initiatives. For individual residents they may refuse to purchase flood insurance or take other measures that have associated costs because they have limited discretional financial resources. They want the insurance but just not have the funds.

Communicating information from the hazards analysis to the public can help shape perceptions of risk and elevate concern for protecting personal property. Further, by acknowledging local environmental risks, the community may initiate strategies that can overcome the individual financial limitations so as to protect the entire community from hazards. Collective action may be advisable when low-income residents who may be renters or homeowners just cannot take individual action. Individual risk assessment, risk management, and impact assessment are all part of using information from a hazards analysis to protect individual citizens and their property.

Community Involvement

The approach described in Figure 1.2 suggests that a broad-based representative methodology be used; this provides for community inputs and provides a base for the development of strategies that address community priorities and concerns. Smit and Wandel (2006) stress that a key outcome of any assessment is to identify adaptation strategies that are feasible and practical in communities. A key recommendation centers on the development of adaptation initiatives that are integrated into other resource management, disaster preparedness, and sustainable development efforts. Samarasinghe and Strickert (2012) suggest a methodology for using qualitative, quantitative, and cognitive mapping to provide insights into public policy formulation from many diverse local stakeholder groups that include both lay and expert insights.

Some agencies and scientists see hazards analysis as a scientific process that includes only the experts. We suggest an alternative approach. Partnering experts in community planning, engineering, modeling, geography, sociology, or other hazards fields with community leaders and members will establish a dialogue relating to risks to reach common goals around generating insights and strategies to build a sustainable community. It operates through a dialogic, hermeneutic approach, similar to fourth-generation evaluation (Lincoln and Guba 1989). An external group of experts can be well positioned to collect the right information, and same local governments time in dealing with unknown areas. Engagement seeks a grassroots understanding of risk, one that is perceived as critical to risk reduction and building local capacity (Center 2000). This report suggests that the engagement process should include the following:

1. Experts (public agency representatives including emergency management and other local agencies along with consultants if used) meet with local residents to explore common goals in a hazards analysis.
2. Identify questions and issues relevant to the residents including the roles of residents or community members and agency experts (outsiders).
3. Develop, through consensus building, common objectives and priorities for the hazards analysis beneficial to both the experts and residents.
4. Describe the hazards analysis process.
5. Develop, through consensus building, an agreed-upon strategy of how the results of the hazards analysis will be shared with the community, organizations, and public officials.
6. Discuss residents' concerns.
7. Initiate the project(s).
8. Present the results to the community for their response.

Butzer (2012) notes the value of community engagement in adaptation to risks and the development of sustainable societies. He acknowledges the intricate interplay of environmental, political, and sociocultural resilience in limiting the damages of the adverse impacts from risks. His model emphasizes resilience in the form of innovation and intensification on a decentralized, protracted, flexible, and broadly based approach. He also stresses the slow pace of risk in the form of the degradation of soils or other biotic resources including deforestation, ground cover removal, soil erosion, or groundwater depletion. He explains that declining resource productivity increases pressure on the environment and may be a precondition of an environmental or economic failure.

Values in Community Engagement

Greenwood and Levin (1998) suggest that this approach is context bound and deals with real-life problems. It is problem focused and joins participants and experts to generate through collaborative discussion. All participants' have meaningful dialogue; diverse points of view are welcomed, and the process leads to action. The key is that actions evolve to address problems associated with hazards, and both the experts and citizens have an increased awareness of options to address problems.

One of the main values in community engagement is community sustainability, characterized by environmental quality, quality of life, disaster resiliency, vibrant economies, and equity as developed through local consensus building (Mileti 1999). Sustainability implies persistence within an ecosystem's carrying capacity (Burby 1998). Systems theory guides the analysis by examining the natural, human, economic, political, and constructed systems. "Hazards researchers and practitioners would do well to take a more systems-based approach …. [it] recognizes multiple and interrelated causal factors, emphasizes process, and is particularly interested in the transitional points at which a system … is open to potential change" (Mileti 1999, pp. 106–107). From this standpoint, hazards are viewed as "complex interactions between natural, social, and technological systems" (Cutter 1993, p. xiv). Those interactions result in vulnerability (Center 2000). To assess those systems and their interactions, the analysis should include an examination of the following:

- *Human social and cultural heritage elements:* This part of the analysis attends to culture, ethnic identity, social institutions (family, faith, economy, education, self-governance), and disaster experiences. Members of the community are encouraged to share family and community photographs and stories related to the human, natural, and physical systems.
- *Protective actions:* (1) Risk communication (warnings); (2) land use and zoning (life and property); (3) participatory community meetings that will involve residents in hazard identification, risk and vulnerability assessment, and planning. Data collection will center on disaster history, vulnerability, and sociobehavioral response. This focus will address interactions between the natural and human systems. We are especially interested in protective actions taken during several phases of disaster, with a particular concentration on social bonds. Kates et al. (2012) comment on the potential outcomes of adaptation by stating that change may be difficult to implement because of uncertainties about risks such as climate change and adaptation benefits, the high costs of transformational actions, and institutional and behavioral actions that tend to maintain existing resource systems and policies.

Implementing transformational adaptation requires effort to initiate it and then to sustain the effort over time.

- *Constructed and physical environment:* This part of the analysis examines how the constructed environment links to the human and natural environments. This focus examines how the community's sense of place mediates their relationship to the social, physical, and built environments. Data from a risk assessment are included such as hurricane wind and storm modeling, riverine flood hazard modeling, earthquake, wildfire, and wind models may be run to examine potential social, economic, and environmental vulnerabilities. This focus addresses the built and natural systems as experienced within the human system.

Critical Thinking: Is the concept of community right to know just a legal obligation or is it based on a broader set of values? The adoption of the community planning right to know act (United States Environmental Protection Agency 1986) changed how local communities conducted hazards analysis and communicated information about hazards to the community. This legislation asserts that the community has a right to know about chemical hazards present in the community. Although this legislation focused on human-caused technological hazards, it was built upon shared societal values—the "right to know" implies other values: transparency, accountability, responsible action, democracy, and active citizen participation. This emphasis on risk communication and sharing of information freely about hazards was changed following September 11, 2001. Since hazards that time, new restrictions have been established relating to access to chemical hazard information "in the interest of security."

Conclusions

In this book, we wish to suggest that a broader approach be used to understand the potential impacts of disasters and that the process can be used for multiple local conditions. Natural hazards have very different impacts throughout the world and the nature and extent of a disaster depends on several factors including:

- Local and regional environments including the landscape, climatic conditions including the probability of an event, how often they occur, and the capacity of the hazard to do harm
- The strength and vitality of the social, economic, and natural environments to withstand and cope with the adverse effects of a hazard
- Response and recovery resources that enable communities, regions, and nations may need to cope and recover from disasters

The capacity of the organization, community, and region to recognize their vulnerability and initiate steps to reduce adverse impacts is critical to the hazards analysis process.

In addition to viewing hazards and disasters as part of a local condition, we wish to stress the use of hazards analysis for mitigation and prevention rather than just for response and recovery or as part of the regulatory permitting hazards process. The results of a hazards analysis can provide information to identify hazard risk management strategies to strengthen social, economic, and environmental systems and enable these systems to withstand the destructive conditions that are inherent in hazards. It may be impossible to reduce the wind or storm surge from a hurricane or the shaking from an earthquake, but we can take steps to build stronger buildings, locate our structures in less-vulnerable areas, and enhance our social structures to cope with displacement, loss of jobs, and critical natural resources.

We link the hazards analysis process to decision making by local community officials, individual citizens, and private and nonprofit organizations. This emphasis on decision making is reflected in the work of Deyle et al. (1998), where hazard assessments provide a factual rational basis for local decision making. Their goal is to achieve safer more sustainable communities through management and informed decisions that are based on estimates of costs and benefits of efforts to reduce risks. The approach suggested in this text builds on the EPA approach of hazards identification, vulnerability assessment, and risk analysis but emphasizes decision making within sociocultural, political, and legal constraints. We stress that the scale of analysis is critical in establishing the type of data that is needed and the degree of precision that may be needed in a community and that data used must be current.

Lindell and Prater (2003) explain that the hazards analysis process is linked to comprehensive emergency management and hazards risk management processes through a disaster impact model. Their view suggests that there are existing conditions that reflect current hazards as well as current social and economic vulnerabilities. In addition, they note that the current physical conditions of the community will impact its resilience. They contend that the social and economic impacts from a disaster are greatly influenced by the community's level of implementing effective mitigation strategies, emergency preparedness, and recovery strategies, as well as the physical characteristics of the disaster and the community's actual response and recovery efforts. This broad view of the hazards analysis process acknowledges the need to understand the nature of the community and develop a broad community profile that includes an examination of local geography, demographics, infrastructure, and response resources. They go further to stress the importance of using hazards analysis in the preparation and monitoring of hazards mitigation strategies, emergency preparedness plans, and response strategies.

The nature and extent of a hazard condition thus influences the potential adverse impacts to the built and human environments. Slow-moving category 4 hurricanes have a greater capacity to cause destruction than a tropical storm. Wildfires that are driven by 30 mph winds have a greater destructive force than fires in low wind conditions. Each of these hazardous conditions has very different impacts on any community depending on local conditions including:

- The character of our built environment such as homes, office building, manufacturing plants, roads, bridges, dams, or levees
- The nature and condition of the natural environment such as wetlands, flood plains, forests, cultivated areas, hills, mountains, or changes in elevation
- The existence of strong and connected families, neighborhood associations, nonprofit groups, and individuals who are engaged in the community

Communities with high unemployment, poverty, excessive crime, and poor education may have great difficulty in coping and recovering from a hazardous event. Communities vary in the resources that may be used to deal with a hazardous event. Some communities may have large numbers of the local population living in poverty and high unemployment, high crime, and limited pubic resources to deal with a disaster event. In addition to social and economic community characteristics, the design of built structures and the nature of natural environments can influence the damage that results from a hazard event. Healthy wetlands and forests, strengthened structures and land that are of higher ground are in a better position to withstand the destructive character of hurricane winds or flooding conditions. Existing conditions may be assessed as the community's capacity to resist damage from disasters.

Measuring a community's resilience has been examined by a growing number of hazard researchers (Bruneau et al. 2003; Tierney and Bruneau 2007; Turner et al. 2003). Vulnerability should be considered not just by exposure to the stresses produced by hazards alone but also the "sensitivity and resilience of the entire sociocultural, economic, and environmental systems experiencing the hazards." Vulnerability assessment must examine coupled human and environmental systems and their linkages within and without the systems that affect their vulnerability. Turner et al. (2003) presents a framework for the assessment of coupled human and environmental systems.

A comprehensive hazards analysis must take this broad approach and acknowledge the human, built, and natural environmental systems and their multiple connections. Hazards analysis must attempt to understand each of these systems and examine their interconnectedness and impact on community resilience.

In this book we stress the need to embrace a broader context for viewing the hazards analysis process. There is a need for more integrative approaches in vulnerability science for understanding and responding to environmental hazards (Turner et al. 2003). Mileti (1999) notes that a new paradigm is needed in dealing with hazards and disasters, one that addresses sustainable hazards reduction. To accomplish this, the following should be addressed:

- *Sustainable culture:* We do not control nature despite our efforts to design levees, dams, or buildings, and in many cases we are the cause of disaster losses. We must understand the nature of hazards and build to reduce losses. The outputs of the hazards analysis process must be used to identify mitigation strategies so as to minimize our vulnerabilities socially, economically, and environmentally.
- *Events, losses, and costs:* Outputs from the hazards analysis process need to characterize our vulnerability and document how disasters have affected our communities.
- *Interactive structure of risk:* The hazards analysis process can characterize our vulnerability and quantify areas that could be affected by a disaster. This process must also provide a broader view so we can see the social, political, economic, and environmental costs to our communities. This broader view of risk allows us to include many different interest groups in making decisions about reducing our vulnerability.
- *Land-use management:* Local decision makers can use outputs of a hazards analysis in land-use plans. Limiting development may contribute to the social, ecological, and economic sustainability of our communities.
- *Engineering codes and standards:* Local government adoption of codes and enforcement process are critical in reducing our vulnerability to disasters. A comprehensive hazards analysis provides critical information to ensure that code enforcement goals are attained.
- *Prediction, forecast, warning, and planning:* A detailed hazards analysis provides a sound basis for ensuring that local communities can offer citizens adequate disaster warning. Procedures for delivering timely warning for disasters can be based on alternative planning scenarios from a hazards analysis.
- *Disaster response and preparedness:* Emergency preparedness plans are prepared on a comprehensive hazards analysis. Policies and operational procedures are driven by the nature of the hazards faced by organizations and communities.
- *Recovery and reconstruction:* Planning for recovery should not begin following a disaster. To be effective, it should be part of a community hazard mitigation plan and include priorities for a community's long-term recovery in the event of a disaster.
- *Insurance:* Insurance is not a prevention strategy, but it can be included as part of a recovery process. The question is how can we use insurance as a means of ensuring that an entity's financial stability is protected.

- *Economic sustainability:* Public, private, and nonprofit organizations must understand the nature of risks facing them and develop strategies to reduce or eliminate losses. A hazards analysis is critical to this decision-making process.

Deyle et al. (1998) stress the application of a community hazards analysis in local or regional decision making and land-use planning. Deyle agrees with the suggestions of Meleti that analysis without action does not address the critical decision that must include a comprehensive understanding of hazards, vulnerability, and risk. Hazards analysis is part of a comprehensive emergency management and risk management process. Hazards analysis is not the goal but is a means toward a goal of promoting social, economic, and environmental sustainability. This emphasis is on hazards risk management and mitigation so as to foster community and environmental sustainability and resiliency.

Discussion Questions

The science associated with hazards is complex and often debated at a local, regional, and national level. Why do we worry when the scientist or policy makers content that risks are minimal or that there are none present?

Why are community hazard analyses necessary? Does the community have a right to know about local hazards?

What is the role of hazards analysis in organizational decision making and public policy at a local, regional, national, or international level? How are disasters and development related? What is the role of a hazards analysis in preventing people from moving into harm's way?

What influences our understanding of risks from natural- and human-caused hazards when we include a discussion of vulnerability and exposure to hazards?

Hazards research shows that there is the potential for significant losses from disasters. How do demographic, economic, political, and environmental systems contribute to vulnerability? How could a better understanding of hazards and their impacts help us to reduce the adverse consequences associated with disasters?

What type of demographic, economic, political, and environmental changes could make a community more resilient or less vulnerable to disasters?

Do you agree that citizens have a right to know about hazards in their community? Under what conditions might a community restrict information about hazards?

How are disasters and community change related? What role does community resilience and adaptation have in the hazards analysis process?

What impact could a disaster have on sensitive natural areas and endangered species?

Which social groups are likely not to receive or not to understand or not to take the warning message seriously? Why?

Are there characteristics of social groups that may make it more difficult for them to be rescued, to receive adequate emergency medical care, to feel comfortable in an emergency shelter? Are there population groups that are likely to suffer to a great extent economically or emotionally in a response as well as in a recovery?

Applications

Take a look at the hazards analysis process outlined in Figures 1.1 and 1.2. Identify examples of how population characteristics, the local economy, the infrastructure in the community, and the natural environment influence the community's vulnerability to natural hazards. What hazards appear at this first look, to be the primary threat for your community and should be addressed through a comprehensive hazards risk management and mitigation strategies?

Websites

Integrated Ecosystem Restoration and Hurricane Protection: Comprehensive Master Plan for a Sustainable Coast (2007). Coastal Protection and Restoration Authority (CPRA) of Louisiana. http://www.lacpra.org/index.cfm?md=pagebuilder&tmp=home&nid=24&pnid=0&pid=28&fmid=0&catid=0&elid=0. The plan includes a comprehensive assessment of the hazards in a coastal environment, impact assessments (social, economic, and ecological), public participating, and recommendations on strategies to protect the social, economic, and environmental assets of the state. This is an outstanding example of a region-wide hazards analysis that included federal, state, and local government agency collaboration.

FEMA (1997) *Multi-Hazard Identification and Risk Assessment: The Cornerstone of the National Mitigation Strategy*. Washington, DC: FEMA. http://www.fema.gov/library/viewRecord.do?id=2214

FEMA: Risk Analysis—Helping Communities Know Their Natural Hazard Risk. http://www.fema.gov/risk-analysis-helping-communities-know-their-natural-hazard-risk

Hazards Analysis of the City of New Orleans completed by the U. S. Army Corps of Engineers (2007). http://nolarisk.usace.army.mil

Technical Guidance for Hazards Analysis Emergency Planning for Extremely Hazardous Substances 12-14-2006. http://www.epa.gov/osweroe1/docs/chem/tech.pdf

Hazardous Materials Emergency Planning Guide (Updated 2001) 02-07-2002 National Response Team-1. The planning, preparedness, and response actions related to oil discharges and hazardous substance releases. http://www.epa.gov/osweroe1/docs/chem/cleanNRT10_12_distiller_complete.pdf

USGS Natural Hazards. http://www.usgs.gov/natural_hazards/

USGS-National Seismic Hazard Mapping Project. http://earthquake.usgs.gov/hazards/

NOAA Risk and Vulnerability Assessment Tool (RVAT). Hazards analysis. An easy-to-use, adaptable, multistep process that includes the hazard identification. http://www.cakex.org/tools/noaa-risk-and-vulnerability-assessment-tool-rvat

References

Abramovitz, J. (2001). *Unnatural Disasters.* Worldwatch Paper 158. Washington, DC: World Watch Institute.

Adger, W. N. (2006). Vulnerability. *Global environmental change, 16*(3), 268–281.

Alexander, D. (2000). *Confronting Catastrophe.* New York: Oxford University Press.

Anders, F., Kimball, S., and Dolan, R. (1989). *Coastal Hazards: National Atlas of the United States.* Reston, VA: U. S. Geological Survey.

Bruneau, M., Chang, S. E., Eguchi, R. T., Lee, G. C., O'Rourke, T. D., Reinhorn, A. M., Shinozuka, M., Tierney, K., Wallace, W. A., and von Winterfeldt, D. (2003). A framework to quantitatively assess and enhance the seismic resilience of communities. *Earthquake Spectra, 19*(4), 733–752.

Bryant, E. A. (2005). *Natural Hazards.* Cambridge, UK: Cambridge University Press.

Bull-Kamanga, L., Diagne, K., Lavell, A., Leon, E., Lerise, F., MacGregor, H., Maskrey, A., et al. (2003). From everyday hazards to disasters: the accumulation of risk in urban areas. *Environment and Urbanization, 15*(1), 193–204.

Burby, R. J. (ed.) (1998). *Cooperating with Nature: Confronting Natural Hazards with Land-Use Planning for Sustainable Communities.* Washington, DC: Joseph Henry Press.

Burton, I., Kates, R., and White, G. (1978). *The Environment as Hazard.* New York, NY: Oxford University Press.

Butzer, K. W. (2012). Collapse, environment, and society. *Proceedings of the National Academy of Sciences, 109*(10), 3632–3639.

Center, H. (2000). *The Hidden Costs of Coastal Hazards: Implications For Risk Assessment and Mitigation.* Washington, DC: Island Press.

Coastal Protection and Restoration Authority of Louisiana (2007). *Integrated Ecosystem Restoration and Hurricane Protection: Louisiana's Comprehensive Master Plan for a Sustainable Coast.* Baton Rouge, LA: Coastal Protection and Restoration Authority of Louisiana. http://www.lacpra.org/index.cfm?md=pagebuilder&tmp=home&nid=24&pnid=0&pid=28&fmid=0&catid=0&elid=0. Accessed September 20, 2013.

Coburn, A. W., Spence, R. J. S., and Pomonis, A. (1994). *Vulnerability and Risk Assessment.* United Nations Development Program. New York, NY: United Nations.

Cook, B. I., Miller, R. L., and Seager, R. (2009). Amplification of the North American "Dust Bowl" drought through human-induced land degradation. *Proceedings of the National Academy of Sciences, 106*(13), 4997–5001.

Crozier, M. J. and Glade, T. (2006). Landslide hazard and risk: issues, concepts and approach. In Glade, T. (ed.), *Landslide Hazard and Risk.* West Sussex: Wiley, pp. 1–40.

Cutter, S. L. (1993). *Living with Risk.* London: Edward Arnold.

Cutter, S. L. (1996).Vulnerability to environmental hazards. *Progress in Human Geography, 20*(4), 529–539.

Cutter, S. L. (ed.) (2001). *American Hazardscapes: The Regionalization of Hazards and Disasters.* Washington, DC: Joseph Henry Press.

Cutter, S. L. (2003). The vulnerability of science and the science of vulnerability. *Annals of the Association of American Geographers, 93*(1), 1–12.

Cutter, S. L., Boruff, B. J., and Shirley, W. L. (2003). Social vulnerability to environmental hazards. *Social Science Quarterly*, *84*(2), 242–261.

Cutter, S. L., Burton, C. G., and Emrich, C. T. (2010). Disaster resilience indicators for benchmarking baseline conditions. *Journal of Homeland Security and Emergency Management*, *7*(1), 1–22.

Cutter, S. L., Mitchell, J. T., and Scott, M. S. (1997). *Handbook for Conducting a GIS-based Hazards Assessment at a County Level.* Columbia, SC: University of South Carolina.

Daniels, R. C., Gornitz, V. M., Mehta, A. J., Lee, S. -C, and Cushman, R. M. (1992). *Adapting to sea-level rise in the U.S. Southeast: The influence of built infrastructure and biophysical factors on the inundation of coastal areas.* Oak Ridge National Laboratory.

Deyle, R. E., French, S. Olshansky, R., and Paterson, R. (1998). Hazard Assessment: The factual basis for planning and mitigation. In R. Burgy (ed.), *Cooperating with Nature: Confronting Natural Hazards with Land-use Planning for Sustainable Communities.* Washington, DC: Joseph Henry Press.

Emergency Management Australia (2000). *Emergency Risk Management: Application's Guide.* Australian Emergency Manual Series, pp. 10–12.

Engle, N. L. (2011). Adaptive capacity and its assessment. *Global Environmental Change*, *21*(2), 647–656.

FEMA (1997). *Multi-Hazard Identification and Risk Assessment: The Cornerstone of the National Mitigation Strategy.* Washington, DC: FEMA. http://www.fema.gov/library/viewRecord.do?id=2214. Accessed September 20, 2013.

Gornitz, V. M. and White, T. W. (1992). *A Coastal Hazards Database for the U. S. East Coast.* Publication No. 3913 and 4101. Oak Ridge, TN: Environmental Sciences Division, Oak Ridge National Laboratory.

Gornitz, V. M., White, T. W., Daniels, R. C., and Birdwell, K. R. (1994). The development of a Coastal Risk Assessment Database: vulnerability to sea level rise in the U. S. Southeast. *Journal of Coastal Research*, Special Issue No. 12: Coastal Hazards: Perception, Susceptibility and Mitigation, pp. 327–338.

Greenwood, D. and Levin, M. (1998). *Introduction to Action Research: Social Research for Social Change.* Thousand Oaks, CA: Sage.

Guba, E. G. and Lincoln, Y. S. (1989). *Fourth Generation Evaluation.* Thousand Oaks, CA: Sage.

Gruntfest, E., Downing, T., and White, G. F. (1978). Big Thompson flood exposes need for better flood reaction system. *Civil Engineering, 78*, 72–73.

Hewitt, K. (1983). The idea of calamity in a technocratic age. In Hewitt, K. (ed.), *Interpretations of Calamity from the Viewpoint of Human Ecology.* Boston: Allen and Unwin.

Hewitt, K. (1997). Technological hazards. In *Regions of Risk: A Geographical Introduction to Disasters.* Essex, UK: Longman.

Hewitt, K. (1998). Excluded perspectives in the social construction of disaster. In E. L. Quarantelli (ed.), *What Is a Disaster? Perspectives on the Question.* New York: Routledge.

Jardine, C. G., and Hrudley, S. (1997). Mixed messages in risk communications. *Risk Analysis*, Vol. 17, No. 4. Pages 489–498.

Kalaugher, E., Bornman, J. F., Clark, A., and Beukes, P. (2013). An integrated biophysical and socio-economic framework for analysis of climate change adaptation strategies: the case of a New Zealand dairy farming system. *Environmental Modelling and Software*, *39*, 176–187.

Kaplan, S. (1997). The words of risk analysis. *Risk Analysis, 17*(4), 407–417.

Kates, R. W., Travis, W. R., and Wilbanks, T. J. (2012). Transformational adaptation when incremental adaptations to climate change are insufficient. *Proceedings of the National Academy of Sciences of the United States of America, 109*(19), 7156–7161.

Kedar, B. Z. (1970). *Again: Arabic Risq, Medieval Latin Risicum, Studi Medievali.* Spoleto, Italy: Centro Italiano Di Studi Sull Alto Medioevo.

Lindell, M. K. and Meier, M. J. (1994). Planning effectiveness: effectiveness of community planning for toxic chemical emergencies. *Journal of the American Planning Association, 60*(2), 222–234.

Lindell, M. K., and Meier, M. J. (1994). Planning effectiveness: Effectiveness of community planning for toxic chemical emergencies. *Journal of the American Planning Association, 60*(2), 222–234.

Lindell, M. K., and Prater, C. S. (2003). Assessing community impacts of natural disasters. *Natural Hazards Review, 4*(4), 176-185.

Long, M. H. and John, J. I. (1993). *Risk-Based Emergency Response.* Paper presented at the ER93 Conference on the Practical Approach to Hazards Substances Accidents. New Brunswick, Canada: St. John.

Mendes-Victor, L. A. and Gonçalves, C. D. (2012). Risks: Vulnerability, Resilience and Adaptation, 2012 World Conference on Disaster Reduction, United Nations Disaster Risk Reduction.

Mileti, D. (1999). *Disasters by Design: A Reassessment of Natural Hazards in the United States.* Washington, DC: National Academies Press.

Miles, S. B. and Chang, S. E. (2006). Modeling community recovery from earthquakes. *Earthquake Spectra*, Vol. 22 (2).

National Research Council. (1983). *Risk Assessment in the Federal Government: Managing the Process.* Washington, DC: National Academy Press.

National Response Team (1987). Hazardous materials emergency planning guide (NRT-1). Washington, DC: National Response Team.

Pearce, L. D. R. (2000). An Integrated Approach For Community Hazard, Impact, Risk and Vulnerability Analysis: HIRV. Doctoral Dissertation, The University of British Columbia

Pelling, M. (2011). *Adaptation to Climate Change: From Resilience to Transformation.* Abingdon, UK: Routledge.

Preuss, J. and Hebenstreit, G. T. (1991). *Integrated Hazard Assessment for a Coastal Community: Grays Harbor.* Reston, VA: US Department of the Interior, Geological Survey.

Samarasinghe, S. and Strickert, G. (2012). Mixed-method integration and advances in fuzzy cognitive maps for computational policy simulations for natural hazard mitigation. *Environmental Modelling and Software, 39*, 188–200.

Schmidt, J., Matcham, I., Reese, S., King, A., Bell, R., Henderson, R., and Heron, D. (2011). Quantitative multi-risk analysis for natural hazards: a framework for multi-risk modeling. *Natural Hazards, 58*(3), 1169–1192.

Shoubridge, J. (2012). *Are we planning a disaster resilient region? An evaluation of official community plans in metro Vancouver.* Master's Thesis, School of Community and Regional Planning, the University of British Columbia, p. 2.

Smit, B. and Wandel, J. (2006). Adaptation, adaptive capacity and vulnerability. *Global Environmental Change, 16*(3), 282–292.

Steinberg, T. (2000). *Acts of God: The Unnatural History of Natural Disasters in America.* New York: Oxford University Press.

Tierney, K. and Bruneau, M. (2007). Conceptualizing and measuring resilience: a key to disaster loss reduction. *TR News, 250,* 14–17.

Tobin, G. and Montz, B. (1997). *Natural Hazards*. New York: Guilford. L., Turner, B. L. II, Kasperson, R. E., Matson, P. A., McCarthy, J. J., Corell, R. W., Christensen, Eckley, N., Kasperson, J. X., Luers, A., Martello, M. L., Polsky, C., Pulsipher, A., and Schiller, A. (2003). A framework for vulnerability analysis in sustainability science. *Proceeding National Academy of Sciences, 100*(4), 8074–8079.

United Nations, Department of Humanitarian Affairs (1992). *Internationally Agreed Glossary of Basic Terms Related to Disaster Management.* (DNA/93/36). Geneva, Switzerland: United Nations.

United States Environmental Protection Agency (1986). *Emergency Planning Community Right to Know Act.* EPCRA, 42 U. S. C. Ann. Section 11001 et seq. Washington, DC: United States Environmental Protection Agency.

United States Environmental Protection Agency (2001). *Hazardous Materials Emergency Planning Guide.* National Response Team-1. Washington, DC: United States Environmental Protection Agency.

Weichselgartner, J. (2001). Disaster mitigation: the concept of vulnerability revisited. *Disaster Prevention and Management, 10*(2), 85–94.

Chapter 2

Hazards Identification

John C. Pine

Objectives

The study of this chapter will enable you to:

1. Define the hazards identification process.
2. Clarify who might be involved in the hazards identification process.
3. Explain the nature of a community profile and how it fits within the hazards analysis process.
4. Explain what mapping community assets and hazards contributes to hazards identification.
5. Explain the problem-solving process and its role in hazards analysis.

Key Terms

100-Year flood
Base flood
Base map
Critical facilities
Flood Fringe
Flood Insurance Rate Map (FIRM)
Flood Insurance Study
Floodplain
Floodway
Hazards profile
Hazards profile

Hydrographs
Primary and secondary effects
Regulatory floodway
Secondary hazards
Sociocultural, economic, and natural capital
Urban and nonurban communities

Issue

How can public, private, and nonprofit organizations characterize hazards affecting their organization or community and use that information to reduce the adverse impacts of disasters?

Hazard risk zones may be identified in local communities and may reveal threats and vulnerabilities that might not have been obvious nor understood. How do you profile hazards and risks without causing alarm? What information should be shared and in what form?

Introduction

Emergency preparedness, response, recovery, and mitigation are based on a careful and comprehensive hazards analysis. Tierney et al. (2001) notes that emergency preparedness and operations plans must be based on a careful examination and understanding of the types of hazards that the community or organization faces. An assessment of the sociocultural, economic, and natural capital in a community or an organization is the initial step of this effort. Community assessment and hazards/risk identification is thus a critical part of the emergency management process and forms the basis for risk analysis and the determination of community adjustments to reduce vulnerabilities (United States Environmental Protection Agency 2001). These adjustments are accomplished through intentional problem solving throughout the hazards analysis process. Critical thinking, analysis, and problem solving are critical skills that are used to see issues more clearly and look beyond just the symptoms of a problem and acknowledge that actions must address problems that make our communities more vulnerable to disasters. Deyle et al. (1998) stress that community decision makers must understand how hazards could affect their community. They suggest that a hazards analysis provides a basis for strategic risk management strategies and hazards mitigation decisions. They point out that hazards identification is the foundation for many decisions associated with emergency management, community planning, and sustainability. Deyle et al. (1998) stresses that risk assessment is an essential foundation for emergency management and which a hazards analysis is based.

Hazards identification is the initial step in the hazards analysis process, includes a description of the community, and provides an analysis through environmental modeling of the nature of the threat. This process stresses the importance of examining both short-term and long-term threats, community engagement, a broad-based multidimensional planning team, and an orientation to action. A key element in the hazards identification process is gathering data about hazards and the inclusion of public, private, and nonprofit partners in the community. Olwig (2012) contends that local understandings of risks cannot be disentangled from global understandings and practices. Global organizations must draw on local agency to influence citizen perceptions of their risks.

This chapter addresses the first part of hazards identification by the nature of the community and sociocultural, economic, and environmental assets. Chapter 3 will build on this by examining environmental modeling and how we can use modeling tools to characterize the nature of hazards. Characterizing the community and hazards through environmental modeling allows us to identify hazards in the community and what problems may be present that can be addressed to reduce vulnerability (Figure 2.1). A critical assumption within this initial step of

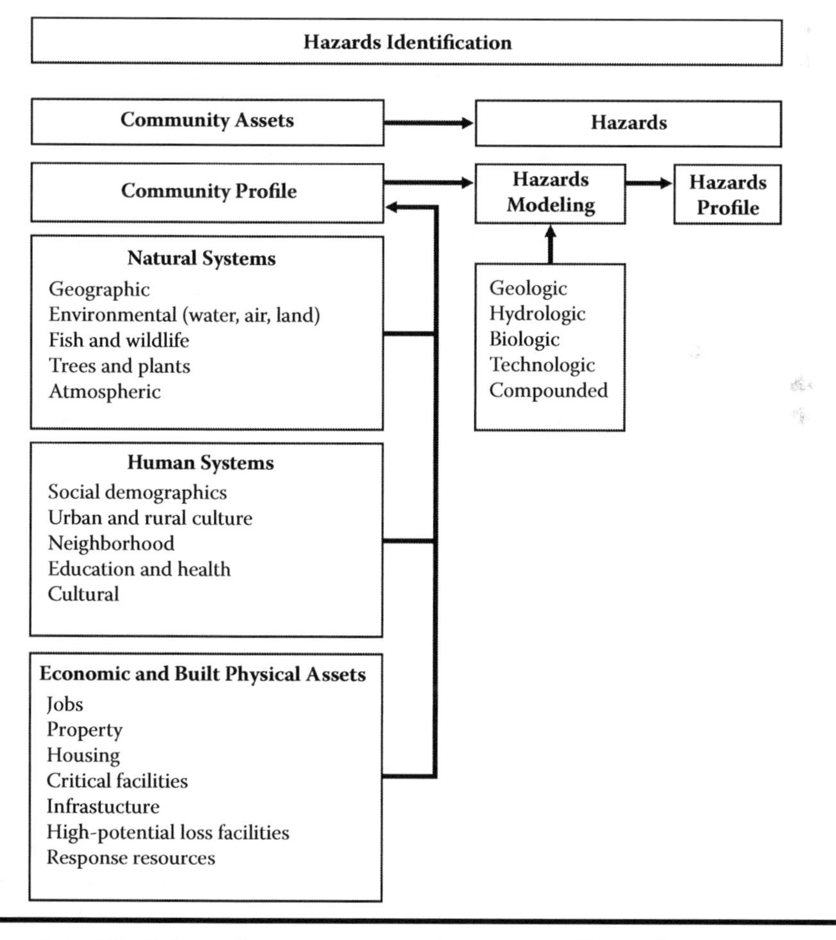

Figure 2.1 The linkages between community assets and hazards.

hazards identification is that effective plans and adaptation strategies must address the unique nature of risks in a community. Response, mitigation, and adaptation strategies should be based on the nature of the hazards event; fires, floods, earthquakes, or windstorms present very different challenges to responders. Further, the nature of the community including its demographics, geography, infrastructure, and critical resources all impact how a community responds and its capacity to recover from a disaster.

Critical Thinking: Hazards identification is an ongoing process that enables communities to establish a culture of hazards risk management, adaptation, and mitigation. It acknowledges that hazards and risks may reveal more fundamental weaknesses in local systems that must be addressed to truly ensure community resiliency and sustainability.

Question: What types of information are critical in understanding community assets and resources?

Hazards Identification Process

The goal of hazards identification is to provide a process that will reveal the most serious hazards or risks in a community or that threaten an organization. This process thus results in a clear description of threats to the human, built, or natural environments. Comprehensive historic data are critical in providing a broad-based description of a community and where it could be vulnerable. The process thus includes an examination of the frequency and impacts of past disasters and the potential for future hazards within the community. We stress that even insignificant hazards could trigger much larger secondary hazards.

It should be noted that hazards are part of our lives, and all communities face risks of many types. Smith (2004) notes that disasters have natural, economic, and human impacts including loss of life and the destruction of homes, businesses, and critical infrastructures as well as damage to sensitive natural areas such as wetlands or water bodies. Disasters result from the interface between natural, the built, and human systems. Clarifying the nature and extent of this interface allows us to determine where the interactions are constructive and when loss or damage might result. Understanding these interactions forms the basis for organizational policy decision making and hazards risk management.

Smith (2004) notes that natural disasters result from the interface between the natural geophysical systems and human systems, both constructed and personal. Figure 2.2 suggests that our community and organizational capital is interrelated. The educational attainment and skills of citizens influence the nature of business enterprises; local economic characteristics can attract and retain skilled workers. Environmental and natural resources can be a great attraction to visitors as well as factors that impact how residents view local natural resources.

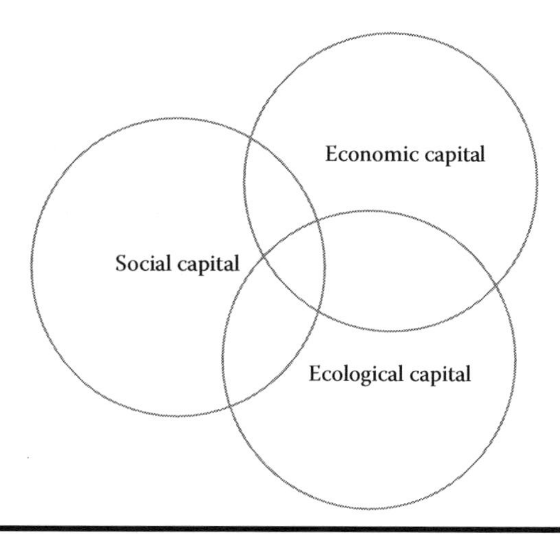

Figure 2.2 Interaction between social, economic, and ecological capital.

Each of these capitals is influenced by public, private, and nonprofit goals and policies. Our capacity to minimize adverse effects of disasters depends on our adaptation and mitigation effort to adjust to the effects of natural events and human actions. Nonstructural mitigation actions include establishment or strengthening building codes, or land-use regulation; structural mitigation initiatives could include the design and construction of our critical infrastructure or protective structures. Our resiliency or capacity to withstand or of recovery from a disaster is influenced by human adaptive actions. Unfortunately, individual and organizational actions can have unintended consequences, which impact social, economic, or environmental capital.

We are not helpless but can take steps to protect our social, economic, and environmental systems. Burton (1993) stresses that the environment is neither benign nor hostile. It is in fact neutral; human action within a hazards zone may either increase or reduce vulnerability. Hazards and risk identification provides a means of examining our natural, human, economic, and constructed systems and serve as a base for actions that may be taken to reduce our vulnerability and enhance our resilience to hazards. The key is that individuals, communities, and private and nonprofit organizations have the best information to deal with hazards. The United Nations (2002) stresses the importance of informed decision making.

Hazards identification, thus, clarifies natural- and human-caused events that threaten a community. This process results in information that reveals a community's capacity to deal with a disaster. It provides an opportunity for us to identify the physical characteristics of buildings, the social characteristics of our community, and local response capabilities. Hazards identification may be directly used in preparedness activities by clarifying hazards zones for response, but Deyle et al. (1998) notes that it can be used for establishing setbacks and zoning classifications.

The comprehensive identification of hazards can support hazards risk management policies and programs and determine benefits of alternative policies to reduce our vulnerability. Hazards analysis thus becomes more of a management tool rather than just one for preparedness.

Organizing a Hazards Identification Team

One does not begin the process of identifying hazards alone; it is done with others from the community or an organization. A broad-based representative group, such as a local emergency responders, health care, public services, and community planners, provides a sound basis for characterizing the community and examining what hazards might affect the community. Team members who have diverse technical skills bring different perspectives on what could happen to the community. Representatives from local agencies such as engineering, building sciences, public safety, emergency medicine, health care, education, community planning, and transportation may have different views on the impacts of disasters in the community.

It would be helpful to form smaller subgroups that can concentrate on one or more of the hazards that could affect the community. Specialized groups enhance our capacity to identify all hazards that could impact the community. For organizations that undertake the hazards analysis process, a diverse team is essential. School districts that desire to reduce vulnerability need teachers, administrators, building maintenance staff, transportation, food preparation, utilities, and security on their hazards analysis team. These teams should also be organized in a cooperative way that allows for information sharing between groups, thus reducing redundancy of effort. For large organizations with numerous sites such as a bank, restaurant, or local school district, there should be consistency among all teams regarding the reporting of findings, the collection of data, and the documentation of sources.

Creating a Community Profile

The goal of the hazards identification and characterization process is to identify and describe hazards that could impact the community or organization. It also provides a basis for future steps in the hazards analysis process. To do this, we need to first look at our community. Just what is at stake in our community from our people, our economy, the environment, and building and infrastructure? When creating a hazards profile, we categorize hazards into groups by type. However, hazards in one category may result in a secondary hazard included in another category. Heavy rains could cause flooding and lead to chemical spills or an avalanche. The sequence that a hazard evolves and the nature and extent of the resulting disaster will influence how we perceive the threat of the hazards to the community. The division of hazards into these categories helps to provide structure for identifying and characterizing risk to a community or organization and helps us to our understanding of the impacts of hazards.

Community Assets

The first step in creating a community or organizational profile is to examine its assets. We examine social, economic, and natural parts of the community and their spatial dimensions that could either increase risk or support effective emergency response, mitigation, preparedness, and recovery activities. There are three principal components of profiling community assets that has been adapted from Federal Emergency Management Agency (2001) guidance including environmental or natural assets, social environment, and the built and economic environments.

1. *Environmental or natural assets:* (Geography; land cover; topography; slope; aspect; lakes; streams, lakes, and waterways; wetlands; watersheds; soils; fault lines; and wild land/urban fire interface, forests, and coastal dunes) These assets may provide natural capital by giving a community beautiful landscapes, resources for recreation, natural flood protection areas, agriculture or timber assets, or natural transportation routes for commerce.
2. *Social and cultural capital:* This includes the capacity of members of the community to anticipate, cope with, resist, and recovery from the impacts of disasters. It includes both social, cultural, and heritage capital. The level of education and their availability to join the workforce; population distribution; population concentrations; degree of poverty; the nature of volunteerism and community engagement; crime rates; educational attainment; availability of health care and critical heathcare rates; and vulnerable population groups—children, non-English speaking, elderly, and disabled.
3. *Built and economic assets:* The nature of our economic capital includes the built and physical environments, as well as building codes and zoning regulations that influence the location and physical assets of a structure. The location and capacity of response resources can impact the vulnerability or residents as well as businesses and could limit a community's capacity to deal with a disaster. The nature and distribution of utilities, transportation, or energy resources will also affect response and recovery efforts. Locations of major employers and financial centers are part of this capital. Our economic capital includes:
    a. Property (land use, types of construction, location of homes, location of businesses, location of manufactured homes, zoning map, and building codes)
    b. Critical facilities (fire stations, ambulance locations, police stations/law enforcement, hospitals, schools, senior centers, day care centers, city hall and other public facilities, prisons and jail facilities, and historic and cultural buildings and areas)

c. Infrastructure (utility lines; water lines; sewer lines; gas lines; pipelines; roads and highways; railroads; airports; waterways; port facilities; bridge locations; communication facilities; transit routes; major energy sources; water purification and treatment plants; landfills; dikes and flood protection structures and facilities)
d. High-potential loss facilities (nuclear power plants; dams; military installations; industrial sites that manufacture and/or store hazardous materials; and gasoline storage and distribution facilities)
e. Response (ambulance districts; fire districts; emergency first-responder districts; early warning systems; emergency operations centers; emergency equipment—fire trucks, ambulances, response vehicles, etc.; hazmat equipment; and evacuation shelters)
f. Commercial and industrial development from financial enterprises to a community's diverse service sector

Profiling community assets is an initial step in understanding hazards in a community for in this process, we establish a baseline for characterizing human, natural, and economic capital. This baseline allows us to see the strengths and weaknesses of our local community. Where we have strong and robust social, economic, and natural assets, we are more likely to respond and recover quickly from a disaster. The capacity of the community to sustain the impact of a disaster with limited damage and recover quickly involves the concept of resilience.

Environmental or Natural Assets and Risks

A community's natural environment plays a critical role in defining hazards that could impact a community's hazards mitigation, response, and recovery strategies. Being located adjacent to a river or in a coastal area could not only increase vulnerability from a hurricane or severe storm but also provide extensive recreational or commercial resources. Communities located in hilly areas could be vulnerable to landslides. Communities located in wooded areas could be vulnerable to wildland fires.

The health and vitality of the community's natural environment is critical to measuring the community's resilience. Clean and plentiful water resources are critical to community sustainability; unfortunately, many communities face declining water resources. A healthy and productive natural environment and ecosystem provides protective buffer for communities. Healthy and productive wetlands provide invaluable flood protection to coastal communities by reducing storm surge and offering broad drainage areas in times of flooding. Healthy forests are less vulnerable to catastrophic wildfires and reduce landslide dangers on slopes. Dunes on coastlines provide buffer from storm surge and winter cold fronts. A damaged, stressed, or unhealthy natural environment can reduce a community's protection from wildfires, floods, storm surge, and in some cases increase the damaging impacts of hazards on the community. A community affected by flooding could in turn have indirect secondary impacts throughout the community. A flood could destroy wetlands by covering them with sediment or erosion, turning productive swamps, marshes, and channels into sterile water feature. Environmental capital may not be obvious on an initial look and thus we need not only to examine the primary effects of hazards but also to determine if there are any secondary or indirect impacts.

Communities that allow productive wetlands to be used for development reduce the natural flood protection and increase the risk of flooding. A forest that is not properly maintained or where human development is unchecked can result in wildfires that destroy valuable forestland. Clear-cutting forests on the sides of hills and mountains increase the chances of landslides. Destruction of coastal dunes reduces vital natural protection to coastal communities from storm surge.

Understanding the direct link between a sustainable and resilient natural environment and a community's vulnerability to hazards is critical to developing effective strategies to preserve and enhance economic, social, or environmental capital. Conducting an inventory of the features of a community's natural environment is an important step in assessing a community's assets. Determine how the community is broken up in terms of geography and how the geographic landscape could influence natural- and human-caused disasters.

The Intergovernmental Panel on Climate Change (IPCC) notes that greenhouse gas emissions, at or above current rates, will very likely result in further warming (2007). This may induce many changes in the global climate system during the twenty-first century and be larger than those observed during the twentieth century. This report suggests that human actions are having unintentional consequences and action must be taken now to avert a disaster on a global scale. Our failure to limit excess carbon dioxide emissions especially from the use of fossil fuels and methane and nitrous oxide will have adverse consequences in the form of increases in global average air and ocean temperatures, widespread melting of snow and ice, and rising global average sea level.

Global climate change has had a significant impact on the frequency and severity of natural disaster events such as hurricanes, floods, tropical cyclones, drought and extreme heat, and fires in recent years (Abramovitz 2001; IPCC 2007). Unprecedented rainfall in some areas of the world in recent years have resulted in larger than usual flooding events. Hurricane Mitch and Hurricane Andrew both attained Category 5 status and resulted in extensive loss of life and damage when they made landfall. Severe droughts have occurred or are ongoing in many areas of the world including the southwestern and southeastern United States. One result of global climate change may be an increase in frequency and severity of natural disaster events around the world. The IPCC notes that extreme events are very likely to change in the magnitude and frequency as a result of global warming. Extended warmer periods are likely to cause increases in water use and evaporative losses and duration of droughts (IPCC 2007). Coastal communities worldwide will be the first to feel the effects of global warming. Unfortunately, low-lying coast already suffering elevated vulnerabilities due to subsidence and erosion will be increasingly threatened by rising sea level and possibly more frequent hurricanes.

Sources of Hazards Data

In compiling information on local environmental risks, local hazards data will characterize which risks pose the greatest potential for loss in a disaster. Chapter 5, Risk Analysis identifies data sources that provide a basis for determining risk. State climate offices and regional National Oceanic and Atmospheric Administration climate centers provide comprehensive historical data on atmospheric events. Hazard models may be used to estimate areas that could be impacted by wind, flood, fire, or hazardous materials risks. Researchers have also acknowledged the value of tree-ring analysis to offer a broader understanding of local risks. Knowledge of past events is important for the assessment of natural hazards. Tree rings have proved to be a reliable tool for describing the past events including flooding, debris flow, landslides, snow avalanche, rockfalls, and extreme droughts (Ballesteros et al. 2011; Lopez Saez et al. 2012; Stoffel and Bollschweiler 2008).

Social Assets

The social environment in a community is represented by its demographic profile. This profile will help emergency planners to determine which segments of the community's population could be impacted by a hazard event. It is essential that local decision makers know which segments of the community will be affected by hazards. What is the community's vulnerability by age, education, access to transportation, ethnicity, degree of home ownership or renters, special needs residents, or single-parent head of household? Does the community have experience in dealing with hazards? Areas of the community that are far from major transportation routes, commercial districts, or less-damaged areas have had limited rebuilding and recovery following Katrina (Pine and Wilson 2007).

Neighborhoods that have high percentages of elderly population and dependent on public assistance may have fewer financial resources to bring to the recovery process. Renters may have less financial investment in a neighborhood but have great attachment to the area and be less willing to relocate when disaster strikes. The centennial census provides information on how long residents have lived in an area and if they are homeowners or renters. Additional information on local residents may be collected from churches, neighborhood or social organizations, clubs, fraternal groups, relief organizations, public workshops, and meetings with local officials. Questions may be addressed such as how many people are in a neighborhood or sector. Which subdivisions are located in high-risk parts of a community? Where do people work, recreate, or gather for civic events? Are there areas for community gatherings and is there a strong bond among local residents? Is there a high percentage of voter turnout for elections? Is there a high degree of civic engagement?

Economic and Constructed Assets

The built environment is a part of the economic capital of a community. It includes residential, commercial, and industrial assets as well as critical sites such as schools and colleges, hospitals and nursing homes, emergency services, day care centers, criminal justice sites, museums and other cultural resources, utility right of ways or sites, and transportation routes and sites from airports, ports, rail yards, or bridges. The built environment also includes political assets such as local building codes and zoning regulations, as well as the extent of local flood control programs and structures.

Many economic activities could be affected by disasters. Major transportation routes can be disrupted limiting supplies needed for commercial enterprises or manufacturers. Shippers are unable to provide needed materials or carry end products thus limiting the capacity of the enterprise to successfully carry out their mission. For many businesses, nonprofits and government agencies, critical fuel supplies, food, and other organizational inputs may be blocked. For some communities, transportation systems (docks, railroads, or bridges) are located near the river to expedite commerce. Although they may have been constructed to a level above anticipated flood levels, flood characteristics change over time as a result of changes in the landscape and land use. New hydrologic modeling might show that higher flood peaks might occur sooner and more frequently as urbanization covers open space with impervious surfaces (roofs, streets, parking lots, etc.).

Infrastructure

The loss of critical infrastructure can cause major direct and secondary disaster impacts in a community, from the destruction of a bridge that disrupts movement of resources and citizens in a community to a major failure at a power station during winter months, where electricity is a life or death factor. Infrastructure would include transportation, communication, and utility networks. Awareness of those infrastructure components, and the mechanisms by which they affect the community both geographically and physically, is a major step toward successful management of hazards.

Critical Facilities

Critical response resources such as police stations, fire and rescue facilities, hospitals, shelters, schools, nursing homes, and other structures of the community are vital parts of the built environment. Information on critical facilities may be obtained from the administrating agencies, boards (drainage, levee, hospital), commissions (planning), departments (public works), or institutions (university departments) at the main office or the building manager, the state office of emergency preparedness, tax assessors, financial institutions, and state/federal agencies. If the community has a hazards mitigation plan, the critical facilities are usually identified in that plan. Taking it to the next step, you can analyze community vulnerability by collecting data on building construction type and quality, age, size, footprint, elevation, building capacity, presence of auxiliary power, and potential evacuation routes. We would also examine the exposure of these critical facilities including the number, types, qualities, and monetary values of various types of property or infrastructure (Schwab et al. 1998).

Economic Activities

Unemployment following a disaster can cause extensive disruption in a community. Following Hurricane Katrina, some colleges including the University of New Orleans and the Louisiana State University Health Sciences Center in the New Orleans area were able to avoid employee layoffs and to shift educational services to Web-based instructional formats and carry on their educational activities. For the medical school, they simply moved their classes outside the disaster area. Internet servers had been relocated to available sites and thus could support adjustments in educational operations. Many schools and colleges were not so fortunate and had no plans for long-term interruption of services. Organizations such as faith-based nonprofit day care centers, health centers, and other service providers were simply shut down resulting in increased unemployment. Today, many of these organizations have made contingency plans to ensure that services are continued following a future disaster.

Sources of information on economic activities in a community may be obtained from the chamber of commerce, the state and local departments of economic development, tax assessors, business associations, planners, university departments, and meetings with local officials. Community zoning information may also be used to characterize parts of a community and the level of activity in manufacturing, retail sales, and general commerce. Determine the location of residential housing and rental properties. Clarify how many housing units are in a specific area of the community, what types of housing units are present, and if the property is insured.

Mapping Community Assets and Hazards

Mapping a geographic representation of the community as a whole provides a framework for examining the nature and extent of a specific hazard and its potential impacts on the community or an organization. Deyle et al. (1998) stresses that a critical component of hazards identification is a hazard map. They indicate areas that are subject to a specific hazard of various intensities and probabilities. Hazards that threatens a community will affect it differently. For instance, although it can be expected that heavy snow will affect the whole community in a uniform manner, avalanches will only be a problem where there are steep slopes or gullies. Maps show the spatial extent of the community and provide a basis for us to see the interactions between hazards and people, structures, infrastructure, and the environment. Many hazard maps are readily available on the Internet, and in public agency records such as local emergency preparedness or hazards mitigation plans. Web maps reflecting flood hazards or wildfire risk may be obtained from national agencies. One should note the date that the map was prepared or how often it is updated.

Hazard maps are used in organizations to help clarify the nature and extent of hazards and form the basis for much organizational decision making associated with preparedness, mitigation, response, and recovery. Maps can show specific areas that require additional studies prior to investment, zoning decisions, or the acquisition of a parcel of land.

Displaying information concerning the nature and characteristics of the natural, social, and built environments provides an excellent means of determining how hazards might impact a community in both emergency response activities and how specific business, public agency and nonprofit sectors might be impacted. Communities that are split by major water features, rail lines and large sites such as colleges, manufacturing, or commercial enterprises could impair movement of critical resources in a disaster response.

A map is a graphic representation of a physical landscape. It may represent social, natural, or built features, such as transportation networks, political boundaries, or provide information on the number of residents by age, education, or employment by sector. Most importantly, maps provide a means of communicating information. All maps should include the following: a title, mapped area, the agency or person who prepared the map, when it was created, the sources of data used, legend showing symbols as displayed in the map, direction, a distance scale, and symbols. Maps provide a frame of reference for displaying a thematic overlay (hazard) on a geographic base map. A map may simply show where something is located or provide data as in the case of the number of students in a school or patients in a hospital.

Figures 2.3 and 2.4 provide examples of choropleth maps, which present information using distinctive color or shading to areas. The percentage of homes in a community or renters can be represented by grouping data by an area such as a city, county, or smaller representation such as a U. S. Census Bureau track or block group. Maps may simply show data such as population density, population by age groups or education, or even the number of injuries or homes damaged. Maps may give a representation as ratios, proportions, or averages that directly or indirectly involve areas.

To truly compare and analyze the risks of a community, it will be important to have risks represented individually on the base map and together on a single aggregate community risk map. Although it will not be impossible to create such a map without a standardized base map, it is much easier and the resulting product is likely to be more accurate if standardization is used.

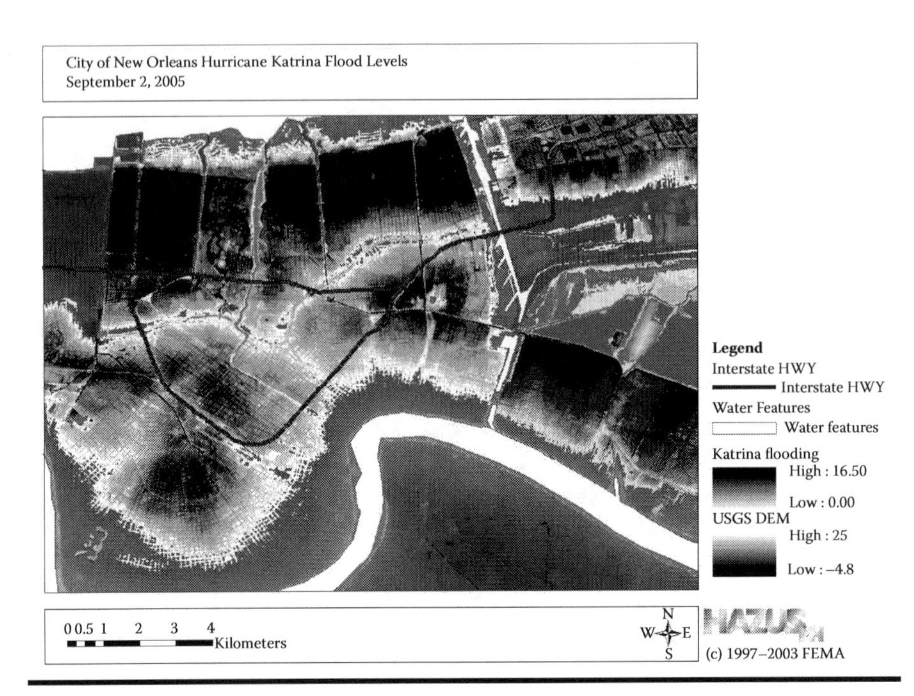

Figure 2.3 Elevation map of city of New Orleans.

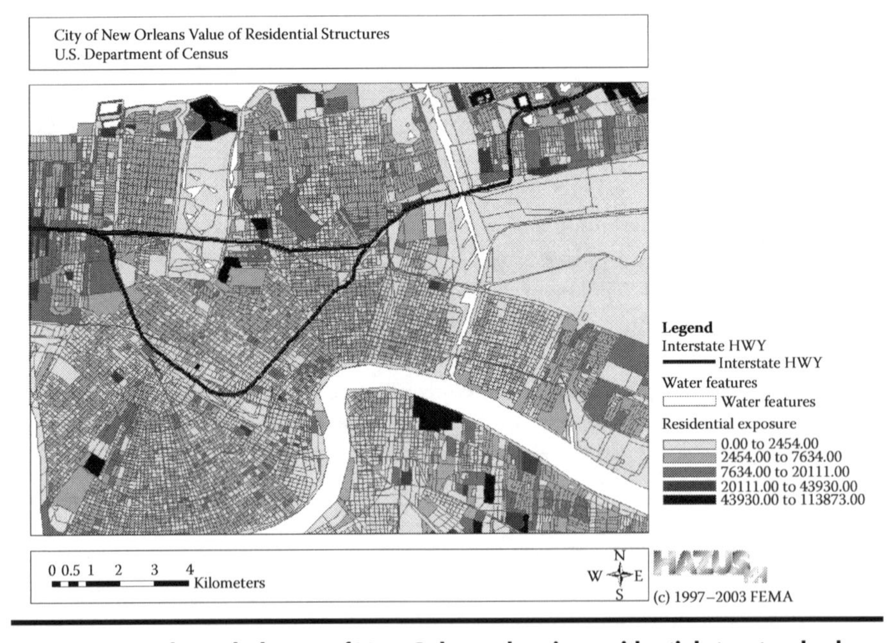

Figure 2.4 A choropleth map of New Orleans showing residential structural values.

Interdependence of Communities

An inventory of community social, economic, and environmental assets provides data related to assets, such as structures, demographics, response capabilities, and environmental conditions. It is also important that information be collected concerning future growth and development plans and how these plans may impact future conditions. It is also critical to recognize that actions taken in one community can impact the vulnerability of surrounding communities to specific hazards. For example, a community that dikes a river will change the natural environment not only in that community but also for communities located down river thereby changing the vulnerability to flood events for both communities.

Mutual aid agreements are an example of how communities and states in a region can join forces to ensure that any one community or state will have the needed response capabilities to effectively respond to a disaster event. Mutual aid agreements can include the provision of equipment and staff from one community or state to another in a time of crisis.

Regional growth and development planning is one effective method for ensuring that all potential impacts on neighboring communities can be measured and considered. Understanding that actions taken in one community ripple out to surrounding communities is the first step in preparing comprehensive hazards risk management strategies for a region.

Identifying Community Problems

There is a critical need for information on how disasters are impacting communities. Disaster costs have been rising and the financial burdens on public, private, and nonprofit organizations as well as for individual citizens has grown to an extent that it threatens the sustainability of many communities (Mileti 1997). Although the number of disasters has not increased the costs associated with them in terms of property damage, injuries and fatalities and human suffering have risen dramatically (Abramovitz 2001). The recent costs associated with response and recovery to Hurricane Katrina were estimated in 2006 at $22 billion for direct property losses, economic losses at $8 billion, and emergency assistance estimated at $20 billion making this the most costly disaster in the United States, and local communities from New Orleans to Biloxi will take years to recover (Kates et al. 2006). We have seen that the recovery will take many years but the devastating social, economic, and environmental impacts motivate us to reexamine how decision making on an individual, family, community, and state level contributes to this continuing rise associated with direct disaster losses. We need to better understand how disasters will impact our communities and more effective problem solving and decision making can reduce losses. The key is quality information on potential damage impacts, and our past practices are not providing the basis for a systematic examination of problems in our communities.

Hazards analysis along with sound risk management and hazards mitigation strategies provide the initial step in reducing losses. Efforts to reduce costs, injuries, fatalities, and indirect damages unfortunately take time to implement and for us to determine if they are achieving the results that we intended. This is why it is so critical that a hazards analysis, plus risk management and hazards mitigation strategies be viewed within the context of comprehensive emergency management. Quality planning and mitigation initiatives are the basis for effective emergency response and recovery; they are interconnected and interdependent. Monitoring and evaluating our efforts on an ongoing basis provides a context to determine if our risk management and hazards mitigation initiatives are having desired results. The fact is that disasters harm people, and it is man who fails to acknowledge the power of nature and take the necessary steps to heed warnings, seek safety, build to withstand storms, and take precautionary actions to avoid losses.

For too long, we have focused on understanding hazards and failed to adequately determine how to minimize our exposure to disasters. The gap between hazards and exposure of many types results in problems associated with human, economic, and ecological vulnerabilities. We need to focus on quality decision making and the adoption of widespread risk management and mitigation initiatives so as to reduce the destructive impacts of disasters. This emphasis on risk assessment and management was given broader recognition by the United Nations International Decade for Natural Disaster Reduction. A worldwide effort was initiated to acknowledge that public policies and individual decisions contribute to disaster losses. Unfortunately, they have not had the desired effects as demonstrated by the great losses from natural disasters. In fact, most state and local governments continue to conduct limited hazards analysis and do not link the results with problem solving and long-range strategies for risk reduction.

Petak (1985) notes that there are significant challenges to those dealing with hazards and that problems are rife with complexity and uncertainty and leaders are not necessarily responsive to them. Further, technical and administrative capacities of organizations are limited in dealing with the complexity of the problems. Clearly, how we go about identifying and dealing with problems that evolve from our analysis of hazards must be carried out in an intentional systematic process. Problem solving always involve risks. To achieve a safer environment, plans and programs must be developed, integrated, and implemented on our ability to understand existing conditions and to predict future consequences.

Problem-Solving Process

Huber (1980) views that organizational decision making lies within the broader problem-solving process. He explains that the first step is to clarify the background on which the problem situation exists, who is involved, how the situation evolved, and eventually a clear statement of the problem. This first step may be viewed as background and problem finding. He suggests that in too many cases, managers fail to adequately understand the roots of the problem and how it has evolved. He contends that many managers spend little effort in attempting to focus on the right problem and simply react to the environment in which they exist. From the background information on the situation, one can develop a clear statement of the problem. A problem can be defined as the difference between the current situation and what is desired. Others have expressed it as a gap between where you are now and where you want to be and you do not know how to get there. Where that difference is great, you have a much larger problem.

The second step is to take the problem statement and determine alternatives for addressing the fundamental issues. This creative part of the problem-solving process is intended to generate many solution options. The result is that one has numerous options in which to pick a strategy for addressing the problem; one option is to combine possible alternatives that have been identified.

The next step is one of choice making and decision making. Huber notes that the decision-making process involves the review of potential solutions to a problem, ranking these options, and then making a choice between them. Decision making then moves to the development of a strategy for implementing the desired solution and then a follow-up methodology should be determined for determining if the selected solution had the results desired.

Critical to dealing with complex problems is the availability of information concerning hazards, impacts, and potential strategies. Wallace and De Balogh (1985) contends that management information systems provide a key resource for effective problem solving.

Huber believes then that decision making lies within a broader problem-solving approach that includes problem finding, defining a problem statement, generating options for addressing the problem, determining a strategy for solving the problem, strategy implementation, and finally monitoring the results of the strategy (Figure 2.5). Did the solution address the problem as desired and are further actions required?

Critical Thinking: Conflicts evolving from problem solving often result from differences in individual and organizational goals and priorities. It is

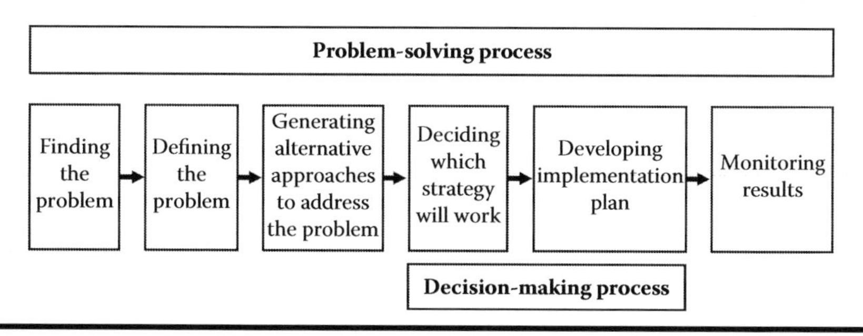

Figure 2.5 The problem-solving process.

helpful to articulate clear statements as to the goals of parties including in the analysis of the problem. In addition to our individual goals, our beliefs and values impact what we believe is important and should be obtained. Having those involved in the decision-making process to clarify their goals, beliefs and values can help move toward a consensus on how to address the problem. How do you see that your values and beliefs might influence your views of problems? How do you approach problems? Do you deal with problems in a systematic way?

We have suggested that an effective hazards analysis involves sound decision making and communication within organizations and with the public, risk management, and hazards mitigation. These activities compliment and extend the outputs of a hazards analysis and provide the basis for including these outcomes in emergency preparedness, response, and recovery initiatives. The effective management of these activities thus becomes a primary goal. Quarantelli (1996) stresses the need for good managing explaining that this is quite different from just managing. He provides a breakdown of the criteria for "good management" by stressing the following:

1. There is a difference between agent and response-generated needs and demands. The first is a tactical need that evolves from a single source; the second is a strategic need that may be identified prior to a disaster and could have long-term implications for the effectiveness of a response. A quality hazards analysis and effective decision making can help identify long-term multihazard response needs.
2. The value of common generic functions is critical to an effective response. Which functions might be included as "generic" should evolve from a comprehensive hazards analysis including the need for housing no matter the source of the disaster.
3. A comprehensive hazards analysis can reveal the nature and type of resources that will be needed for most any disaster. Quality decision making can result in the effective coordination of resources that the right people are able to help in an efficient manner.
4. Organizing resources effectively is based on a clear understanding of what is needed, and the hazards analysis provides a sound basis for how a disaster could impact the community or organization. Getting the right people to the right place is not to be taken for granted but must be managed effectively.
5. Most communication issues that are associated with problems in disaster response relate to communication. Quarantelli stresses that the problems are not in the technology but in what is communicated. The key is that we should stress the content of our communication and not be distracted with just how we deal with others. This is a key insight and stresses the need to communicate clearly the results of our hazards analysis.

6. The need for quality decision making is stressed throughout the emergency management process. Efforts to institutionalize effective decision-making processes provide a basis for ensuring that a culture of sound decision making is used when there is a disaster.
7. The development of a comprehensive hazards analysis includes many people from throughout an organization and from outside interests. Effective coordination then becomes a fundamental element of good management.
8. Good management also recognizes the contribution that emergency groups and behavior have on successful responses to disasters. An organizational culture that encourages emergency behavior by individuals and groups must be nurtured, and the hazards analysis process is an excellent opportunity to instill this as a positive organization value.
9. Effective management is viewed by Quarantelli to anticipate the needs of individual citizens for information. Working with the media is not only needed in emergency response but also especially in communicating the results of a hazards analysis.
10. A quality hazards analysis provides essential information for preparedness and response operations. It should not be viewed as a book to be placed on a shelf but a basis for not only good planning but also emergency response. Understanding our hazards and clearly articulating the adverse impacts of a disaster provide a solid basis for operations in a disaster.

Quarantelli has suggested that good management does not just happen but is the result of intentional actions that impact the effectiveness of any response operation. He provides specific recommendations to help us move toward effective management. A comprehensive hazards analysis is a key factor in establishing and maintaining good management.

Problem Solving in a Nonstructured Environment

Attempting to examine environmental hazards where risk can never be guaranteed and the ultimate solution is never clear presents a situation that is classified as nonstructured (Radford 1981). Decision making that is nonstructured includes the following characteristics:

1. There is a lack of complete information associated with hazards, the environment, and potential strategies that could be considered to reduce the adverse impacts of a disaster.
2. We can quantify the number and frequency of past disasters and their characteristics, but sufficient information to provide a valid estimate of potential disasters is not available. Appropriate measures just are not available to give us a high degree of certainty on the frequency of disasters.

3. Decisions that result from a hazards analysis attempt to address multiple objectives. Unfortunately, these objectives may be in conflict and difficult to quantify. For example, we may want to reduce the damages from a disaster but local resources are limited and increased taxes or fees place a greater burden on some than others.
4. Multiple participants who represent alternative positions and power bases make decision making complex and ill structured.

A critical strategy for dealing with problems in a nonstructured situation is to ensure that the methodology for examining the problem is clear and stated in advance. This process should provide opportunities for numerous alternative positions to be raised and priorities to be examined. Ensuring that a representative from alternative positions is critical in ensuring that all parties support the outcomes of a decision.

Decision Traps

Russo and Schoemaker (1990) provide helpful observations for improving the decision-making process by outlining decision traps that pose barriers to effecting resolution of problems.

1. Failing to take the time to adequately consider the problem and prematurely reach a conclusion without putting the issues into some context. The key is to adequately express what are the issues, who is involved, what is at stake, and potential difficulties that could be encountered if the problem is not addressed.
2. Too often, we do not take the time to clearly define the problem and spend too much effort addressing the wrong issues. The key here is to frame the situation so that the fundamental problem is stated in a clear manner. The frame could be influenced by specific goals of the agencies involved as well as legal or ethical issues.
3. Attempting to understand hazards and their impacts is complex, and we must admit that we do not have all the answers and that we have doubts concerning the potential consequences of some solutions. We need to be clear about the assumptions that we bring to the decision-making process and how our own values, beliefs, and knowledge influence our analysis.
4. For many, we often look for ways to simplifying complex situations and as a result fail to appreciate the broad nature of the problem or the consequences of our actions. Maintaining an appreciation of the powerful force of hazards is a constructive position from which we view the problem-solving process. We need to ensure that where we need additional viewpoints and position or knowledge, which we get it.

Perception of Risks by Citizens

Siegrist and Gutscher (2006) surveyed both experts and citizens to determine their perception and preparedness to flood risks. Results of their survey revealed that citizens in high-risk zones had similar perceptions of potential hazards as the experts but that citizens in "no-risk" areas had lower perceptions of risk than those who lived in areas with higher levels of designated risk. They also found that many people overestimated their risk and thus were more afraid of a disaster than was justified by the facts. Overall, people underestimate risks and failed to show prevention behavior when needed; they were not prepared for disaster events. One would conclude from this study that communities face significant challenges in conveying appropriate risk of hazards in a community and that risk communication is a critical element in ensuring that communities appreciate the hazards that are faced and understand what appropriate preparedness steps would be appropriate.

In a related study, Lindell and Hwang (2008) studied the effect of hazards proximity, experience, risk information sources, and household characteristics on risk perception and action. They found that perceptions of personal risk of flood and hurricane hazards were related to the adoption of flood and wind mitigation measures, with the purchase of flood insurance. They also found that experience with flood and hurricane disasters influence citizen decisions to purchase flood insurance. In addition, they found that there was no evidence of a direct effect of information sources on hazard adjustments such as the purchase of flood insurance. Information from news, the news media, and the Internet was unrelated to risk exposure. They concluded that perceived personal risk and direct hazard experience were significant factor to influencing preparedness and mitigation actions. A key outcome of this study suggests that proximity to a hazard and personal experience have a great influence on how people perceive and act on risks. Communities interested in communicating local risks to citizens must appreciate the need to build on past disaster experiences of the public and the nature and extent of local hazards. The study determined that those engaged in community preparedness and mitigation activities should target their audiences and use specific messages that communicate personal risks. The key is that hazards adjustments are available.

Siegrist and Gutscher (2008) also found that personal experience with a disaster perceive risks at a much greater level. People who have not experienced losses due to a disaster strongly underestimate the negative effects associated with a disaster event. Local officials interested in helping citizens to identify local risks and hazards need to appreciate that many citizens who have not experienced a disaster might underestimate the risks that are present in their community. We should not assume that when we communicate risks that citizens appreciate just what could happen.

Conclusions

Current hazards analysis processes and decision-making approaches tend to put a great deal of power in the hands of technical experts and professional administrators who are not directly accountable to the public. Elected officials must, therefore, be included as representatives of the public and actively engage in the process of exercising value judgments, which will lead to agenda setting, resource allocations, staffing, training, and, ultimately, the effective implementation of a program designed to mitigate against, prepare for, respond to, and recover from disasters when and if they should occur.

The determination of an acceptable level of risk is ultimately a policy that is based on judgment and not just facts. We can attempt to quantify uncertainty, but in the end we must confront our limits in dealing with natural hazards. Our decisions combine facts and values, and problems require a framework that can facilitate a broad view of the risks associated with floods, earthquakes, wind, or other hazards. Further, our decisions will be exposed in a public arena and require open discussion of our analysis. Our process, therefore, must be broad based and one that deals with complex uncertain conditions. We must create and sustain a culture of quality disaster planning that includes a comprehensive hazards analysis, sound decision making, and both risk management and hazards mitigation strategies that address community and organizational vulnerabilities.

Discussion Questions

If an organization or community has limited resources, how can they prepare a community profile of hazards that could affect them?

Why would it be a good idea to perform different risk identification methods to identify the hazards in the community?

How do the nature and the health of a community's natural environment influence the social and build systems?

Why is it important to use a single, standardized map or mapping system to display a hazards profile?

What are the relationships between communities within a region and how could this impact hazards risk management planning?

Applications

Using the community profile format outlined in this chapter, identify the social, economic, natural, and built systems in your community and their assets. From these assets, what do you see are the risks that are associated with them and how losses associated with these assets might impact your community?

The Community Profile provides an excellent basis for examining strengths and weaknesses of a community. Using the Community Profile process:

1. What do you see are the opportunities presented by the strengths of your community?
2. What do you see are the weaknesses of the community?
3. What do you see are the problems that result from these weaknesses? What could be the factors that may have caused these problems?

Websites

Population Data

Community Data is an excellent source for community information. http://www.nationalrelocation.com

Community Analysis (ESRI): http://www.esri.com/software/arcgis /community-analyst

Environmental Scorecard: A community's chemical pollution information site. http://www.scorecard.org

Good site for statistics for many communities:

http://www.demographicreports.com/

http://www.bestplaces.net

U. S. Census Data Source: http://www.census.gov and http://factfinder.census.gov

Information on Cities, Counties, Towns, and Communities: http://www.epodunk.com/

Health Data

Health Data: This site is dedicated to making high-value health data more accessible to entrepreneurs, researchers, and policy makers in the hopes of better health outcomes for all. In a recent article, Todd Park, United States Chief Technology Officer, captured the essence of what the Health Data Initiative is all about and why our efforts here are so important. http://www. Health Data.gov

Health Landscape: Allows for the creation of maps and table of health status in a community including populations at risk, health outcomes, and distribution of health interventions. http://www.healthlandscape.org/index.cfm

Environmental Data

Information related to the Environment: http://cfpub.epa.gov/surf/locate/index.cfm

Agriculture: U. S. Department of Agriculture—Economic Research Service (ERS).

The Economic Research Service provides key indicators, outlook analysis, and a wealth of data on the U.S. food and agricultural system. Along with information on farming practices, structure, and performance, ERS produces data on diverse topics such as farm and rural households, commodity markets, food marketing, agricultural trade, diet and health, food safety, food and nutrition assistance programs, natural resources and the environment, and the rural economy.

http://www.usda.gov/wps/portal/usda/usdahome?navid=DATA_STATISTICS

Economic Data on Food, Agriculture, and the Rural Economy
Natural Resources, Environment, and Conservation
Rural America
Food Consumption (per capita) Data System
Farm Income

Foreign Agricultural Service (FAS): FAS maintains a global agricultural market intelligence and commodity-reporting service to provide U. S. farmers and traders with information on world agricultural production and trade for use in adjusting to changes in world demand for U. S. agricultural products. Reporting includes data on foreign government policies, analysis of supply and demand conditions, commercial trade relationships, and market opportunities. In addition to survey data, crop condition assessment relies heavily on computer-aided analyses of satellite, meteorological, agricultural, and related data.

National Agricultural Statistics Service (NASS): NASS collects, summarizes, analyzes, and publishes agricultural production and marketing data on a wide range of items, including number of farms and land in farms, acreage, yield, production, stocks of grains, and numerous commodities. The Census of Agriculture is conducted every 5 years to collect information on the number of farms; land use; production expenses; value of land, buildings, and farm products; farm size; characteristics of farm operators; market value of agricultural production sold; acreage of major crops; inventory of livestock and poultry; and farm irrigation practices.

Rural Development: Rural Business-Cooperative Service collects, summarizes, analyzes, and publishes data from annual surveys of U. S. farmer, rancher, and fishery cooperatives. The data are published in RBS Service Reports. The data are also available in Excel spreadsheets by state and type of cooperative. The Cooperative Directory presents a listing of those cooperatives willing to be listed by state. The directory contains contact information, type of cooperative, and products sold, and it is updated monthly.

Transportation

State Transportation Statistics: State Transportation Statistics are a series of reports highlighting major federal databases and other national sources related to each state's infrastructure, safety, freight movement and passenger travel, vehicles, economy and finance, and energy and the environment. Along with tables generated for each state, the reports describe databases and give information on access, formats, and contact points.

http://www.rita.dot.gov/bts/sites/rita.dot.gov.bts/files/publications/state_transportation_statistics/index.html

Energy

Data sources available for estimating local energy consumption: http://www.eia.gov/consumption/residential/index.cfm

Residential Energy Consumption Survey (RECS): Administered by the U. S. Energy Information Administration (USEIA), the RECS combines information on housing units, energy usage patterns, and household demographics with data from energy suppliers to estimate residential energy usage and costs. The latest RECS was collected in 2009 and includes data from a representative sample of 12,083 households.

Data are available in an Excel format for 4 census regions of the United States, 9 subdivisions, and 16 individual states. Using representative household characteristics as a point of comparison, residential energy consumption characteristics can be extrapolated from state or regional data to create estimates for local communities.

Commercial Building Energy Consumption (CBECS) table: In contrast to the RECS, the CBECS collects energy usage and cost data from commercial buildings only. The latest CBECS data are available for 2003 and organized by census region and subdivision.

Manufacturing Energy Consumption Survey (MECS): The latest MECS, administered in 2006 by the USEIA, collected data for all 50 states and the District of Columbia on energy use among manufacturing establishments. The 2006 MECS surveyed approximately 15,500 establishments, providing estimates for 50 industry sectors and 21 industry subsectors listed in the North American Industry Classification System.

Transportation Energy Data Book: Oak Ridge National Laboratory (ORNL) publishes a collection of data on energy consumption in the transportation sector. Although the Data Book is comprehensive in scope, it is a secondary source in the sense that data are compiled from various sources and not collected by ORNL.

Transportation Energy Consumption Survey: Using household travel data-collected U. S. Department of Transportation's National Household Travel Survey, the USEIA has derived information on transportation energy consumption and expenditures.

(Energy consumption data can often be localized by estimating figures based on regional travel data collected by a metropolitan planning organization.)

Business Statistics

- Bureau of the Census: http://www.census.gov—The Census Bureau site will lead you to the full range of popular and obscure census data series. The site has a comprehensive A-to-Z listing of data subjects, as well as American FactFinder and Censtats, query-based means for accessing data for your area from a variety of census series.
- Bureau of Labor Statistics: http://www.bls.gov—Bureau of Labor Statistics (BLS) has a wealth of information available through its website. BLS jobs, wages, unemployment, occupation, and prices data series are available through a much improved query-based system. Also see Economy at a Glance for an integrated set of BLS data for states and metro areas.
- Bureau of Economic Analysis: The Bureau of Economic Analysis (BEA) makes its gross state product, Regional Economic Information System (REIS), and foreign direct investment data available on its website. You can also use this site to access BEA's national income account data and its publication of record, the Survey of Current Business.
- FedStats: http://www.fedstats.gov—The FedStats website, maintained by the Federal Interagency Council on Statistical Policy, provides clickable maps to obtain state and local data profiles drawn from multiple federal statistical agencies, with links to the original data sources. It also offers links to the websites of over 70 federal statistical agencies. You can visit this website if you want to get a quick overview of what federal agencies provide what kinds of data.
- Geospatial and Statistical Data Center, University of Virginia: http://www.library.virginia.edu—The University of Virginia's Geospatial and Statistical Data Center provides a query-based system for obtaining a wide variety of federal data by state and area, including REIS, county business patterns, and historical census data.
- Geography Network, ESRI: http://www.econdata.net—Through the Geography Network, you can access a wide range of geographic content, including live maps, downloadable data, and more advanced services, from hundreds of providers around the United States and the globe.
- Regional Economic Conditions, Federal Deposit Insurance Corporation (FDIC): http://www.econdata.net—The Regional Economic Conditions site produced by the FDIC is high value added. Not only does it provide access to employment, income, housing, and real estate data for states, counties, and metropolitan area but also it offers tools to build maps, tables, and charts.
- State of the Cities Data Systems, Department of Housing and Urban Development (HUD): HUD's Office of Policy Development and Research has worked with federal statistical agencies to produce special data runs on a number of economic performance indicators for metro areas, including

demographics, employment, and crime. The unique aspect of this site is that data are disaggregated by central cities and suburbs.

- Economagic.com: http://www.economagic.com—Economagic.com gives you easy access to more than 100,000 data series including state, metro, and county employment data compiled by federal statistical agencies. The site will create spreadsheet files of data online as well as graphing data in your Internet browser. Registered users can generate forecasts from historical data.
- State Economic Data Sources, Association of University Business and Economic Researchers (AUBER): The Association of University Business and Economic Researchers is one of those organizations we suggest everybody get to know. Your local AUBER member is often an insightful observer and invaluable resource on your regional economy. Every state has its own experts and specialized data collections. The fastest way to find them is AUBER's state-by-state directory of resources.

Mapping Resources

National Atlas: http://www.nationalatlas.gov.

Maps that capture and depict the patterns, conditions, and trends of American life. Maps that supplement interesting articles. Maps that tell their own stories. Maps that cover all of the United States or just your area of interest. Maps that are accurate and reliable from more than 20 federal organizations. Maps about America's people, heritage, and resources. Maps that will help you, your children, your colleagues, and your friends understand the United States and its place in the world. This is **nationalatlas.gov**™, and it shows us where we are. It allows you to use your imagination and, by probing and questioning, to choose the facts that fit your needs as you explore the American story.

Layers include the following: agriculture, environment, biology, geology, people, boundaries, government, transportation, climate, history, and water. The data we use to generate maps can be downloaded for free. Vector files are available in Shapefile format. Geostatistical data are in DBF format. Our images are GeoTIFF files. We bundle and compress National Atlas Data to facilitate delivery on the World Wide Web, so you will need decompression software.

Google Earth Pro: http://www.googleearth.com

References

Abramovitz, J. (2001). *Unnatural Disasters*. Worldwatch Paper 158. Washington, DC: Worldwatch Institute.

Ballesteros, J. A., Bodoque, J. M., Díez-Herrero, A., Sanchez-Silva, M., and Stoffel, M. (2011). Calibration of floodplain roughness and estimation of flood discharge based on tree-ring evidence and hydraulic modeling. *Journal of Hydrology, 403*(1), 103–115.

Burton, I. (1993). *The Environment as Hazard.* New York: Guilford Press.

Deyle, R. E., French, S. P., Olshansky, R. B. and Patterson, R. G. (1998). Hazard assessment: the factual basis for planning and mitigation. In R. J. Burby (ed.), *Cooperating with Nature: Confronting Natural Hazards with Land-Use Planning for Sustainable Communities.* Washington, DC: Joseph Henry Press.

Federal Emergency Management Agency (2001). *Understanding Your Risks: Identifying Hazards and Estimating Losses.* Washington, DC: Federal Emergency Management Agency.

Huber, G. P. (1980). *Managerial Decision Making.* Glenview, IL: Scott, Foresman.

Intergovernmental Panel on Climate Change (IPCC) (2007). *Impacts, adaptation and vulnerability: Working Group II contribution to the Intergovernmental Panel on Climate Change.* Geneva Switzerland. http://www.gtp89.dial.pipex.com/chpt.htm.

Kates, R. W., Colten, C. E., Laska, S., and Leatherman, S. P. (2006). Reconstruction of New Orleans after Hurricane Katrina: a research perspective. *Proceedings of the National Academy of Sciences, 103,* 14653–14660.

Lindell, M. K. and Hwang, S. N. (2008). Households' perceived personal risk and responses in a multi-hazard environment. *Risk Analysis, 28*(2), 539–556.

Lopez Saez, J., Corona, C., Stoffel, M., Schoeneich, P., and Berger, F. (2012). Probability maps of landslide reactivation derived from tree-ring records: Pra Bellon landslide, southern French Alps. *Geomorphology, 138*(1), 189–202.

Mileti, D. (1997). *Designing Disasters: Determining Our Future Vulnerability,* Natural Hazards Observer. Volume XXII, Number 1, September 1997.

Olwig, M. F. (2012). Multi-sited resilience: the mutual construction of "local" and "global" understandings and practices of adaptation and innovation. *Applied Geography, 33,* 112–118.

Petak, W. J. (1985). Emergency management: a challenge for public administration. *Public Administration Review, 45,* 3–7.

Pine, J. C. and Wilson, H. (2007). Community at risk: an examination of four neighborhoods flooded from Hurricane Katrina in New Orleans. *Journal of Race and Society, 3*(1), 7–27.

Quarantelli, E. L. (1996). *Ten Criteria for Evaluating the Management of Community Disasters.* Preliminary Paper No. 241, Newark, DE: Disaster Research Center, University of Delaware.

Radford, K. J. (1981). *Modern Managerial Decision Making.* Reston, VA: Reston Publishing Company, Inc.

Russo, J. E., and Schoemaker, P. J. H. (1990). *Decision Traps. Ten Barriers to Brilliant Decision Making and How to Overcome Them.* New York: Simon & Schuster.

Schwab, J., Topping, K. C., Eadie, C. C., Deyle, R. E., and Smith, R. A. (1998). Hazard identification and risk assessment. In *Planning for Post-Disaster Recovery and Reconstruction.* Planning Advisory Service Report No. 483/484. Washington, DC: American Planning Association.

Siegrist, M. and Gutscher, H. (2006). Flooding risks: a comparison of lay people's perceptions and expert's assessments in Switzerland. *Risk Analysis, 26*(4), 971–979.

Siegrist, M., and Gutscher, H. (2008). Natural hazards and motivation for mitigation behavior: People cannot predict the affect evoked by a severe flood. *Risk Analysis, 28*(3), 771–778.
Smith, K. (2004). *Environmental Hazards: Assessing Risk and Reducing Disaster*. London: Routledge Press.
Stoffel, M. and Bollschweiler, M. (2008). Tree-ring analysis in natural hazards research? An overview. *Natural Hazards and Earth System Science, 8*(2), 187–202.
Tierney, K. J., Lindell, M. K., Perry, R. W. (2001). *Facing the Unexpected: Disaster Preparedness and Response in the United States*. Washington, DC: Joseph Henry Press.
United Nations (2002). *Living with Risk: A Global Review of Disaster Reduction Initiatives*. New York, NY: The Inter-Agency Secretariat of the International Strategy for Disaster Reduction.
United States Environmental Protection Agency (2001). *Technical Guidance for Hazards Analysis: Emergency Planning for Extremely Hazardous Substances*. NRT-1. Washington, DC: United States Environmental Protection Agency.
Wallace, W. A. and De Balogh, F. (1985). Decision support systems for disaster management. *Public Administration Review*, Vol. 45, Special Issue, Emergency Management: A Challenge for Public Administration, pp. 134–146.

Chapter 3

Modeling Natural- and Human-Caused Hazards

John C. Pine

Objectives

The study of this chapter will enable you to:

1. Clarify the role of hazard models in the hazards analysis process.
2. Identify the nature and types of hazard models.
3. Explain the criteria that may be applied to assess a hazard model.
4. Explain the advantages and disadvantages of hazard models.
5. Explain the purpose and the elements of a hazard profile.

Key Terms

ALOHA
Deductive reasoning
Deterministic models
Dynamic models
Experimental design
Flood discharge values
Hazard risk vulnerability zone
HAZUS-MH
HECRAS
Hypothesis

Models
Model validity and reliability
River gage stations
Statistical models
Uncertainty

Issue

Social, economic, and environmental hazard models are based on theoretical frameworks, sets of assumptions, and influenced by interrelated dynamics that can change over time. Users of models must understand the purpose and limitations of hazard models and how they can be used in decision making, public policy, and the development of hazards risk management and hazard mitigation strategies.

Role of Hazard Modeling in Hazards Analysis

The U.S. Army Corps of Engineers (USACE), the U.S. Environmental Protection Agency (EPA), National Oceanographic and Atmospheric Administration (NOAA), the U.S. Geological Survey (USGS), the Federal Emergency Management Agency (FEMA), the U.S. Defense Department, the National Institutes of Health, the Centers for Disease Control, and the Department of Homeland Security have used hazard models for many years to clarify the nature and extent of tropical cyclones, inland flooding, wind, fire, earthquake, explosions, radiological and nuclear hazards, landslides, chemical releases, volcano hazards, and the spread of disease. These agency applications of models provide examples of the use of hazard models in community hazards analysis to reduce community vulnerability. More importantly, they serve as a basis for hazard mitigation and community preparedness programs.

A model is a simplified representation of a physical phenomena (Brimicombe 2003; Drager et al. 1993), and in the case of a hazard, it simulates the nature and extent of a potential or real-time disaster event. We use models as a tool to represent natural events and for determining how a specific storm, chemical incident, fire, landslide, tornado, or threat of disease could affect a geographic area. Sophisticated computational models are based on complex mathematical formulas and assumptions and are thus designed to simulate an event. Models are quantitative and attempt to reflect the dynamics of physical, economic, natural, and social systems and processes. Models can be reflected in regression lines predicting an output and based on input variables and mathematical formulas that use complex processes within a computer program.

Chorley and Haggett (1968) suggest that models provide many uses within a scientific context and help us to

- Visualize complex processes and interactions that add to our understanding of social, economic, and physical events.
- They are tools that we can use for teaching and learning.
- Describe a social, economic, or physical phenomenon or process.
- Compare and contrast events, situations, processes, and dynamics of complex systems.
- Collect and manipulate data.
- Explore or construct new theories, concepts, or dynamics.

Most of the hazard models that we use today are based on deductive reasoning. That is, one starts with specific observations of the natural, social, or economic environment and suggest a theory that is based on a hypothesis. An experimental design is determined and based on real data and results in a predicted outcome or phenomena. Model results either provide outcomes that may serve as a basis for changing assumptions or correct errors in our expected outcomes.

As part of the emergency management and disaster science community today, we are able to take advantage of computer technology advancements to use hazard models to characterize hazards and identify potential vulnerability of our social, economic, and natural systems. Many hazard models address a broad range of possible disaster events and allow us to characterize the social, economic, and environmental outcomes.

Critical Thinking: The key to models is the development of personnel (internal staff or external contractors) to set up the hazard simulations and to interpret the results. Models have grown in capacity to simulate very complex dynamics. What may be needed are resources to help design and implement hazard model simulations. Scientific support from local universities or consulting organizations could set up models for a jurisdiction and help in adjusting the model inputs for various hazard scenarios. An interdepartmental team could also be assembled from local agencies such as public works, planning, geographic information systems (GIS), engineering, public health and health care, and emergency service agencies to explore how the results could impact the community (Pine et al. 2005). What hazard models are being used in your community or region? Who is involved in setting them up and using them?

An Example of a Hazard Model

The Tennessee Valley Authority and the USACE have also been leaders in the initiative to characterize the nature of hazards using hazard models and maps. Congress authorized the National Flood Insurance Program (NFIP) in 1968 with the enactment of the National Flood Insurance Act, which was administered by the U.S. Department of Housing and Urban Development (FEMA 1997). Flood insurance rate maps (FIRMs) were prepared for communities throughout the United States and based on hydrologic modeling for drainage basins. The Corps of Engineers also use models to understand the economic and social impacts of levees and dams or proposed community or regional mitigation projects.

HECRAS (Hydrologic Engineering Centers River Analysis System) is a riverine model developed by the USACE and used to describe and simulate inland flooding events. The HECRAS model may have been run for a local community as part of a FEMA flood study. FEMA contractors use this widely accepted flood model in preparing or revising FIRMs. Obtaining the HECRAS model file from FEMA or their contractor allows a local jurisdiction to use well-regarded hazard models in a local hazards analysis. Traditionally, FEMA asks local public works, engineering, or planning well-regarded work with engineering consultants to ensure that revisions to local flood maps represent current conditions.

FEMA saw the need of engaging local communities in understanding flood hazards and the need for resources to support the development of hazard mitigation plans. HAZUS-MH Flood (Hazards United States—Multi Hazard Flood) module was developed by FEMA to enhance flood mitigation planning and emergency response programs. The HECRAS model outputs may be used within HAZUS-MH to characterize alternative flooding events and to determine potential social and economic impacts.

Nature and Types of Models

Mathematical models come in different forms such as statistical, dynamic, or combination (statistical and dynamic together). Statistical models are used to predict or forecast future events by using data from the past. These models compare current hazard characteristics with historical data of similar events. Historical records may cover many parts of the continental United States and include data for over 100 years. Note that data collection methods have changed over time, and our understanding of extreme weather or geologic events is far more detailed today than prior to the application of sensitive direct and remote sensing technology.

Dynamic Models

Dynamic models use real-time data to forecast extreme climatic events. For example, a dynamic model might take current atmospheric conditions such as wind, temperature, pressure, and humidity observations to forecast a specific event. This type of model is very useful where we have extensive data on the nature of the environment. This is more likely the case for numerous data sources along coastal areas of the United States and water features in inland areas. The use of powerful computers with real-time hazard data collection has led to great improvements in dynamic models.

Combination models can take advantage of both dynamic and statistical approaches. For areas of the world where precise data measurements are not available, combination models can take a more global perspective and provide good predictions of hazard events on a regional basis.

Deterministic

Deterministic models are based on relationships that may be part of an environmental condition. For example, a digital elevation model (DEM) provides a description of an area on the earth's surface. The points or contours are related to nearby geographic features and may be used to determine the flow of water on the earth's surface by examining the relationship spatially between contours or points. An interesting dynamic that is seen in this type of deterministic model is that location matters. Tobler (1970) explains that the "first law of geography" is that "everything is related to everything else, but near things are more related than distant things."

Hydrologic models use DEM files to characterize water flow dynamics and the relationships between elevation and the spread of water over the landscape. Models such as HECRAS are influenced by numerous factors such as slope, soil types, land use, surface roughness, and water feature characteristics. The data inputs reflect local conditions that are derived from empirically based measurements. The data inputs reflect specific geographic locations and thus limited to a local context. Data are fed into a model and relationships emerge usually in the form of rules. As a result, we are able to represent and examine the relationships between very complex dynamic physical processes over a landscape.

Probabilistic

In 1967, the U.S. Water Resources Council (WRC) published Bulletin 15, *A Uniform Technique for Determining Flood Flow Frequencies* (Benson 1968; U.S. Water Resources Council, Hydrology Committee 1967). The techniques used to determine flood flow frequencies were adopted by the WRC for use in all federal planning involving water and related land resources. This bulletin has been updated several times with the latest version in 1982. Practically, all government agencies undertaking floodplain mapping studies use flood flow frequencies as a basis for their efforts [Bulletin 17B Interagency Advisory Committee on Water Data (IACWD) 1982]. Flood flow frequencies (Bulletin 17B IACWD 1982) from this national initiative are used to determine flood discharges for evaluating flood hazards for the NFIP. Flood discharge values are a critical element in preparing FIRMs. Corps of Engineers models such as HEC1 requires flood discharge data that may come from the flood frequencies discharge values in this bulletin.

Statistical probabilistic models such as HEC1 have been used in the NFIP for many years. The HEC FFA (Hydrologic Engineering Center Flood Frequency Analysis) model was developed by the Corps of Engineers in 1995 to perform a flood-frequency analysis. It performs flood-frequency analysis based on the guidelines delineated in Bulletin 17B, published by the Interagency Advisory Committee on Water Data in 1982. The model estimates flood flows having a given recurrence intervals or probabilities; these calculations are needed for floodplain management efforts and the design of hydraulic structures. The program estimates annual peak flows on recurrence intervals from 2 to 500 years. It characterizes the magnitude and frequency of annual peak flows for water features.

Most hazard models determine a risk vulnerability zone for a specific hazard and suggest that individuals in the risk zone are vulnerable to harm. Flood models could suggest that residential, commercial, and industrial property could be at risk from flooding if structures are located in an area near a water feature. To determine if specific structures would actually be flooded, additional information would be required concerning the precise location of the structure and the ground elevation of the structure. If this type of data were not available, then the model would not be able to determine the extent of flooding for a single building in the flood zone. It might flood or the water might not reach the first flood elevation of the structure.

Models Used in Hazards Analysis

Some hazard modeling programs, however, go beyond determining the vulnerability of individuals and property. FEMA and the Defense Threat Reduction Agency collaborated on a multihazard program Consequence Assessment Tool Set (CATS) that uses hazard modeling to clarify the risks associated with earthquakes, tropical cyclones, hazardous material releases, and risks from explosive, radiological, or nuclear hazards. The CATS suite of models displays hazard model outputs in the form of risk zones for use in understanding the potential impacts of disasters including building damage, injuries, and fatalities for some hazards. As a result, it can be classified as a consequence assessment tool for its attempts to model the consequences of an event rather than just showing who might be vulnerable.

REALITY CHECK: The Army Corps of Engineers completed a risk assessment of potential flooding for the City of New Orleans in an effort to understand the risk of potential flooding in neighborhoods throughout the city. Figure 3.1 provides a map showing the depth of water in regions of New Orleans. This figure allows a homeowner or business representative a way of identifying the nature and extend of possible flooding.

HAZUS-MH Model

In 1997, FEMA issued the first release of Hazards United States (HAZUS) for modeling earthquakes in the United States. In January of 2004, FEMA released HAZUS-MH and broadened the types of modeling that could be carried out at the community or regional levels. The most recent release of HAZUS allows for modeling

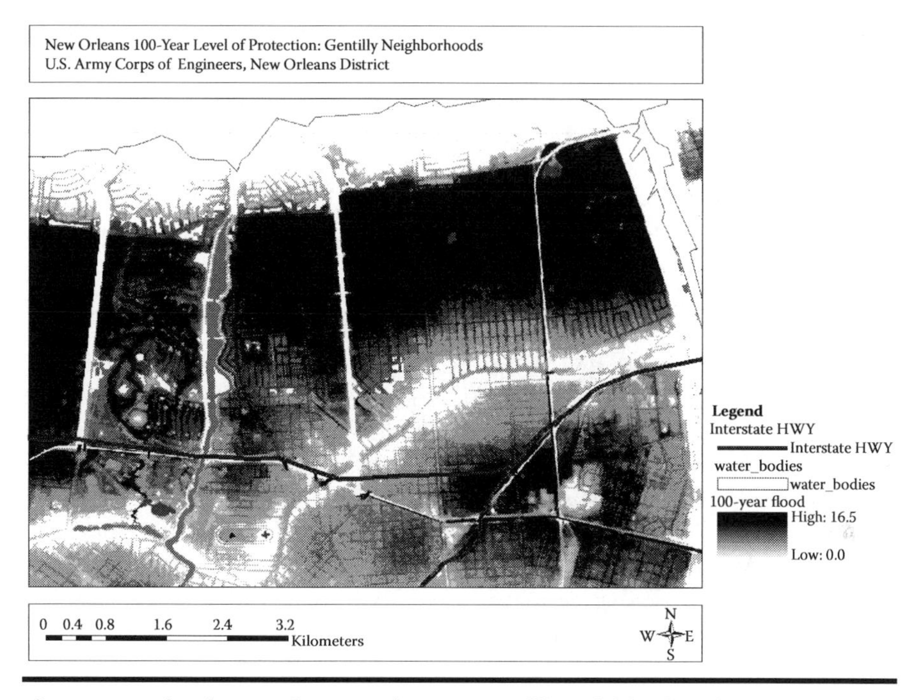

Figure 3.1 Flood map of New Orleans—Gentilly neighborhood.

of not only earthquake risks but also riverine and coastal flooding, wind hazards, and releases of hazardous materials using ALOHA (Areal Locations of Hazardous Atmospheres), dispersion modeling package developed by NOAA and EPA.

The HAZUS-MH mapping and modeling software uses the power of GIS and hazard modeling to estimate associated social and economic losses and to characterize the nature and extent of flood, wind, and coastal hazards. HAZUS-MH supports emergency management by enhancing local capacity for determining the potential damage from inland and coastal flooding, hurricane winds, earthquakes, and chemical hazard events (FEMA 2003). Local, state, and federal officials can improve community emergency preparedness, response, recovery, and mitigation activities by enhancing the ability to characterize the economic and social consequences from flood, wind, and coastal hazards (O'Connor and Costa 2003).

Officials at all levels of government have long recognized the need to more accurately estimate the escalating costs associated with natural hazards (FEMA 1997). The Hazard Mitigation Act of 2000 requires that local jurisdictions complete a comprehensive hazards analysis as a part of their hazard mitigation plan to qualify for FEMA mitigation funds. HAZUS-MH provides needed tools to estimate the adverse economic impact of a flood, wind, and coastal hazards in a community.

HAZUS-MH is just one of the utilities that are available to communities and organizations to characterize risks associated with natural hazards. Allowing local communities and organizations the opportunity to model natural hazards and control the nature of disaster scenarios used in the planning and mitigation process builds modeling capacity at the local government level. Modeling programs for clarifying the nature of natural hazards has long been available to the higher-education research community, federal agencies, and their research laboratories associated with USGS, EPA, NOAA, FEMA, National Aeronautics and Space Administration, and the Army Corps of Engineers. It was not until 1988 that NOAA and EPA developed and released ALOHA for use by local jurisdictions in emergency planning and response. ALOHA has proven that local communities have the capacity and interest in using hazard modeling in local emergency planning, response, recovery, and mitigation activities.

Today, with the use of HAZUS-MH and other user-friendly fire, landslide, volcano modeling programs, local communities and organizations can develop the capacity to use modeling within their organizations rather than remain dependent on engineers and environmental scientists from our research institutions, state or federal agencies, or private consulting companies. Communities have the opportunity to develop in-house capacity for hazard modeling and mapping.

HAZUS-MH Flood provides basic and advanced analysis for flood hazards and their impacts. The basic analysis uses USGS DEM surface grids and discharge frequency values from either the National Flood Frequency Program (Jennings et al. 1994) or, when available, USGS gage stations. The advanced analysis uses either USGS DEM surface grids or higher-resolution DEMs from LIDAR (light detection and ranging).

Advanced flood modeling in HAZUS-MH uses hydraulic analysis from the USGS HECRAS. As is required for a basic analysis, users conducting an advanced analysis must identify a flood study area and obtain a USGS DEM for the area. A USGS website link within HAZUS-MH provides the pathway to a USGS mosaic of 30-m DEM and 10-m quads specific to the study area. The mosaic file is in the form of a GRID file and reflects the surface elevations throughout the study region.

Outputs from the advanced analysis using HEC-RAS are the same as the basic analysis; the depth grid, however, is determined from an engineering hydraulic analysis rather than the general statistical discharge estimates reflected in the National Flood Frequency Program or values from the USGS River gage system. Depth is determined for a specific flooding event by comparing the flood elevation along a water feature with the land surface elevations as denoted in the GRID file. Flood elevations for specific cross sections of the water feature are determined using HEC-RAS.

The initial input into a community's hazard mitigation or emergency preparedness program may be from HAZUS-MH basic flood analysis. This type of general analysis renders a foundation for an assessment of the nature and extent of flooding in a study area. The damage calculations reflected in the basic flood analysis help form a general comparison between regions in the study area. This basic analysis establishes a basis for prioritizing future analyses using the advance features of HAZUS-MH and the HEC-RAS. Local jurisdictions may use advanced flood analysis capabilities of HAZUS-MH by incorporating previous HEC-RAS outputs into the program. Time constraints are a limiting factor because setting up each HEC-RAS study area requires georeferencing the cross sections of peak water elevations. In regard to clearly stated limitations, the HAZUS-MH documentation states that a level 1 "basic analysis" is a generalization of the flood hazard in a local jurisdiction. The hydraulic analysis is not specific to each part of the study area but is derived from the National Flood Frequency Program. USGS regression equations and gage records are used to determine discharge frequency curves. The depth grid, which is the output from the HAZUS-MH basic flood analysis, is a much more detailed illustration of flood range depths than what is viewed on a FIRM. Level 1 analysis is the simplest type of analysis requiring minimum input by the user. However, the flood estimates are crude and are only appropriate for initial loss estimates to determine where more detailed analyses are warranted. Some refer to this type of analysis as "screening." Further studies using the HAZUS-MH advanced Level 3 analysis are required for specific decision making at the local level.

Evacuation Transportation Modeling

Planning for the evacuation of populations at risk from hazards is a key element of the preparedness process and an outcome of the hazards analysis process. Murray-Tuite and Wolshon (2013) note that population evacuations of over 1000 are common in the United States with an evacuation occurring once every 3 weeks. Evacuations associated with many hazards may be needed virtually anywhere and often carried out with little advanced warning time. The major components of roadway transportation evacuation modeling include forecasting evacuation travel demand, the distribution of evacuation demand on road networks on a regional basis including alternative destinations, assignment of evacuees to alternative modes of transportation, and testing alternative management strategies to assess the capacity of evacuation networks. Murray-Tuite and Wolshon (2013) stress the importance of effective communication of evacuation orders. They note that disaster warnings are a social process involving the interaction of warning sources and receivers. Recognizing that who needs to evacuate and when varies over a geographic area and suggests the use of evacuation zones. Timing the notice to a specific area can reduce congestion. Evacuation models, thus, focus initially on estimating the number of households that will evacuate and then on their departure time.

Evacuation demand modeling is a key part of this modeling effort of large or small scenarios. Just who will leave and when is a critical part of simulating this complex situation. Murray-Tuite and Wolshon (2013) note that social networks have a critical role in individual decision making and that the use of cell phones and the Internet have enhanced individual decisions. Evacuation modeling thus links decision making associated with the social sciences with engineering functions involving road capacity. A key outcome of evacuation modeling reveals that maximum flows under actual evacuation conditions differ from nonemergency periods (Wolshon and McArdle 2009).

Postevacuation surveys have been very useful in validating the assumptions that are critical to modeling of individual choices in the determination to leave, when and what route is selected. The postevent surveys combined with agency monitoring of evacuations has been very helpful in validating the assumptions built into the models. Surveys also revealed the need to consider special care facilities such as hospitals or special care facilities for the elderly, disabled, children, or those with significant handicaps. A final consideration that has evolved in more recent modeling efforts is the increasingly importance of pets to household decision making. Murray-Tuite and Wolshon (2013) suggest that future developments in evacuation modeling will include automobile-based information systems that take advantage of new technologies. Agencies may be able to use models and observed data to communicate with evacuees as the event unfolds.

Modeling Community Resilience

Models not only have the capacity to represent the nature and potential impact of various hazard events but also can be extremely useful in helping us to understand the nature of our communities and just how it might respond and recover from a disaster. Regional economic resilience to disasters has been modeled to anticipate the types of challenges that might face a community following a disaster and appropriate strategies to overcome issues presented (Rose and Liau 2005). Computable general equilibrium (CGE) analysis is an excellent approach to disaster impact analysis because it is able to model the behavioral response to input shortages and changing market conditions. A key element of this type of modeling is to acknowledge that utility lifeline supply disruptions can have significant impacts on regional economic activity in the aftermath of a disaster.

CGE models represent a regional economic focus and are used to understand economic impacts and in determining alternative public policies. These are also characterized as multimarket simulation models that are founded on optimizing the behavior of individual consumers and firms, subject to economic and resource constraints. Rose and Liao (2005) note that they are excellent for analyzing the impacts of natural hazards and alternative policies from public, private, and nonprofit sectors.

Park et al. (2011) note that most of the analysis of the impacts of disasters focuses on the direct and indirect impacts of disasters on the economy. Economic impacts, however, may vary widely because of the characteristics of the community impacted. The key is to understand the loss of production flow of goods and services associated with damage to capital stock including suppliers and customers.

Characterizing community recovery may also be modeled by using agent-based models that examine homeowner dynamic interactions following a disaster. This type of modeling simulates the dynamics of postdisaster recovery and homeowner decision-making behavior during reconstruction (Nejat and Damnjanovic 2012).

Communicating Risks from Models

The results from a hazards analysis should be formatted for decision makers in several forms. The reports generated should be explicit and easy to read. However, the user should be ready to raise questions about the hazards and associated risks. The analysis should present information in an orderly arrangement and in a form that assists the decision maker in hazard mitigation. It is critical that local government officials, businesses managers, and nonprofit organizations understand the nature and extent of hazards. A GIS provides a tool for understanding specific risks. Today, many hazard models are linked to GIS tools to display risk zones or identify population areas and infrastructure (roads, bridges, utilities, or industrial areas) at risk.

Critical Thinking: Using the highest quality data for a model will provide outputs that are a more accurate indication of potential damage impacts. Emergency managers and others who use the results of hazard models should discuss the quality of the data with those who are running the model to ensure that the model outputs are used in a manner that was intended by those who created them. Given the potential limitations of the data, current modeling technology allows the user to predict close approximation to the real event. Review a metadata file by going to the following site: http://atlas.lsu.edu/rasterdown.htm. Look for DOQQ images and select one of the sets of images (they offer three dates to select from). From the Downloader, search for the New Orleans East DOQQ. Download one of the four Quads—they will be in an image format. A sample is provided below. Included with the files is a "Metadata" file. Review it and see what information is included.

Bunya et al. (1968) note that coastal Louisiana and Mississippi are very prone to large hurricanes because of their location in the north of the Gulf of Mexico. They note that between 1941 and 2008 this area was impacted by 16 major hurricanes including Audrey in 1957, Betsy in 1965, Andrew in 1992, Katrina and Rita in 2005, and Gustav and Ike in 2008. A team of researchers led by Luettich and Westerink developed a complex wind-driven coastal surge model to represent large hurricanes. Luettich et al. (1992) describe their early efforts to model the complex elements of hurricane storm surge and explain their Advanced Circulation (ADCIRC) hydrodynamic model. Luettich et al. (1992) explains the nature of this hurricane storm surge model at: http://www.youtube.com/watch?v=umnchBR3yZQ).

The validity of environmental models is explored by comparing model results based on atmospheric conditions as storms progressed in a geographic region over time with actual measurements of recorded events. Bunya et al. (2010) made a comparison of storms in the Gulf of Mexico in 2005 (Hurricane Katrina and Rita) to clarify any errors in the ADCIRC model. The study verified that the model provided an accurate representation of the hurricanes. The ability of model accuracy was attributed to the use of high-resolution topography and bathymetry data. Rapid advances in observational systems such as LIDAR (Light Detection and Ranging), satellite-based and airborne Doppler radar, and airborne microwave radiometers allowed for accurate data reflecting coastal storm characteristics. Advances in high-performance parallel computing also allows the model to function at a faster pace and provide outputs for use by public agencies and provide enterprises that have assets in the storm's path (Dietrich et al. 2011).

A recent output from the ADCIRC storm surge model is shown in Figure 3.2. The data reflected in the ADCIRC model represents a compilation of peak storm surge over time. The large area impacted by Hurricane Sandy in 2012 is reflected in this figure and provides public agencies, business, and nonprofit agencies with information to identify risks for future storms. The capacity of the ADCIRC model to represent large geographic areas during a storm is a significant benefit to both emergency planners and responders. Such data outputs allow for the assessment of hazards from a local to regional scale. The detailed characterization of coastal storms from the ADCIRC model provides an accurate representation of a disaster that impacted most of the eastern seaboard of the United States.

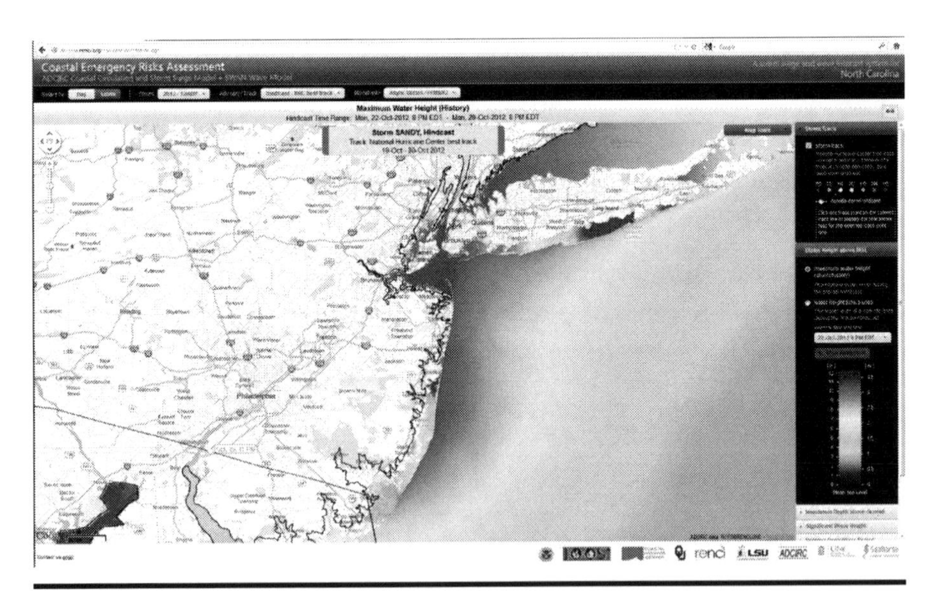

Figure 3.2 Hurricane Sandy storm surge as captured by the ADCIRC (Advanced Circulation) Coastal Circulation and Storm Surge Model. (Image from the Coastal Emergency Risks Assessment website, http://nc-cera.renci.org; provided by the Coastal Hazards Center, the University of North Carolina, Chapel Hill.)

Assessing Hazard Models

Validity

Hazard models are key tools for understanding potential risks to communities. It must be acknowledged, however, that no model is perfect for models are simplifications of reality. Our goal is to obtain what is described as a good fit between the model outputs and what can be observed or process validity. In some cases, we are able to test a model by running a simulation of an actual disaster. Validity is thus a key element in determining the effectiveness of an environmental hazard model. Model validity is determined by examining if the outputs provide the same results with the same inputs or model reliability (Brimicombe 2003). These criteria are extremely useful for determining which hazard models would be appropriate for our use in characterizing hazards and their impacts. As a beginning, our criteria will include the quality of the model outputs, timeliness of model functioning, and completeness of model results.

Quality

Quality concerns the overall accuracy of the model in describing the nature and extent of a specific risk under specific conditions. A critical concern is model validity or determining if the model results accurately represent the potential damaging impacts of the hazard event on the physical environment. Finally, we like to know if the model is truly accurate and often are able to compare a model scenario to a disaster. Any differences must be understood and included in guidance provided to users of the model.

Quality also is related to the replication of results when similar scenarios are repeated. Does the model give the same results when replicated and when the model is run for a large or small area? Is the model representing environmental phenomena giving the same results when replicated? We want consistent results in similar situations. This means that one would obtain the same outputs each time the same parameter values are used in the model. An interesting test for an environmental model is to make a slight adjustment to one input variable to determine if the outputs are the same. You want to see how the model reacts to very small changes in model parameters and then see if the results from the model change.

Availability of Model Documentation

A quality model is one that has extensive documentation and any strengths and limitations explained to potential users. The key here is that model limitations are stated in a clear straightforward manner and appropriate warnings are provided to the user when attempting to learn how to use the system.

Data Accuracy, Resolution, and Availability

Quality data are critical in providing accurate results for planning, mitigation, and decision making. Data issues have become more important today especially since we are able to obtain higher-quality data for use in models. We do understand that models are a representation of complex phenomena and the outputs from models include uncertainty. We must acknowledge that we cannot be 100% correct. Users of our model results must appreciate the inherent errors in data, calculations, and visual display of the results.

Critical Thinking: Intrinsic uncertainty involves potential errors that occur in our data from our collection methods or the manner in which we processed the data. This inherent uncertainty may evolve from the age of the data set or changes in the study area from new housing or commercial developments, or land-use changes that affect hydrological dynamics and potential damage impacts to the built environment. It is thus critical that model users appreciate the purpose and limits of the environmental model and how the output may and should be used.

The date that the data were created could be important to the use of model results. High-resolution images of populated areas that are experiencing rapid growth could provide very different representations of development. The communities that have been experiencing rapid growth could have large subdivisions where just a few years earlier none were present. The difference could be twofold. First as developers worked the area to be developed, they might have changed the elevation of the landscape so as to alter drainage patterns and influence the rate of drainage into a water feature. Second, adding roads or commercial areas with paved parking could have also been established and thus increased the level of water draining into a water feature.

The resolution of the data can greatly influence model outputs. As an example, many hydrological riverine models use DEM files as a basis to represent peak water levels for streams, rivers, and other water features. The resolutions of the DEM files available from the USGS are available in a 30-m grid. Today, higher-resolution DEMs may be obtained using laser technology (LIDAR) and establish a 6-m grid file. The higher-resolution 6-m DEM has the capacity to show areas that may be impacted by floodwaters when compared to model outputs for the same geographic area using a 30-m DEM. The higher-resolution 5-m grid DEM may, thus, show greater variations of contours and flood elevations simply because more data points were used in the flood hazard model.

The geographic resolution of data used as input in a hazard model can have great influence on model outputs. Figure 3.3 shows a high-resolution DEM files obtained from LIDAR and the older version of the (USGS) DEMs. In the past, most USGS elevation contour data were based on a 30-m resolution data format; LIDAR is a new technology that measures the contour of the earth's surface. The new version of the DEM is formatted at a higher-resolution grid. For flood hazards, the higher-resolution LIDAR DEM reveals areas of the landscape that could be impacted by flooding. The lower-resolution 30-m grid DEMs are not able to show the level of

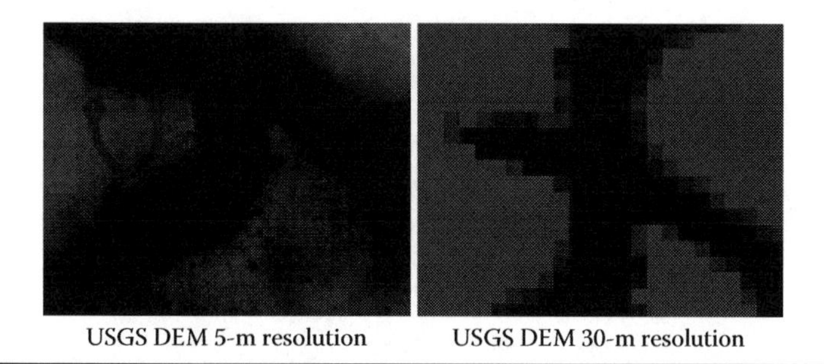

Figure 3.3 U.S. Geological Survey's (USGS) digital elevation model files at 5-m and 30-m resolutions.

detail in potential flooding as with the higher-resolution LIDAR DEMs. Figure 3.3 "DEM files at 30-m and 6-m resolutions" provides an illustration of the differences in the two data sets. One can see greater changes in the LIDAR DEM files when compared to the lower-resolution 30-m files.

Critical Thinking: When one compares alternative geographic digital elevation data sets for use in hazard models, we can see the impact that higher-resolution data can make a great impact on model outputs. Users of hazard models want the best possible results for project development, permitting, and planning. What other local data might be used in hazard models and provides more accurate information to guide emergency planning, mitigation, and recovery efforts?

Postdisaster modeling to verify the accuracy of a disaster event is a common practice for public agencies. Obtaining data that were collected by gages at the time of the disaster provides a critical basis for determining if the program function as the developers stated it would. Agencies such as FEMA have sponsored the development of data clearinghouses for major disaster events (Warren Mills et al. 2008).

The clear presentation of hazard risk vulnerability zones is a key part of conveying flood zones, earthquake hazard area, or other hazard risk zones. See Chapter 7 for a more detailed discussion of issues in risk communication.

Coupling Models with GIS

Today, models do not stand-alone and are in many cases coupled with GIS technology. This combines the modeling program with the display characteristics of a GIS. Parks (1993) noted that environmental modeling tools lacked any spatial data handling and manipulation tools as offered by GIS. But it goes further, for GIS today has a role to play on several fronts. It should not be too surprising that hydrological and hydrogeological models were helping to guide the shift from one-dimensional to two-dimensional approaches given their need to understand the high-sensitive spatial configuration and characteristics of the natural landscape (McDonnell 1996). Many files needed for analysis benefit from the ability of GIS to format data sets or change their characteristics so that environmental models may more easily use them. For example, in hydrological riverine modeling, a DEM file is invaluable for determining land elevations around water features. These elevations can be used when water levels rise where the water goes along the banks of main channels and tributaries. But in addition, the elevations can demonstrate where floodwaters go when flows are constricted by impediments in the channel such as bridges or culverts of just the banks of the water feature. These elevations can show backwater flooding that moves into areas not directly along a water feature but are in the end subject to floodwaters from the water feature.

Brandmeyer and Karim (2000) established a typology for categorizing how GIS and environmental models interface. The most simplistic relationship is one of "one-way data transfer" that provides a linkage between the GIS and an environmental model. HAZUS-MH Flood provides an illustration of this type of linkage where a model is run independently and then linked to HAZUS-MH. The GIS within HAZUS-MH takes the hazard description and constructs a hazard zone map reflecting the simulated disaster scenario. In the case of a flood hazard, values of flood elevations within HECRAS provide a depth grid and flood boundary for a specific flood event. In this example, if changes are to be made in flood conditions or dynamics, they are made in HECRAS and then the new flood model outputs are linked with HAZUS-MH.

A more complex relationship is described in a loose coupling interface. In this category, there is a two-way interchange between the model and the GIS allowing for data exchange and change. Processing of environmental data may be made in a GIS using spatial analysis tools and then the data are moved to the model as a data input.

A shared coupling design links shared data sets for the GIS and the model (Kara-Zaitri 1996). HAZUS-MH includes a utility to allow the GIS to display residential, commercial, and industrial building data by census block. In addition, a consequence assessment or damage estimate is determined by comparing the census building data with a flood grid, wind field grid, or other type of hazard grid file (coastal flooding or earthquake). Building damage estimates are thus calculated using a common data set of local building inventories that may be revised and updated over time.

A joined coupling design may also be established where both the modeling and GIS use common data sets but integration occurs in common script language for both the modeling and GIS (Goodchild et al. 1993). Newer versions of hydrological models have been developed so that as environmental conditions change data inputs may be used to revise hazard outcomes and GIS displays. The highest level of integration is one where the modeling and GIS are combined in a common user interface either on the same computer or by way of a network connection. Many functions are joined and shared within the programs including data management, spatial data processing, model building and management, model execution, and finally visualization of model outputs in a GIS.

Critical Thinking: Hazards are very complex phenomena and may include inputs such as wind velocity, surface roughness, air temperature, stream flow, and geographic surface features. Physical geographic or atmospheric features impact the effects of natural events and may be included in hazard models in the form of mathematical algorithms or formulas. When using hazard models, it is critical to understand how the model is constructed and what data are required. Technical documentation is provided for many hazard models and provides users the necessary information for clarifying how the model was constructed and should be used.

Some hazard models are developed and use data sets that are available globally or for communities in the United States. The United States Census Bureau distributes demographic data that may be used to assess social vulnerability from a neighborhood to a regional level. Similar data sets may be available for other countries. HAZUS-MH uses data obtained from the U.S. Census Bureau to determine the number of residential structures, their value, and when they were constructed. This data allow hazard models to determine an estimate of the number of people who might be impacted from a flood earthquake or wind hazard. Although the data are updated on a 10-year basis, it do provide a good basis for predicting the consequences of disasters (Meyer 2004). Meyer (2004) showed that residential housing counts and values were very accurate but that the commercial and industrial building data in HAZUS-MH were not as accurate as the residential data. HAZUS-MH does provide options for importing and editing the building inventory data and thus may open the use of this hazard model to other communities outside the United States.

Even with the limits of the technology, modeling still provides the best estimate of potential impacts from natural- or man-made disasters. The outputs from models may provide the basis for determining vulnerability zones to floods, landslides, wildfires, earthquakes, or wind hazards and may be used in various emergency response plans and procedures.

Brimicombe (2003) makes an interesting observation in comparing the GIS and environmental modeling communities. He notes that the GIS community has worked to establish data standards and open GIS access across networks, applications, or platforms. In contrast, the modeling community is larger in size and more diverse in representing hazards. We can anticipate that in the future more modeling programs will be linked across networks, applications, and platforms.

A further use of a GIS is that we can use it as a tool to adapt data for use in displaying the model outputs and complete analytical processes to display model results. The GIS can change a DEM grid file so that it may be used in flood modeling. The adapted file is used with channel cross sections that express elevation measures along the banks of a water feature. The TIN (triangulated irregular network), cross sections, and flow conditions all are used to determine the elevation of the water along the water feature. This high water calculation is then measured against the DEM (digital elevation model) to demonstrate the anticipated location and depth of flooding along the water feature. Analytical tools built into a GIS are thus a critical element of the modeling process and far more than just a display tool for demonstrating where a hazardous condition will be seen. Brimicombe (2003) notes that early environmental models did not link outputs with GIS and with these tools we could move from a one-dimensional output to one that would be two dimensional or even three dimensional. Coupling environmental models with GIS tools and capabilities is a major breakthrough for simulating hazards and their outputs.

Critical Thinking: Using high-quality data in a hazard model provides outputs that are a more accurate reflection of local damage impacts. Emergency managers and others who use the results of hazard models should discuss the quality of the data with those who are running the model to ensure that the use of model is consistent with the data used in the program. Given the potential limitations of the data used in a model, current modeling technology allows the user to predict close approximation to the real event. What barriers inhibit a full understanding of environmental hazard model outputs by users?

Our ability to understand the complex relationships between model elements is provided by statistical tools that can prioritize or characterize which parameters influence the simulated model outputs. Brimicombe (2003) notes that when we increase the "number of parameters at smaller units, we raise the level of uncertainty in model outputs and makes validation of the outputs almost intractable (p. 165)."

Static versus Adaptable Outputs

A key criticism of many models is that they reflect a static environment and do not reflect changes such as weather conditions. The U.S. EPA program ALOHA (Areal Locations of Hazardous Atmospheres) has provided for many years the option of user input for weather conditions or direct input from weather sensors. As winds change in direction or velocity or temperatures vary, the model receives the changing conditions from the sensor, models the results, and displays the outcome of the dispersion of hazardous chemicals either in a text format or on a GIS. A real-time dispersion modeling program is a great asset in an emergency response situation.

Uses of Model Outputs

A key element of a quality model concerns its potential use by decision makers. When the model is completed, the results must be formatted for use by decision makers. Will decision makers who use the model results be able to easily apply the results for its intended use? Is there too much information or confusing results from the hazard model? Does overload result from the model outputs?

Timeliness

The timeliness criterion is sensitive to concerns that many day-to-day decisions must be made in a limited time frame or are time sensitive. Many models run on desktop or laptop computers and can be run in just a few minutes. Other programs such as the hurricane storm surge models may take several hours to run and require the power of a server. Decisions on how to respond to situations must be made quickly. Timely information has several components: is the information provided when it is needed for decision-making. Is the information from the model updated as needed? When conditions change such as more concrete in the drainage basin displacing more water, is information provided as often as needed or at an appropriate frequency?

We have learned in many models that critical variables in the model may have data that is out of date. Many flood studies result in changes in FIRMs and are updated as needed. In many cases, flooding changes because of development in the river basin. Figure 3.4 provides an illustration of development in a rapidly growing community. Despite efforts to provide retention ponds in new subdivisions, flooding could occur. The images provide a contrast in a small area of this growing community in South Louisiana prior to the rapid growth following Hurricane Katrina in 2005.

Completeness

The results of the model must be complete to be of value to decision makers. Is the scope of the information sufficient to allow the decision maker to make an accurate assessment of the situation and to arrive at a suitable decision? Does the decision maker have access not only to current information but also to past history? Are the results of the model presented to the decision maker in a concise form, but with sufficient detail to provide the decision maker with enough depth and breadth for the current situation? Is sufficient relevant information provided to the decision maker without information overload?

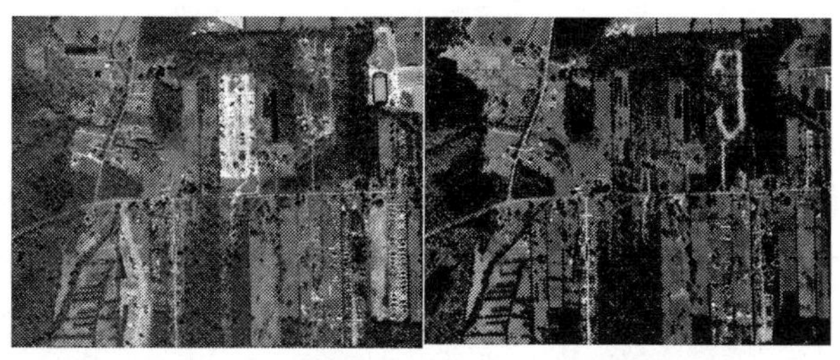

USGS DOQQ 2004–St. Gabriel USGS DOQQ 1998–St. Gabriel

Figure 3.4 Development in a rapidly growing community.

The outputs from a model may be complete given the intended use of the results. As an example, a hazard mitigation study prepared for a local jurisdiction should reflect hazards for the area. The basic analysis provided by HAZUS-MH provides sufficient information to allow decision makers to make an accurate assessment of the risks in their jurisdiction and arrive at suitable decisions. The model outputs are adequate for identifying general risk zones, but an advanced analysis using hydraulic modeling results such as HECRAS provides more accurate results for decision makers. The more advanced models such as HECRAS are suitable for determining base flood elevations and the risk of flooding for specific buildings and infrastructure. This advanced flood modeling capability takes considerable time and is not possible unless the more advanced flood modeling capability are available to the jurisdiction. A local jurisdiction can set a goal to obtain the detailed hydraulic analysis for their area and input the data into HAZUS-MH.

A balance between conciseness and detailed results should be determined by the user. A general analysis may be obtained for understanding the range of potential losses, but more detailed information can be obtained from a model that requires extensive local data and processing time. Figure 3.5 provides an example of a clear assessment of flood risks.

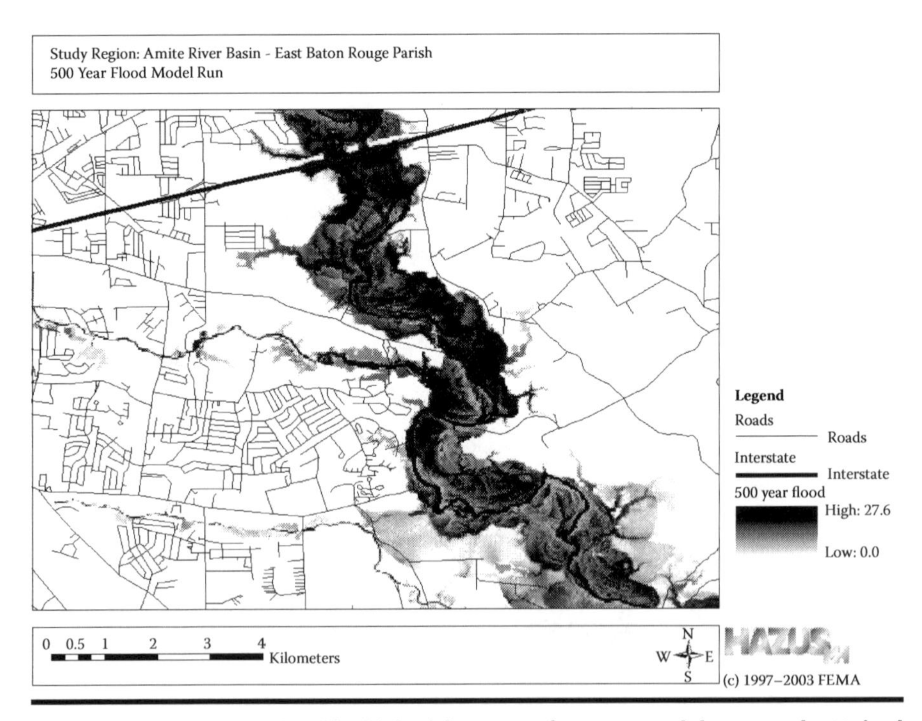

Figure 3.5 Easy to identify high-risk zones from a model (Hazards United States—Multi Hazard [HAZUS-MH]).

Advantages and Disadvantages of Hazard Models

Hazard models that are based on comprehensive data sets may provide an accurate representation of a complex environmental dynamic. If this data are up-to-date, accurate, and in a format that may be used by modelers, then the modeling outputs are more likely to be received in a constructive manner. If users have severe reservations concerning data quality that would be used by a hazard model, then they will resist the application of the model in community policy making.

Hazard models are based on a set of assumptions that should be conveyed to the model user and a part of the outputs from the model. These assumptions could involve decisions by model developers that a geographic area mapped was flat, in a rural area, and that weather conditions did not vary over the study zone. Unfortunately, many hazard models include assumptions that are not fully understood by users (Goodchild et al. 1993).

Kirkwood (1994) suggests that environmental hazard models may not provide an accurate representation of the risk to local citizens. He suggests that when we fail to communicate clearly the nature of environmental risks, public officials and citizens alike can have a false sense of security, or if risk is overestimated and consequently cause fear. Unfortunately, environmental models require a people to interpret the results.

Environmental data used in modeling hazards should come with an associated meta-data file. Meta-data provided information about the data set and include its purpose, an abstract, when it was developed, by whom, any geographic parameters, sponsoring agency, how it is distributed, contact information for anyone who has questions, and use constraints.

Critical Thinking: Agencies have spent considerable time in the development of meta-data files that provide insights in the use of hazard models. What is the value of reviewing a meta-data file?

Model Limitations

Most models have limitations that impact their use and application. For example, the U.S. EPA and the NOAA developed ALOHA, an air dispersion model that is widely used by local emergency responders. In the initial setup of this hazard model, the user is warned that the ALOHA (Ariel Locations of Hazardous Substances) model should not be used with chemicals that are a mixture of hazardous substances, particulates, or incidents that exceed a 1-hour duration. The user documentation for ALOHA stresses that the model provides an approximation of the risk zone for an area that could have property damage, injuries, or fatalities. The user of any hazard model must understand how the assumptions contained within the model affect outputs and how variations of data input could impact results. Errors in data input by the users of hazard models can lead to distortions of the hazard vulnerability zone so that the hazard zone outputs do not reflect the potential danger in the simulated hazard. It is critical that data inputs reflect the hazard scenario and the best data available that is to be used in the model.

Hazard Profiles

Once hazards have been identified and mapped in a community, it is possible to create a hazard profile for the community or organization. A hazard profile provides a clear and concise picture of the local context, and influence of each hazard, as well as a general description of the hazard for reference, uses FEMA (2001). This information will prove vital in developing a hazard mitigation strategy and emergency response plan, and to form a basis for recovery in the event of a disaster. Many hazards have different names, so it is important that the risk statement clearly identifies what is meant by the hazard being profiled. For instance, the hazard "storms" could easily mean windstorms, snowstorms, hurricanes, torrential rainfall, windstorms, or other hazards. By providing a name of the hazard and other hazard identifiers that may also be considered in the risk statement, some confusion will likely have been eliminated. There are many measurement and rating mechanisms for hazards that may have changed over time or may be extremely useful in determining the local context of a hazard. Those conducting the assessment determine the amount of information provided in these descriptions. However, it is important that both minimum and maximum requirements and expectations are established before the process begins to ensure consistency in reporting, which will alleviate some confusion and ensure that risk perception effects are kept to a minimum.

Sources of Hazard Information for the Hazard Profile

1. Research the disaster and emergency history of the community (newspapers, town/city government records, the Internet, public library "local history" section, local historical societies, and older members of the community, local incident reports).
2. Review existing plans (regional and state transportation; FEMA hazard mitigation plans; environmental, dam, or public works reports; land-use plans; capital improvement plans; building codes; land development regulations; and flood ordinances).
3. Interview local residents, risk managers, community leaders, academics, and other municipal and private sector staff who regularly perform risk management tasks; floodplain managers; public works departments; engineering, planning, and zoning; and transportation departments, fire department, police department, emergency management office staff, and local businesses personnel. Do a search on FEMA's website for state hazard mitigation officers.

4. Perform site visits to public or private facilities that serve as a known source of risk for the community (chemical processing operations, transportation carriers, and utilities).
5. Examine a local map of your jurisdiction and note major transportation routes (rail, motor carrier, and water), medical facilities, schools, commercial and industrial locations, apartment complexes, and residential neighborhoods. Examine the relationship between major water features including wetlands and drainage areas. A sample FEMA FIRM is provided in Figure 3.6. What are the critical transportation links in the community? How could they be impacted by a major disaster? How might a single hazard event (flooding) trigger secondary impacts (chemical release)?

Hazard profiles categorize the nature of a potential hazard event and include an analysis that is part of local hazard mitigation. Including a hazard profile on each community's risks should ensure that the process clarifies the risks that exist to a community. The profile includes the following:

1. A description of the hazard that could impact a community.
2. The potential magnitude that the hazard could have.
3. The frequency of occurrence.

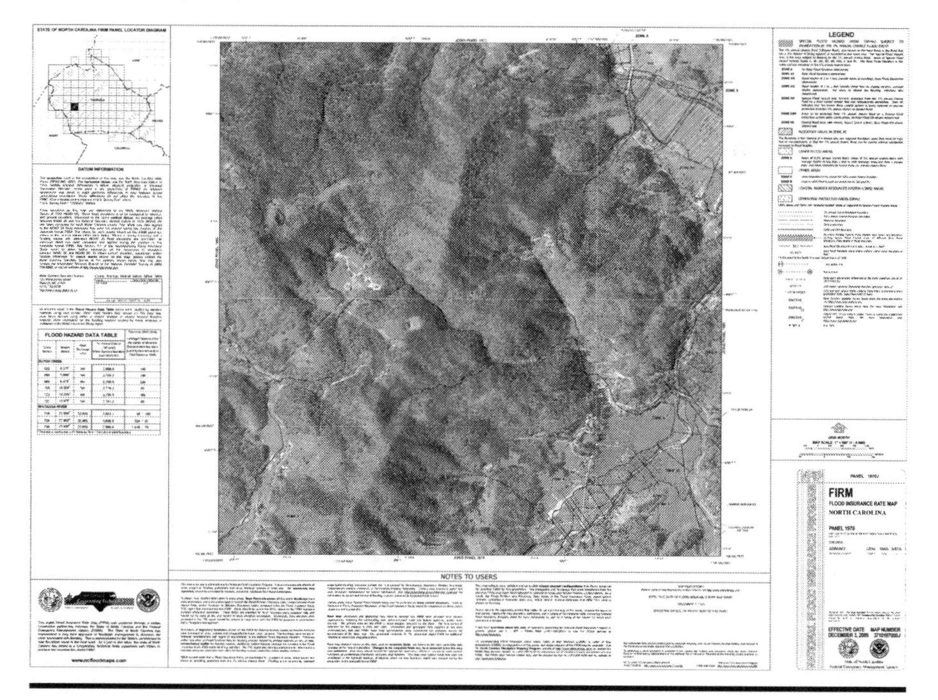

Figure 3.6 Federal Emergency Management Agency's digital flood insurance rate map (FIRM). Watauga County CID No. 370251, Panel 1970J North Carolina (2009).

4. Is there is a seasonal pattern to the hazard?
5. The duration of the hazard.
6. The hazard's potential speed of onset.
7. Warning systems that are available.
8. The location and spatial extent of the potential hazard event.

A sample of a hazard profile is included in the "Websites" section of this chapter.

Description of Hazard

Organize the hazard profile by first identifying the type of hazard that could impact the community or organization. Include both natural- and human-caused hazards. Natural hazards include floods, droughts, extreme heat, extreme cold, hurricanes, thunderstorms and lightning, tornadoes, severe snowstorms and blizzards, ice storms, land subsidence, or expansive soils. Natural hazards could also include disease and poisoning. Human-caused hazards include major transportation accidents, hazardous materials spills, widespread power failures, water or sewer failures, telecommunication disruption, computer system destruction, gas line breach, intentional destruction, laboratory accidents involving biological hazards, building collapse, or fires.

Magnitude

Based on the hazard maps developed above, this may be a single figure or a range of possibilities. The magnitude and possible intensities will be important in the analysis step, as they will help the team to determine the possible consequences of each hazard and to determine what mitigation measures would be adequate. If a chart detailing the characteristics of a hazard as a function of magnitude or potential intensity was provided in the general description of the hazard, then it is easy to see how the local expected magnitude and potential intensity would be highly useful to the Hazards Risk Management team.

Frequency of Occurrence

Historical incidences of the hazard should be displayed in a standardized format, either as a spreadsheet, a chart, or a list. If the hazard is one that happens regularly, it may be better to indicate that fact and list only the major events that have occurred. This is often true with floods and snowstorms, for example. The predicted frequency of the hazard should be provided as well. Oftentimes mitigation measures, development, changes in the environment, or other reasons can cause the average annual frequency of a hazard to rise or fall. It may also be helpful to include any comments or reasons why the frequency has changed or is expected to change in the future.

Seasonal Pattern

This is simply a description of the time of year that a hazard is most likely to appear, if such a pattern exists. Knowing seasonal patterns of hazards allows the Hazards Risk Management team to analyze interactions between hazards that could occur simultaneously, and devote less time to those that are not likely to occur at the same time. For instance, there is little chance that a town in South Carolina would need to plan for the simultaneous occurrence of a blizzard and a heat wave. Although this may seem obvious, this type of information is useful in determining the adequacy of emergency management assets and for other logistical and planning issues.

Duration

For hazards that have occurred frequently in the past, it will be possible to give an accurate estimation of the duration of the hazard based on previous response efforts. However, for disasters that rarely occur or have never occurred, such as a nuclear accident or a specific type of hazardous material spill, it will be necessary to estimate the chance of the incident. These estimations should be based on the description of the possible hazard, the vulnerability of the community, the response capability of the jurisdictions emergency response organizations, the response capability provided through mutual assistance agreements with neighboring communities, and anticipated state and federal assistance. This estimate will be measured in days rather than hours or minutes, but will be very useful in subsequent steps that analyze the possible consequences of hazards.

Speed of Onset

The speed of onset of a hazard can help planners in the mitigation phase of Hazards Risk Management determine what actions are possible, impossible, and vital given the amount of predisaster time they are likely to have. For a hurricane, there may be several days before the disaster where people could be evacuated and protection measures could be applied to structures and other physical objects, for instance. For a tsunami, there may be 10 minutes to several hours of advance warning, which allows for an immediate evacuation, but little structural work. However, for an earthquake, there is often little or no warning at all, so all preparations must be made with no knowledge of when the next event will occur.

For each of these, the public education and communication systems that can be planned will be drastically different. Warning systems and evacuation plans must reflect the availability or lack of time within which action can be taken. If responders can be readied before the disaster, the speed of response will be increased significantly. For all of these reasons, and many more, it is easy to understand why knowing the speed of onset of a hazard is vital in planning.

Availability of Warnings

This information is indirectly related to the speed of onset of a hazard, but is also independent in some ways. Each hazard is distinct and has certain characteristics that either do or do not lend themselves to prediction through the use of technology or other means. Even hazards that can have a fast onset, such as a volcano, can be predicted with some degree of confidence (though not always), whereas some hazards with slower onset times, such as biological terrorism, cannot be predicted accurately at all. Lastly, some provide no advance warning at all, such as the case of a chemical accident.

Even if advance knowledge of a disaster is possible, the capabilities of the local warning system further determine the possibility of adequately informing the public about an impending disaster. Local warning systems are more than the physical alarms, sirens, or announcements; they are also the ability of the public to receive, understand, and act on the warnings they receive. These are all factors that must be considered when determining the availability of warnings.

Location and Spatial Extent

For most hazards, the map that was generated above will be sufficient and highly informative in the analysis step of the process. However, if there are some individual areas or regions within the community that need special mention, and likewise, special consideration, this factor should be included as a separate comment in addition to the map. This helps to ensure that those special areas are not overlooked in subsequent processes. This information is also likely to be found on the maps that were developed in the previous step. However, for some hazards there may be special additional comments or facts that need to be added separately from the visual representation provided by the map.

Conclusions

Hazard models provide users with an exceptional tool in the hazards analysis process. Over the years, hazard models have been constructed for ease of use and to provide the user quality outputs. The models generate information on where to expect losses in the community and considerable information on the nature of these losses. Difficulties in using models may be overcome as users become more skilled in using the power of GIS. Training scenarios can be prepared to help users anticipate common problems in using a model.

A key element in using hazard models is determining the validity of the outputs and that they may be used correctly in developing emergency preparedness or hazard mitigation plans. Too often, we see that our scientific tools such as hazard model outputs are not used correctly and that users fail to understand the limitations of the models and how they should be used.

Discussion Questions

Many people contend that the public does not often understand detailed scientific results from complex models. Why is there such a lack of communication between the scientific community and the public?

Why are some hazard models so difficult to use at the local government level? What is the extent of the use of programs such as HAZUS-MH at the local government level?

Many models combine hazard risk zone outputs with visualization tools such as GIS. How does this tool help in understanding environmental hazard models?

Those who develop and use hazard models stress that the outputs may not reflect real-world conditions. What are the benefits of using models if we are unable to be completely assured that the results are accurate?

Although many communities and organizations have an all-hazards approach for managing disasters, why is it important to understand how specific hazards could impact a public entity, business, or nonprofit organization?

FEMA has created HAZUS-MH as a powerful tool for the assessment of local disaster losses. The tool allows the user to execute a local analysis in a reasonable period of time and estimate losses to the jurisdiction. It provides a basis for examining the economic impact of different hazard events and use loss estimates to establish hazard mitigation priorities. HAZUS-MH is thus a powerful hazard mitigation tool. What contributions does this hazard assessment tool provide to local decision makers? See http://www.fema.gov/media-library/assets/documents/5231?id=1985 for a description of HAZUS-MH.

Applications

National Flood Insurance Program

Review an NFIP map for your jurisdiction and note when the map was published. For these maps, not all geographic areas are modeled using riverine modeling programs or coastal hurricane storm surge programs. Areas are marked with a reference when the flood zone is based on an environmental hazard model. Other areas are noted on which a flood zone is provided but that it was not based on model results. For flood zones that are not based on flood model results, how might the map risk zones be interpreted?

http://www.floodsmart.gov/floodsmart/static/landing1.jsp?WT.mc_id=FEMA_Google1&WT.srch=1

NOAA and the U.S. EPA developed a dispersion modeling program for use in emergency response and preparedness activities. They wanted a program that would not only be easy to use and provide quick results but also be reliable and give results that reflected the conditions provided by the user. To download ALOHA, go to the U.S. EPA website and search for ALOHA. Using the following criteria, evaluate this model for the following:

Does the model provide quality results?
Is the model easy to use and provide timely results?
Are the results of the model complete and usable for decision making in emergency preparedness and response efforts?
Is the data concerning weather conditions, chemical characteristics, and the accurate, at an appropriate resolution, current and available to the user?

Hazard Profile Sample for a Tornado

The following provides an example of a profile for a tornado hazard. Review the description of the elements of the profile that was described in this chapter. From this sample profile, construct a local hazard profile for a risk that faces your jurisdiction.

Hazard Profile: Tornado

Sample

Hazard: Tornado (twister, funnel cloud)

Description of hazard: A tornado is a rapidly rotating vortex or funnel of air extending groundward from a cumulonimbus cloud. When the lower tip of the vortex touches the earth, the tornado becomes a force of destruction. Approximately 1000 tornadoes are spawned by severe thunderstorms each year.

Tornado damage severity is measured on the Fujita Tornado Scale. The Fujita Scale assigns numerical values based on wind speeds and categorizes tornadoes from F0 to F5. Tornadoes are related to larger vortex formations, and therefore often form in convective cells such as thunderstorms or in the right forward quadrant of a hurricane, far from the hurricane eye. The strength and number of tornadoes are not related to the strength of the hurricane that generates them. Often, the weakest hurricanes produce the most tornadoes. Large fires can also generate tornadoes.

The path and width of a single tornado generally is less than 6 miles. The path length of a single tornado can range from a few hundred meters to dozens of kilometers. A tornado typically moves at speeds between 30 and 125 mph. The average lifespan of a tornado rarely exceeds 30 minutes (FEMA 1997).

Potential magnitude: Limited (10%–25% of area affected).

Frequency of occurrence: Likely between 10% and 100% probability in the next year or at least one chance in the next 10 years. Wayne City lies within a state that experiences an average of 50–100 tornadoes per year. Although only some of the tornadoes directly affect the university, any event occurring within the city had the potential to strike the campus.

Seasonal pattern: All year, though more common in the spring, summer, and autumn months.

Probable duration: Between 1 and 30 minutes for the tornado and up to 24 hours for the response.

Potential speed of onset: None to minimal (less than 3 hours).

Identify existing warning mechanisms: The National Weather Service (NWS) issues tornado watches and warnings. The City uses a system of sirens indicating a tornado watch or warning. The University uses a fire alarm system to warn students of any warning. The University also uses the university Web page and university building public announcement systems to warn residents.

Identify location and spatial extent of potential event: Entire campus is at risk. The University has identified designated tornado shelters where students are instructed to go during a tornado warning.

For additional information on hazards useful in the development of risk statements and hazard profiles, the following resources are provided.

Websites

NOAA Climate Prediction Center

NWS, NOAA Climate Prediction Center Hazard Assessment Maps. The Climate Prediction Center (CPC) staff climate and weather observations and data collected by NOAA and other international partners and then use that data in models to assess the meaning, significance, and current status of hazards and how likely these hazards will result in future climate impacts. Their findings are issued as assessments, advisories, special outlook discussions, and bulletins. http://www.cpc.ncep.noaa.gov/products/predictions/threats/threats.php

Areas of interest include the following:

- Global Ocean Assessment
- Drought Information (United States)
- Climate Assessments
- Weather and Crop Bulletin

Avalanches

http://www.avalanche.org

Coastal storms: FIRMs also cover coastal regions and provide detailed information concerning coastal flooding or risk from the water-related hazards of coastal storms. The Saffir–Simpson Scale can be used to categorize storms and their wind. Storm surge can be shown on the FIRM and range from 4 to more than 18 feet above normal sea levels. Inland wind measurements, provided by NOAA, can be used to characterize the maximum sustained surface wind that is possible as storms move inland. Historical data on hurricanes can also be used to characterize future storm events. Data on past storms are provided by NOAA and provide information on tropical cyclone tracks, and their associated wind speeds, from 1927 until the present day.

http://www.fema.gov/hazards/hurricanes/

http://www.aoml.noaa.gov/hrd/tcfaq/G12.html

http://www.nhc.noaa.gov

http://www.nhc.noaa.gov/pastall.shtml

http://www.nhc.noaa.gov/pastall.html (Historical data on hurricanes)

Dam Safety

FEMA provide a good guide for assessing the safety of dams.

http://www.fema.gov/dam-safety-0

Drought

Both the USGS and the National Drought Mitigation Center at the University of Nebraska-Lincoln provide excellent resources to understand drought hazards.

Water Use in the United States: http://water.usgs.gov/watuse/

National Drought Mitigation Center: http://www.drought.unl.edu

Heat Hazards: (NOAA CPC)

http://www.cpc.noaa.gov/products/predictions

Earthquakes

The USGS provides maps detailing peak ground accelerations (PGAs), which measure the potential strength of ground movements that could result from an earthquake. The maps provide the values of the 10% and 2% statistical incidence of such PGAs occurring during the coming 50-year period.

USGS Geologic Hazards Science Center: http://geohazards.cr.usgs.gov

USGS Earthquake Hazards Program: http://earthquake.usgs.gov

National and Regional Seismic Hazard Maps: The USGS provides seismic hazard assessments for the United States and areas around the world. http://earthquake.usgs.gov/research/hazmaps/

Flooding

The FEMA NFIP Flood Insurance Studies provide extensive information for local communities concerning potential for flooding. Today, these maps are displayed on high-resolution images that show many local features and reveal vulnerability. These maps show not only water or hydrographic features but also coastlines and extensive risk information from base flood elevations for homes, businesses, and other sites. The maps also show geographic elevation reference marks and in many cases elevation contours. Note the location of flood maps at the conclusion of this chapter under Internet resources.

Flood Smart for understanding your risk to riverine flooding risk: http://www.floodsmart.gov

FEMA's Map Services Center

http://msc.fema.gov/webapp/wcs/stores/servlet/CategoryDisplay?catalogId=10001&storeId=10001&categoryId=12001&langId=&userType=G&type=1

National Flood Insurance Program: Mapping http://www.fema.gov/mit/tsd/
Natural Hazards: http://www.usgs.gov/themes/flood.html

WaterWatch: http://water.usgs.gov/waterwatch/

Water Resources: http://water.usgs.gov/nwc/

http://www.earthsat.com/wx/flooding/floodthreat.html

Landslides

The best predictor of future landslides is past landslides because they tend to occur in the same places (FEMA 2001). However, because of the complexity of geotechnical factors that must be used to determine landslide risk, there can be no general assumptions made based on blanket designations (such as for wind speeds) or on elevation (as with floods), for example. The expertise of geologists and engineers, who must conduct surveys of soil conditions, slopes, drainage, climate, prevalence of earthquakes and volcanic eruptions, flora cover, erosion, industry-induced vibrations, construction, and other alterations to terrain, among many other factors. The primary elements that help understand landslides include the following: (1) areas that appear to have failed due to landslides, including debris flows and cut-and-fill failures, (2) areas that have the potential for landslides by correlating some of the principal factors that contribute to landslides such as steep slopes, geologic units that lose strength when saturated, and poorly drained rock or soil—with the past distribution of landslides, and (3) areas where landslides have occurred in the past are likely to occur now and could occur in the future.

http://landslides.usgs.gov

National Severe Storms Laboratory: Lightning http://www.nssl.noaa.gov

Snow

NOAA Snow Analysis: http://www.noaanews.noaa.gov/stories/s300e.htm

Tornadoes

Communities measure their tornado risk as a factor of the maximum possible wind speed they may experience, should a tornado occur. Although tornadoes are highly unpredictable and can occur just about anywhere in the world, there are areas that have historically been more prevalent. This high prevalence exists in the midwest, southeast, and southwest regions of the United States. Locate the design wind speed for your area and the level of wind resistance that should be factored into the construction of community shelters as determined by the American Society of Civil Engineers. Historic tornado data may be reviewed to determine the exact historical frequency of tornadoes in their jurisdiction.

http://www.noaawatch.gov/themes/severe.php

Tsunamis

Tsunamic hazard maps identify inundation zones and can reveal low-lying areas in a community that are at risk. Most maps have been produced at the state level, and considerable local mapping has been done as well. Tsunamic maps are available from the West Coast/Alaska Tsunami Warning Center of NOAA/NWS Pacific Tsunami Warning Center; Oregon Department of Geology and Mineral Industries, University of Washington, and the Pacific Marine Environmental Laboratory.

http://www.pmel.noaa.gov/tsunami

http://www.usgs.gov/themes/coast.html

University of Washington: http://earthweb.ess.washington.edu/tsunami/

http://www.pmel.noaa.gov

Wildfires

Wildfires are influenced by the availability of fuel, topography, and weather. Fuel availability may be characterized using the Urban Wildland Interface Code (2000)

Fuel Model Key. Since fire spreads much faster as slopes become steeper, topography maps can reveal areas that have a high probability of fire. Critical weather elements, such as relative humidity and high wind, can be helpful in predicting a community's risk from wildfires. This information can be obtained from local and state fire marshals, state foresters, the NWS, or from the NOAA.

http://www.noaawatch.gov/themes/fire.php

http://www.spc.noaa.gov/products/fire_wx/

http://fs.fed.us/links/maps.shtml/

NOAA Fire Hazard Mapping: http://www.ssd.noaa.gov/PS/FIRE/hms.html

References

Benson, M. A. (1968). Uniform flood-frequency estimating methods for federal agencies. *Water Resources Research, 4*(5), 891–908.

Brandmeyer, J. E. and Karim, H. A. (2000). Coupling methodologies for environmental models. *Environmental Modeling and Software, 15*(5), 479–488.

Brimicombe, A. (2003). *GIS, Environmental Modeling and Engineering.* New York: Taylor & Francis.

Bunya, S., Dietrich, J. C., Westerink, J. J., Ebersole, B. A., Smith, J. M., Atkinson, J. H., Jensen, R., et al. (2010). A high-resolution coupled riverine flow, tide, wind, wind wave, and storm surge model for southern Louisiana and Mississippi. Part I: model development and validation. *Monthly Weather Review, 138*(2), 345–377.

Chorley, R. J. and Haggett, P. (eds.) (2013). *Socio-Economic Models in Geography.* Routledge.

Dietrich, J. C., Zijlema, M., Westerink, J. J., Holthuijsen, L. H., Dawson, C., Luettich, R. A. Jr., Jensen R. E., et al. (2011). Modeling hurricane waves and storm surge using integrally-coupled, scalable computations. *Coastal Engineering, 58*(1), 45–65.

Drager, K. H., Lovas, G. G., Wiklund, J., and Soma, H. (1993). Objectives of modeling evacuation from buildings during accidents: some path-model scenarios. *Journal of Contingencies and Crisis Management, 1*(4), 207–214.

Goodchild, M. F., Parks, B. L., and Steyaert, L. T. (1993). *Environmental Modeling with GIS.* New York: Oxford University Press.

FEMA (1997). *Multi Hazard Identification and Risk Assessment: A Cornerstone of the National Mitigation Strategy.* Washington, DC: FEMA.

FEMA (2001). *Understanding Your Risks: Identifying Hazards and Estimating Losses.* Washington, DC: FEMA.

FEMA (2003). *HAZUS-MH Flood Model: Technical Manual.* Washington, DC: FEMA.

Interagency Advisory Committee on Water Data (1982). *Guidelines for Determining Flood Flow Frequency.* Bulletin No. 17. Bulletin of the Hydrology Subcommittee. Reston, VA: Office of Water Data Coordination, U. S. Geological Survey,.

Jennings, M. E., Thomas, W. O. Jr., and Riggs, H. C. (1994). *Nationwide Summary of U. S. Geological Survey Regional Regression Equations for Estimating Magnitude and Frequency of Floods for Ungaged Sites, 1993.* U. S. Geological Survey Water-Resources Investigations Report 94–4002. Reston, VA: U. S. Geological Survey, p. 196.

Kara-Zaitri, C. (1996). Disaster prevention and limitation: state of the art; tools and technologies. *Disaster Prevention and Management, 5*(1), 30–39.

Kirkwood, A. S. (1994). Why do we worry when scientists say there is no risk? *Disaster Prevention and Management, 3*(2), 15–22.

Luettich, R. A. Jr., Westerink, J. J., and Scheffner, N. W. (1992). *ADCIRC: An Advanced Three-Dimensional Circulation Model for Shelves, Coasts, and Estuaries.* Report 1. Theory and Methodology of ADCIRC-2DDI and ADCIRC-3DL (No. CERC-TR-DRP-92-6). Vicksburg, MS: Coastal Engineering Research Center.

McDonnell, R. A. (1996). Including the spatial dimension: using geographical information systems in hydrology. *Progress in Physical Geography, 20,* 159–177.

Meyer, J. (2004). *Comparative analysis between different flood assessment technologies in HAZUS-MH.* Ph.D. Dissertation, Faculty of the Louisiana State University and Agricultural and Mechanical College in partial fulfillment of the Requirements for the degree of Master of Science in The Department of Environmental Studies by Jennifer Meyer, B. A., Louisiana State University.

Murray-Tuite, P. and Wolshon, B. (2013). Evacuation transportation modeling: an overview of research, development, and practice. *Transportation Research Part C: Emerging Technologies, 27,* 25–45.

Nejat, A. and Damnjanovic, I. (2012). Agent-based modeling of behavioral housing recovery following disasters. *Computer-Aided Civil and Infrastructure Engineering, 27*(10), 748–763.

O'Connor J. E. and Costa J. E. (2003). *Large Floods in the United States: Where They Happen and Why.* Reston, VA: U.S. Department of Interior.

Park, J., Cho, J., and Rose, A. (2011). Modeling a major source of economic resilience to disasters: Recapturing lost production. *Natural Hazards, 58*(1), 163–182.

Parks, B. O. (1993). The need for integration. In Goodchild, M. F., Parks, B. O., and Steyaert, L. T. E. (eds.), *GIS and Environmental Modeling: Progress in Research Issues.* New York: Wiley, pp. 31–34.

Pine, J. C., Mashriqui, H., Pedro, S., and Meyer, J. (2005). Hazard mitigation planning utilizing HAZUS-MH flood and wind hazards. *Journal of Emergency Management, 3*(2), 11–17.

Rose, A. and Liao, S. Y. (2005). Modeling regional economic resilience to disasters: a computable general equilibrium analysis of water service disruptions. *Journal of Regional Science, 45*(1), 75–112.

Tobler, W. R. (1970). A computer movie simulating urban growth in the Detroit region. *Economic Geography, 46,* 234–240.

Warren Mills, J., Curtis, A., Pine, J. C., Kennedy, B., Jones, F., Ramani, R., and Bausch, D. (2008). The clearinghouse concept: a model for geospatial data centralization and dissemination in a disaster. *Disasters, 32*(3), 467–479.

U.S. Water Resources Council, Hydrology Committee (1967). *A Uniform Technique for Determining Flood Flow Frequencies.* Bulletin No. 15. Washington, DC.: USWRC.

Wolshon, B. and McArdle, B. (2009). Temporospatial analysis of Hurricane Katrina regional evacuation traffic patterns. *Journal of Infrastructure Systems, 15*(1), 12–20.

Chapter 4

Spatial Analysis

John C. Pine

Objectives

The study of this chapter will enable you to:

1. Define spatial analysis and explain how it is used in hazards analysis.
2. Explain the types of spatial analysis.
3. Describe how to visualize data using the results of a spatial analysis.

Key Terms

Accuracy
Buffering
Choropleth maps
Error
Geospatial data
Hydraulic analysis
Hydrologic data
Hypothesis
Metadata
Methodology
Precision
Reliability

Spatial analysis
Statistical analysis
Transformations

Issue

What tools are available to examine the spatial and temporal nature of hazards, risks, and disasters and their impacts?

Introduction

Dr. John Snow unraveled the causes of cholera in the mid-ninth century in London by recording on a map the incidence of cholera. He was able to observe from his map the relationship of a public water pump in the center of the cholera outbreak. Although his use of maps to track cholera outbreak did not prove the cause, it raised a question as to the relationship between drinking water and the outbreak of cholera. Stronger evidence was obtained to confirm his contention when the water supply was cutoff and the outbreak subsided (Gilbert 1958).

This illustration shows critical elements of the productive use of spatial analysis in a hazards analysis. First, John Snow collected accurate health data and made accurate georeference placement of this data on a map. He also noted on his map, other related items such as public water pumps. The scale of the map was of a small area within London and provided an appropriate view of the area in which to test his hypothesis. More importantly, Snow simply used his analysis of spatial data to raise a hypothesis that benefited from further study. He thus used information from his analysis, had a sound basis for choosing his data sources and how he would use this data in forming a hypothesis. His methodology was goal directed and determined the scope of his analysis. The key to spatial analysis is clearly stating what we intend to accomplish and determining a methodology that is suitable to achieve the desired results.

Definition of Spatial Analysis

Spatial analysis is a set of tools and methods that are used to examine the relationships between social, cultural, economic, ecological, and constructed phenomena. Bailey and Gatrell (1995) suggest that in broad terms, spatial data analysis is a quantitative study of phenomena that is located in space. In spatial data analysis, we explore the spatial relationships, associations, correlations, or spatial persistence of observable phenomena to determine areas of homogeneity often considered as "hot spots." For our purpose in examining hazards and their impacts, spatial analysis provides a means of understanding the nature of hazards and their social, economic, or ecological impacts. Spatial analysis is the center of how geographic information systems (GIS) are used in transforming and manipulating geographic data. It provides methods that may be used to support organizational decision making by government agencies, businesses, and nonprofits (Longley et al. 2005). The methods and tools provided by spatial analysis thus give us a means of turning raw data, such as what John Snow collected, into useful information. In the case of understanding natural hazards, we can enhance our understanding of the nature of hazards and their impacts by using spatial analysis. The results of our analysis can also help us to better communicate within organizations and with the public. Spatial analysis adds meaning, content, and value to our quest to better understand hazards and their impacts.

Spatial analysis is more than just a fast computer and expensive digital data. It is the formation of a hypothesis or question that may use geospatial data in expanding our understanding of how the physical environment interfaces with our social, economic, and natural environments. Geospatial data such as what was used by Snow reflect information that is associated with a spatial context. Statistical methods are used in our analytical methods to understand the relationships between spatial data elements. A statistical analysis might simply show cases of cholera within a one-mile radius of a point on the map. Alternative statistical methods allow us to examine associations that may reveal answers to our questions or hypothesis. Through spatial analysis, we are able to reveal patterns and processes that otherwise might not have been observed and confirm or disprove our hazard-related hypothesis (Anselin et al. 2006).

The association between (Turner et al. 2003) hazard events or exposure and human or ecological vulnerability is a key framework for spatial data analysis and fundamental in our quest to build and maintain sustainable systems. This suggests that there may be complex associations between natural and human systems that could be explained by spatial analysis. The relationship may be examined when two phenomena have locational data that may be compared. One set could have environmental data that reflects water quality, air temperature, and group elevation. The second used in the analysis could have health information, household income, or housing characteristics as well as locational data. The spatial analysis process thus uses a common locational relationship and examines possible linkages between the attribute data. The key is to acknowledge that any association is likely to be complex and our analysis subject to differences in the scale of our analysis, variation over time, the nature of the diverse systems that are the focus of the analysis, and the identification of potential causal relationships between and within different social and natural systems.

Fischer (1996) notes that spatial data analytical techniques perform a variety of functions within a GIS and are important for the types of questions and concerns that policy makers address in private, public, and nonprofit organizations. He further stresses that using geographic spatial relationships provide a framework for understanding the meaning of data. Spatial analysis evolved in the early 1960s as part of quantitative geography and the application of statistical processes in examining spatial relationships of points, lines, and area surfaces. A spatial temporal perspective was added to allow us to examine these relationships over time.

Geospatial Data Set

Geospatial data associated with hazards come in many forms and enable us to characterize both the nature and extent of the hazard event and the many elements that help shape or characterize the hazard. Geospatial data might reflect the characteristics of individuals or households, distribution points for food or water, and population density.

For an analysis of flooding events, it is helpful to have high-resolution elevation data to describe the broad geographic area that makes up a river basin, subbasin, or drainage area. Further, we need to characterize the size and shape of water features that make up the river basin and water flows over time (discharge) for a specific water feature. Other factors that influence flooding in a river basin include the size of the drainage area or the amount of impermeable services (paved roads or parking lots and residential structures, commercial buildings, or industrial sites). Flooding threats in a river basin may change over time if property near water features is changed from a natural landscape to one that has new subdivisions, commercial development, or major changes in roads or parking lots. Rain may flow more quickly into a water feather as a result of changes in the development of the landscape.

Riverine flood models use discharge values, soil types, land-use data, and elevation data in characterizing flooding events in a river basin or drainage area. Flood modeling programs use a variety of spatial analysis tools to determine the nature and extent of a flooding event for a specific geographic area. The accuracy of these data, which provide the input into the model, influences the validity of the modeling outputs.

Critical Thinking: Spatial analysis is dependent on the identification of accurate timely data and appropriate tools for manipulating the data to ultimately show where floodwaters will go over time and the depth of the water in a spatial context. The methodology that we establish must include the identification and selection of an appropriate data set that can support the results of our analysis.

Riverine flood modeling addresses the question of just how deep will the water be at a given time and location. The map shown in Figure 4.1 provides an illustration of the use of spatial analysis to show the anticipated depth of water for a 100-year flooding event in a drainage area. The HAZUS-MH (Hazards United States—Multi Hazard) Flood model developed by Federal Emergency Management Agency (FEMA) provides the means of using many types of data to characterize

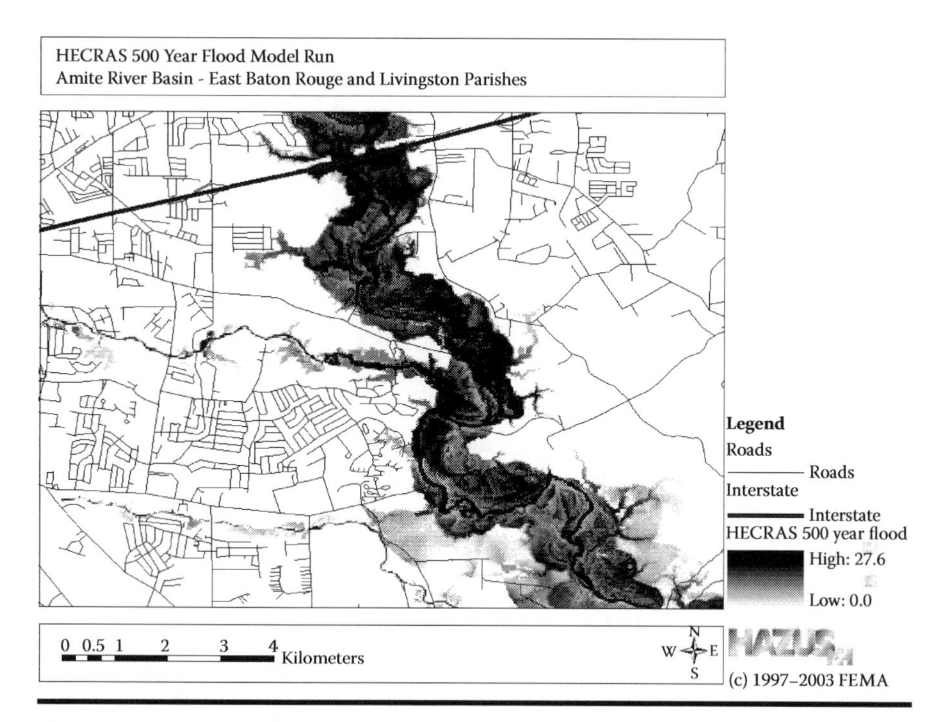

Figure 4.1 Riverine flood modeling results within HAZUS-MH (Hazards United States—Multi Hazard) flood.

riverine flooding events for a specific drainage area. The analysis of flooding in an area requires the following information:

1. Hydrologic data (just how much water may be in the water feature). Hydrology is the science that deals with the properties, distribution, discharge, and circulation of water on the surface of the land, in the soil and underlying rocks, and in the atmosphere. It also refers to the flow and behavior of water features.
2. Hydraulic analysis is a form of spatial analysis that determines flood elevations for a specific flooding event at a location on a water feature. Hydraulic data, thus, reflects anticipated areas to be flooded and the depth of flooding within these areas. Flooding depth and extent is determined by comparing the "modeled" flood elevations along a water feature with land contours [digital elevation model (DEM) land elevations]. A hydraulic model such as HECRAS (Hydrologic Engineering Centers River Analysis System) is used by FEMA and the U. S. Army Corps of Engineers to prepare community flood maps for the National Flood Insurance Program. How will the water move and flow in the drainage area? What will be the depth of the water?
3. High-resolution land contour data such as LIDAR (light detection and ranging) provides detailed data associated with land elevations.

4. Spatial modeling tools such as HECRAS calculate the depth of water along the water feature. GIS tools may be used to determine the banks of the water feature and in the deepest areas of the water bed.
5. Location of bridges or culverts that might limit or constrict the flow of the water.

The development of geographic information data centers has provided key data to state and local agencies along with businesses and nonprofit agencies interested in examining potential hazards. These data centers provide Internet access to public agency geospatial data that is critical in conducting a hazards analysis in a geographic area. Execute a search on the Internet to identify the location of data sets that could be used in a local community. Determine if you can find layers reflecting water features including streams, rivers, or lakes. Find road layers that reflect local, state, or national highways. See if you can locate data providers who can provide demographic information for a town, city, or county.

Spatial Data Quality

We should acknowledge that any data set will not be 100% accurate. Errors and uncertainty are inherent in a spatial data set (Openshaw 1996) and should be noted in information associated with a specific data set. Information associated with a geospatial data layer is known as metadata and helpful in understanding when and who created the data set, the purpose of the data and how it might be used, and contact information if users have questions.

Critical Thinking: To what degree does our data set accurately represent our environment (social, economic, ecological, and built environments)? Understanding the limitations of the data set is critical in formulating a sound methodology for our hazards analysis. What special problems are present in data sets? How does the availability of data influence our methodology that we use in our hazards analysis?

Many users of hazards analysis inherently trust computer outputs especially in a complex environmental hazards analysis. We should acknowledge that the computer model is just a tool that includes assumptions about the environment and the relationships between its variables. We should be very clear as to the limitations of the data inputs and the assumptions that the model makes in simulating a complex environmental hazard.

Those that use the outputs from a hazards analysis need to appreciate the uncertainly that is inherent in spatial data sets and the consequences of using these data sets in our analysis. There are clear limitations in any data set used in a hazards analysis; clearly expressing these limitations is critical for an appropriate application of the hazards analysis outputs in decision making. Goodchild (1993) stresses that GIS layers have inherent errors that may be obvious to GIS specialists but not understood or appreciated by those from other disciplines. The key is that one should not ignore inherent errors that are just part of a geospatial data set. Errors may occur in either the source of the data or in the processing steps of the GIS.

Hazards analysis combines the use of spatial analysis and environmental modeling to reveal characteristic hazards, risks, and potential disaster impacts. The linkages and integration between spatial analysis and environmental modeling are evolving and exist, which must be understood in completing a hazards analysis. Clarifying how spatial analysis tools and environmental models are used must be explained in our approach to a hazards analysis. Explaining our approach including data sources and analysis tools is reflected in methodology for a hazards analysis.

Data are collected within a specific context. Metadata files associated with a data set describes the process of the data collection, the purpose of the data set, and time lines for data collection, processing, assessment, and distribution. Understanding why and how the data were collected must be part of the methodology for our hazards analysis. Any conflicts that are identified with the scope and purpose of the data set and our use of the data must be explained.

We stress that our spatial analysis approach or methodology must include an examination of our metadata files that document who established the data set, when, the intended use, and date of outputs rarely address the accuracy of the data set. The metadata will provide us the information to explain why our selected data are suitable for what we hope to accomplish in our hazards analysis. This may be because it is just too costly to assess the spatial error in the data or because of the complexity of completing such an assessment.

Key terms associated with spatial data quality include error, accuracy, precision, and reliability. Error is any deviation of an observation and computation from what exists or what is perceived as truth (Brimicombe 2003). Accuracy is the degree of fit between our observation or computation with reality. Precision is the degree of consistency between our observations and what exists in the natural, social, or built environments. Reliability involves our confidence in the fit between our data and our intended application of the data in the hazards analysis process. For our purposes, the persons responsible for establishing a methodology for a hazards analysis have the responsibility to articulate what data sets are being used in our analysis, and why we believe that they are appropriate for our use. Our judgment as to the reliability of the data sets is critical in ensuring that users of the hazards analysis have confidence that their decisions are sound and can be supported by our methodology.

Uncertainty is an inherent element of data used in a hazards analysis and is associated with the ways that we obtain data sets, use them, store and manipulate them, and present the results of our analysis as information in support of organizational decisions. The outputs from our hazards analysis are, thus, dependent on data quality and model quality (including any spatial, statistical, or GIS tools that we use) (Burrough and Frank 1996).

A few illustrations can help demonstrate the importance of understanding the purpose of and intended use of a data set. We should understand who collected the data, when it was collected, and how and when it was disseminated. Many community and organizational hazards analysis efforts use U.S. Census road, water feature, community boundary, and point files. The U. S. Geologic Survey prepared these files many years ago and for much of the United States, these files have not been updated or corrected. The Census Bureau partnered with the U. S. Geological Survey (USGS) to add and edit data to the lines (roads, rail lines, and water features), points (community features such as schools, churches, or public buildings), and polygons (lakes and political boundaries). In editing and updating these files, the files now may accurately reflect correct street names, water features, or landmarks in a community. Understanding that these road and address files have been updated is critical in completing a quality hazards analysis.

A second illustration is seen in road files edited by many local governmental emergency communication districts. These units have taken the Census Bureau road files and aligned them over very-high-resolution digital images of their community. For many communities, high-resolution images of a half-foot resolution provide a basis for ensuring that a road feature or a school location is highly accurate. Prior to these corrections were being made, the Census map files so often used in a hazards analysis have extensive errors in the name of a specific feature and its location. It is not uncommon that a road or other feature may be off by as much as 100 ft. when observed on a high-resolution image of a community. Unfortunately, easy-to-use GIS programs were not available to local communities when the Census Bureau created the map files that have been used as part of the Centennial Census. Errors thus could be present in either geographic representation of the object or because of errors in the attributes reflected in the data (i.e., the road name or feature name is incorrect).

We should not avoid using Census map files in our hazards analysis, but insist that the metadata be reviewed. Errors in the road or water feature files must be fully understood and explained in our methodology in the hazards analysis.

Figure 4.2 provides a comparison between common community road files and edited road files both displayed over a high-resolution image. The image was taken after Hurricane Katrina in January 2006. These road files had been edited by the New Orleans Regional Planning Commission GIS unit for the City of New Orleans Planning Department years before Hurricane Katrina struck south Louisiana. These edited street files have been a long-standing asset to local and regional hazards analysis efforts in the public, private, and nonprofit sectors in the New Orleans area. The unedited Census road files on the left in Figure 4.2 are the type of road files that are available from many sources and commonly used by local jurisdictions as part of their base map. The edited files have been corrected using high-resolution images such as the ones above. The edited road files provide a highly accurate basis for spatial analysis. High-resolution photos were not available when the USGS created the road and street files. Many communities have edited the Census road and street files so that they more accurately reflect the local landscape when imposed over high-resolution images. Users of data such as Census road files must appreciate

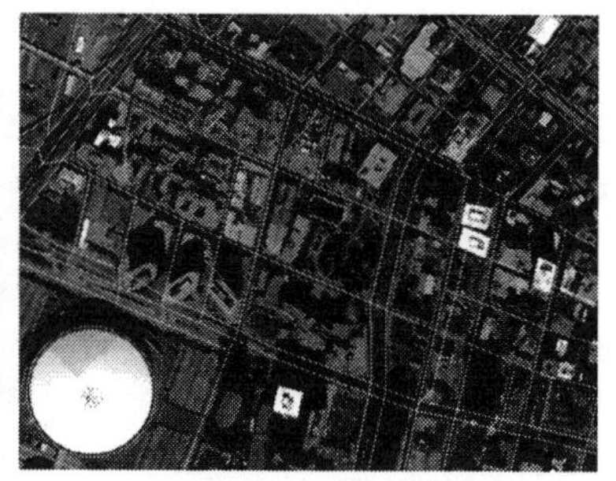

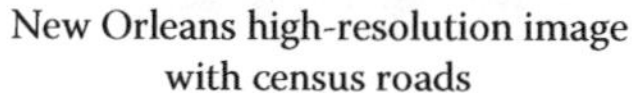

New Orleans high-resolution image with census roads

New Orleans high-resolution image with edited roads

Figure 4.2 Comparison of census road files and edited files.

that errors may exist in the files and if they are an appropriate basis for analysis of hazards at the community level.

Critical Thinking: If a local community is using unedited Census road and street files along with Census population data at the Block or Block Group resolution, could the data sets be used without potential errors distorting the results of the analysis?

Many errors that are inherent in data sets used in a hazards analysis occur because of changes over time. Changes in water features, land use, or landscapes may occur naturally or because of human interventions. We must examine any data set that is part of our hazards analysis to understand if changes have occurred and that these are noted in our methodology.

It should be noted here, as we discuss data quality, that users of environmental models assess the quality of their outputs by comparing the results of simulated disasters with actual events. Comparing the results of a hazards analysis provides a basis for adapting models and improve their predictive capacity for future studies.

Types of Spatial Analysis

Queries

How many people, commercial businesses, or residential homes might be impacted by a flood or storm surge? How many roads or bridges are in the area with the deepest flooding? How many structures are in the high-wind zone of a hurricane? How many renters or homeowners may be displaced by a flooding event? What is the average income of population of a community directly impacted by a hurricane? How many employees are affected by businesses in a flood zone?

Spatial analysis can address these questions and access to major transportation routes by renters, households below the poverty level, households with no automobiles and access to public transportation routes, or households with handicapped individuals over the age of 65 and shelters. Spatial analysis provides a means of comparing renters and homeowners and access to evacuation routes, evacuation access points, or shelters. Figure 4.3 shows the percent of renters by Census Block Group level in New Orleans. The analysis could help determine if renters might be more vulnerable than homeowners if an evacuation was ordered. Further, the analysis would be able to show which Block Group areas for either renters or homeowners are at higher risk by living further from an evacuation route, pickup point, or shelter. With this information, emergency management staff could target specific areas of the community for a contingency plan to ensure that all residents would be safe in a emergency.

Using Spatial Analysis to Answer Questions

Hurricane Katrina flooded many communities in the greater New Orleans area. What was the area flooded in the City of New Orleans? How did this change as rescue efforts progressed and pumps were used to remove the water? What is the average residential parcel or lot size in flood areas of the city? Using land-use classification data for the City of New Orleans, how much commercial or industrial property was flooded? How much public property for parks and open space was flooded? How much of the city's poor neighborhoods were flooded as compared to more wealthy areas?

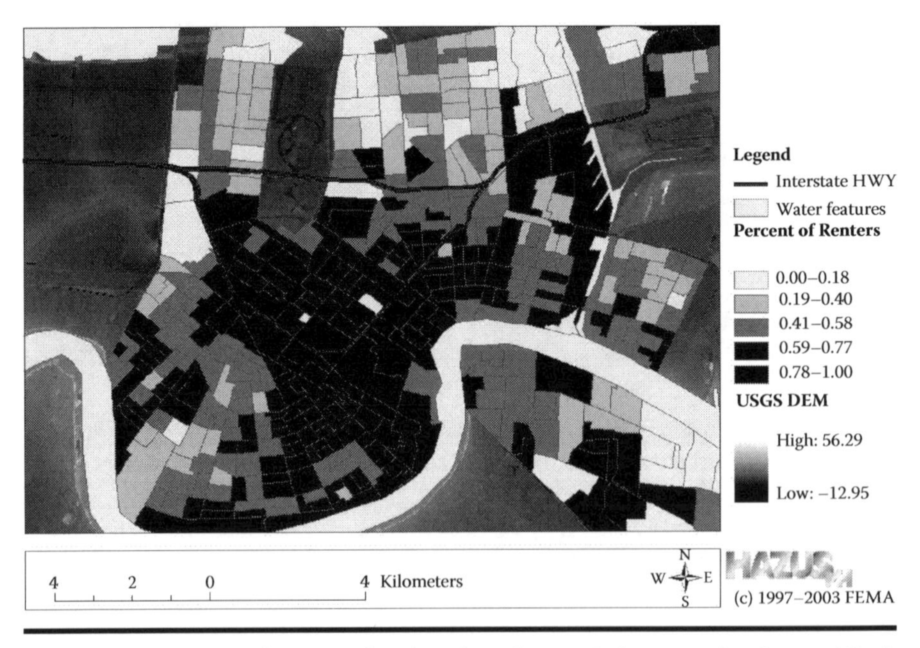

Figure 4.3 Percent of renters for the City of New Orleans at the Census Block Group level.

With the flood depth grid shown in Figure 4.4 for the City of New Orleans during Hurricane Katrina, one could use spatial analysis to determine if a higher percentage of households in flooded areas had incomes below the poverty level, had no access to an automobile, were handicapped, were renters, or had a single head of household with children below the age of 18. Pedro (2006) examined these questions in her Master's of Science thesis using the flood depth levels from a hurricane simulation for 2005 and determined that there were no differences between households on these characteristics when comparing flooded and nonflooded areas of the City of New Orleans. She also addressed the hypothesis that the simulated flooding did not have a disproportionate impact on the percentage of households in Census Block Groups who were African-American, below the poverty level, renters, or who did not have an automobile. With this analysis, she was able to pinpoint areas of the City of New Orleans where the depth of flooding might be very high and a greater percentage of residents would not have access to an automobile, were below the poverty level, and had a single head of household with children under the age of 18.

Kosar et al. (2011) provide a good example of the type of analysis that can be performed. Their study examined the association between disaster vulnerability and recovery within the context of community resilience. They established a spatial recovery index to assess the level of recovery and community resilience for a region in Austria. The study identified key spatial indicators of recovery and a spatial recovery index.

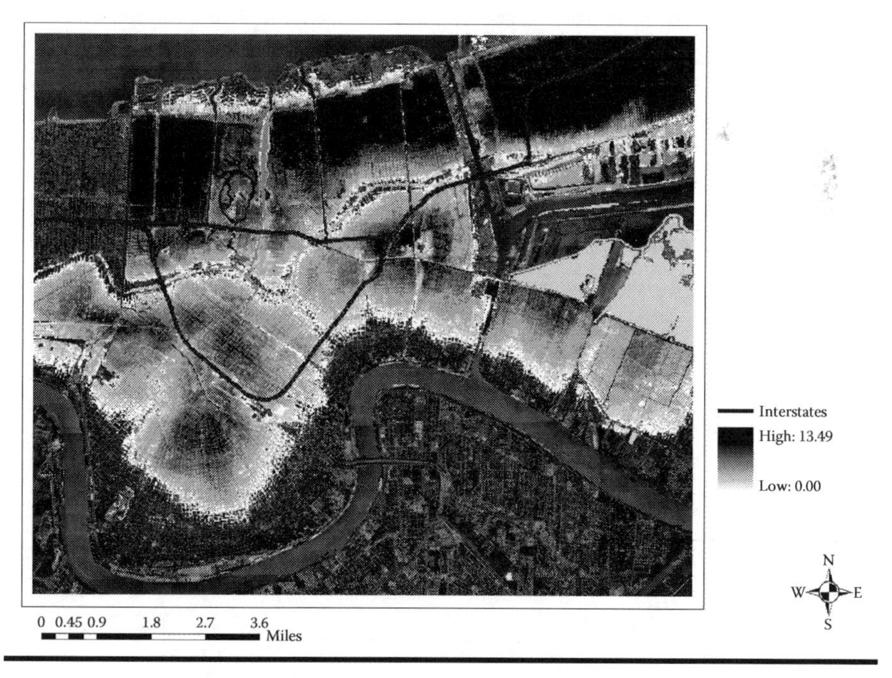

Figure 4.4 City of New Orleans flooding following hurricane Katrina. (Image courtesy of the National Oceanographic and Atmospheric Administration.)

Spatial analysis provides us with a set of tools in which we can explore questions concerning potential vulnerability and damage to hazards. It can provide information that be used to take precautionary measures or to further explore if some neighborhoods were more vulnerable than others and assistance with evacuations or sheltering was needed (Anselin and Getis 2010).

Transformations

These analysis tools allow the user to transform GIS data sets to reveal relationships and dynamics of the physical environment (Fekete 2012). Examples include buffering a point, line, or area to highlight potential change. If a new school were to be built in a specific location, what is the population in a two-mile area? If a commercial area were to be flooded, what other enterprises in a three-mile area could handle the additional business? If a rail line was damaged as a result of an earthquake, how many industrial enterprises within a 10-mile area could be impacted?

Buffering

Buffering was used by Pine et al. (2002) to determine if African-Americans were in closer proximity to 13 large chemical processing sites in Iberville Parish in Louisiana. The question centered on whether African-American residents were not closer to chemical processing operations than non-African-Americans. The buffer zones were determined using dispersion modeling programs from 13 sites, and the number of African-Americans was calculated for each risk zone from Census 2000 data. The buffering spatial analysis tool was helpful in examining claims of disparate impact of chemical releases for African-Americans in a community. The study showed that African-Americans did have a greater chance of living closer to 1 of the 13 chemical processing operations than non-African-American residents.

Spatial interpolation is used to help estimate potential flooding along a water feature where hydrologic modeling programs determine the depth and extent of flooding at various locations along a stream, bayou, or river. Riverine flood models use precise elevation measurements along water features as part of the flooding program. Spatial interpolation is used to estimate the depth and extent of flooding between survey points. Many hydrological flooding efforts include field survey cross sections along water features as a basis for determining flood depths. The depth of flooding between the cross sections is interpolated spatially. The model thus produces a smooth flood zone and depth of flooding for the area impacted using this spatial analysis tool.

Descriptive Summaries

Data sets that reflect unique elements of a disaster provide opportunities for understanding potential relationships between a disaster and associated human characteristics. Following Hurricane Katrina, 911 emergency calls (point locations) were examined to see if the calls were clustered in some way to suggest the vulnerability of the caller, areas at extreme risk, or the locations of safe refuge. One issue that was examined centered on a possible relationship between clusters of 911 emergency calls and water depth. Spatial analysis was used to identify hot spots, where there was a high number of emergency calls for assistance. Further analysis examined the depth of the water at these areas and social characteristics of the area.

Optimization Techniques

Spatial analysis is also used in site selection and transportation routing to help locate the ideal setting for an emergency shelter, medical clinic, or police substation, the shortest evacuation route of multilane roads and highways. Evacuation routes that are scenario specific can be developed to aid community planners in evacuating large populations from a metropolitan area. State departments of transportations have used these tools to mark major evacuation routes as aids to move citizens from vulnerable areas, that is, either from hurricanes, earthquakes, or riverine flooding.

Hypothesis Testing

This type of spatial analysis was used in New Orleans to anticipate the rate of return to specific neighborhoods following the flooding from Hurricane Katrina. A statistically valid sample of household surveys was conducted to determine a family's capacity and willingness to return to the city when the residence was flooded. The address of the respondent was obtained in the survey along with the resident's perception of the level of damage to their home and neighborhood. Independent surveys of residential structures conducted by FEMA, the City of New Orleans, and the Louisiana Road Home Program provided an independent perspective on individual property damage. Household sentiment to return was then compared to their perception of the level of damage to their home and the independent property damage assessment. Spatial analysis was used to use the results of the surveys to infer if other residents would return in a given time period. This type of analysis provided a basis for testing a set of hypothesis relating to the willingness to return to a specific structure or neighborhood.

Spatial analysis was used to examine a hypothesis concerning social vulnerability and flood depth following Hurricane Katrina. The question centered on an association between risk zones measured by depth of floodwaters and social vulnerability. This study examined the relationship between risk flood depth (risk) and population characteristics (vulnerability) including race, income, disability, home ownership, single family member and head of household, and household assess to an automobile. The analysis revealed that African-Americans had the strongest association with deeper floodwaters when compared with other population characteristics for the Orleans Parish (Curtis et al. 2010).

Since two different data sets were used in the study, one had to be converted to a common type and scale. Flood depth values from a grid file were selected and then averaged for each Census Block Group. Since land elevations in New Orleans varied only slightly within a Block Group, this conversion produced good flood estimates for each Block Group. It is critical that the methodology used in a hazards analysis fully explore any potential problems with geospatial data that are used and explain the source of any data and how it may have been adapted for the spatial analysis (Rosenberg and Anderson 2011).

Spatial Data Visualization

A critical part of the hazards analysis process is displaying the results on our analysis. Hazards are geospatially oriented and thus being able to show the results of our analysis are a key element in supporting individual, organizational, and community decision making. No matter if the results of an analysis come from a modeling program such as HAZUS-MH [earthquake, flood, coastal hazards, or ALOHA (Areal Locations of Hazardous Atmospheres)] display geospatial data is a critical means of conveying information to users of our final hazards analysis or as we work with the data using spatial analysis techniques. GIS allows us to interactively examine map layers to reveal information from the landscape in a variety of ways. We can add different spatial layers, such as where people live, transportation routes, quarantine areas, key facilities, or infrastructure, and display them over high-resolution images of a community.

Visualization of the results of a hazards analysis and the use of spatial analysis and mapping tools can help us to

1. Identify patterns within complex data sets or multiple data sets of related data.
2. Make sense of large data sets.
3. Appreciate that local geospatial features change over time.
4. Geospatial features may be similar, or interact more frequently, within smaller geographic scales.
5. May provide a means of conveying complex information without oversimplification of the data.
6. Can give an EOC (Emergency Operations Committee) at a local, state, or regional level critical information on the nature of hazards and their potential impacts.

Both Figures 4.3 and 4.4 offer illustrations of how we visualize hazard vulnerability in a simulation or exercise, or an actual disaster response. On examination of the map of the City of New Orleans in Figure 4.3, one sees that there is a great variation in the percentage of people who are renters. Two patterns may be seen to suggest that the higher percentage of Census Block Groups are in the Central and Mid-City neighborhoods while those areas on the urban fringe of New Orleans have the lowest percentage of renters. A test of this theory and the association between income and percent of renters can be determined quantitatively using spatial analysis. The map shown in Figure 4.4 provides us with a broad view of the community and a basis for testing a hypothesis using spatial analysis.

Additional hypothesis could be identified using broad views of the community raising questions as to the relationship between ground elevation and household income, the relationship between major transportation routes and rental housing, or the association between housing values and community recreation areas (parks). A look at Figure 4.3 shows that City Park is located in the top of the image and that limited rental units are available near the community recreation area. A spatial analysis could address the hypothesis to determine if there is a clear association between large community recreation areas and block groups with low percentages of households of renters.

Figure 4.4 was used extensively by emergency responders at the local, state, and federal levels following the flooding from Hurricane Katrina. As the water depth changed as the result of pumping from the City of New Orleans, one could provide information to emergency responders concerning the use of major transportation routes. Spatial analysis was used to determine the best rescue routes throughout the City of New Orleans.

For any hazard, there is both a spatial and temporal dimension. The spatial dimension has various scales (local to international) depending on what the hazard is. The temporal dimension also has multiple scales (minutes to months).

Mapping data related to a hazard uses many different data sets and types. What are the best ways to show hazards from wind, flooding, storm surge, earthquake, landslide, drought, or other disaster?

Every map is a graphic representation or a model of reality or milieu.

1. A map may represent economic or cultural features, such as neighborhoods, settlement patterns, political–administrative boundaries, and so on.
2. A map represents physical features, such as elevation, water features, land cover, and so on.

A map can also display mental abstractions that are not physically present on the geographical landscape. An example is to map people's attitude (quality of life for or against gun control, etc.).

Choropleth Maps

A choropleth map is defined by the International Cartographic Association as "a method of cartographic representation, which employs distinctive color or shading to areas other than the feature boundaries. These are usually statistical or administrative areas." Making a choropleth map starts with the collection of data by a specific geographic area. An areal symbolization scheme is then devised for these values, and the symbols are applied to those areas on the map whose data fall into the symbol classes. The selection of symbol classes is based on a classification method.

It is important that choropleth maps show relative data in contrast to absolute data. Relative data includes densities (e.g., population density—people per square mile), percentages (e.g., percentage of people 65 years and older), and rates (e.g., number of homicides per 100,000 people). The following three decisions have to be made when compiling any choropleth map:

1. Number of classes: A trade-off exists between too many and too few classes. Too many classes make the choropleth map complex and difficult to perceive and understand by the map-reader. Too few classes results in too much information loss.
2. Type of classification method. This includes equal steps that could be intervals such as 0–10, >10–20, >20–30, and so on. A second one involves natural breaks in the data, which is reflected in the sorted data. Quantiles could be used as a classification method using equal number of observations in each class. Finally, a standard deviation can be used that includes the average deviation of the data values from the mean (average) of the data set. This approach measures the variability in the data and is a good relative measurement tool.
3. Color or areal symbolization scheme can be used, which simply shows the different areas (ZIP codes, incorporated areas, or districts) in various colors.

Critical Thinking: Figure 4.5 provides an example of how we can display the same information using different classification methods. The maps show that the manner in which we classify the data will influence how we view the information and the conclusions that we draw. Which of the four maps provide the best spatial and temporal perspectives?

An example of a proportional symbol map is the common dot map. Common dot mapping involves the selection of an appropriate point symbol to represent each discrete element of a geographically distributed phenomenon. The symbol form does not change, but its number changes from place to place in proportion to the number of objects being represented. Design decisions involve the placement of dots and the selection of dot value and dot size. Figure 4.6 shows data in a jurisdiction that could have resulted from an analysis of spatial data. The dot symbols reflect data for a specific geographic boundary.

A proportional symbol map uses a form (circle, square, or triangle) and varies its size from place to place, in proportion to the quantities it represents. The map in Figure 4.7 shows a quantitative distribution by examining the pattern of differently sized symbols. Proportional point symbol mapping is selected when data occur at points or when data are aggregated at points representing areas as illustrated in the Figure 4.6.

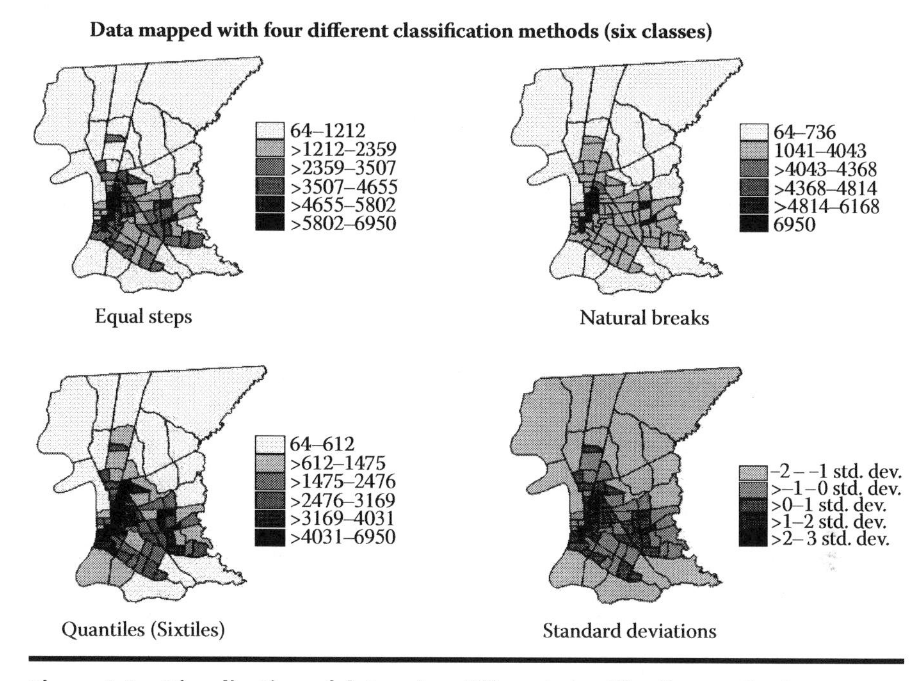

Figure 4.5 Visualization of data using different classification methods.

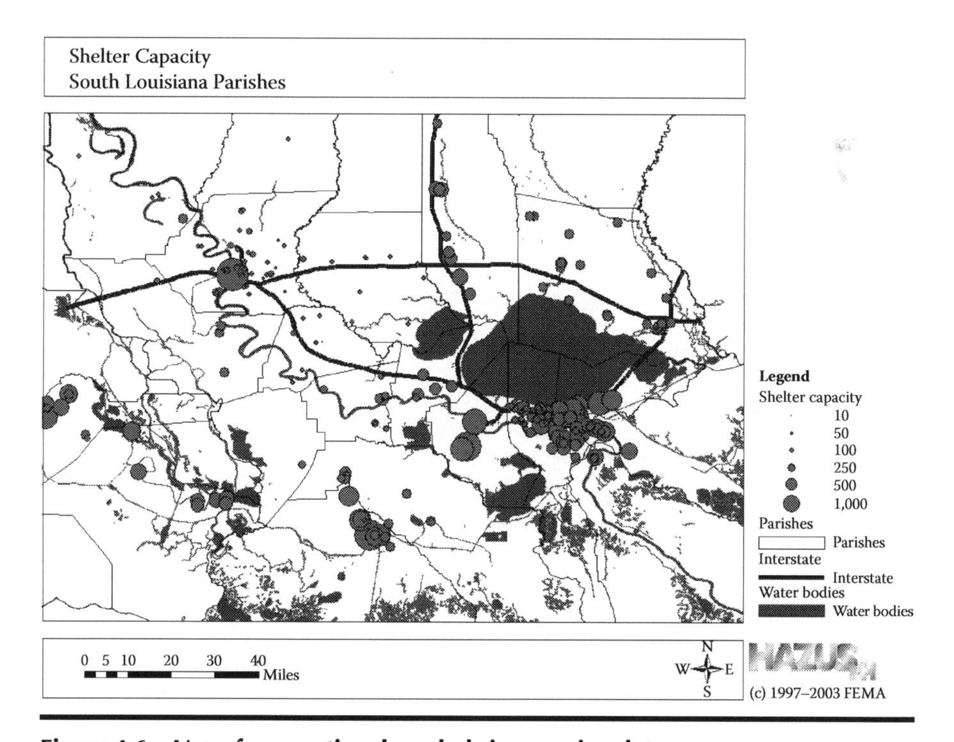

Figure 4.6 Use of proportional symbols in mapping data.

Conclusions

Brimicombe (2003) stresses that suitability of a data set for our use in a hazards analysis centers on its "fitness for use." Rather than focus on errors, he encourages us to view our use of data in a wider examination of uncertainty. We should examine the quality of the data and explain that the data set is an appropriate application for our use. We thus examine the quality of the data set and how we plan on using the data in our methodology. In the end, we want to ensure that our data fits our methodology and is an appropriate use of the data set in our analysis. A fitness-for-use test thus includes an evaluation of the data quality and an explanation of any limitations of the data as it is used in our hazard models or spatial analysis techniques. The key is that this is a managed process that accounts for limitations in our data. Our methodology in completing the hazards analysis should clearly explain the steps that we are taking in the spatial analysis including the source of our data and our analysis of this data.

We stress that our methodology must include an examination of our metadata files which document who established the data set, when, the intended use, and date of outputs rarely address the accuracy of the data set. The metadata will provide us the information to explain why our selected data are suitable for what we hope to accomplish in our hazards analysis. This may be because it is just too costly to assess the spatial error in the data or because of the complexity of completing such an assessment.

Discussion Questions

Many efforts to examine environmental hazards require the use of DEM data. The USGS has published for many years DEM data for the United States in different scales (1:30 m or 1:20 m). More recently, many state and federal agencies have created higher-resolution DEM data using LIDAR remote sensing technology (1:5 m resolution). What difference does using the higher-resolution DEM make for an environmental hazards analysis?

Given that any data set will not be 100% accurate, why is it so important that errors be examined and explained in a hazards analysis?

What do the terms error, accuracy, precision, and reliability mean, and why are they so critical to the use of data in a hazards analysis?

How might the results of a spatial analysis be visualized so as to communicate the nature of hazard risks and help anyone understand the outputs of a hazards analysis?

Applications

Metadata provides information about a geospatial data set to guide the user in determining how best to use the data in a hazards analysis or other application. Metadata files are provided for many data sets as illustrated by the Atlas Internet site at Louisiana State University. Go to this site, select the LIDAR data set, and download for any geographic area. Read the file and determine how the data were obtained, who collected the data, when was it collected and made available to the public, what is the resolution of the data, and what is the intended purpose of the data? The file notes that no data quality tests were performed on this data. What might that mean to the user of the data?

Map a disaster using a common dot map approach for the same study area using different dot sizes and dot values and observe the differences between the different dot maps. Use any mapping or GIS software package to perform this exercise.

Websites

Census TIGER Line Data: http://www.census.gov/geo/www/tiger/tgrshp.html
Federal Statistics: http://www.fedstats.gov/
One Stop for Data: http://gos2.geodata.gov/wps/portal/gos

References

Anselin, L. and Getis, A. (2010). Spatial statistical analysis and geographic information systems, pp. 35–47. In L. Anselin and S. J. Rey (eds.), *Perspectives on Spatial Data Analysis.* Springer-Verlag Berlin Heidelberg.

Anselin, L., Syabri, I., and Kho, Y. (2006). GeoDa: an introduction to spatial data analysis. *Geographical Analysis, 38*(1), 5–22.

Bailey, T. C. and Gatrell, A. C. (1995). *Interactive Spatial Data Analysis* (Vol. 413), Essex: Longman Scientific & Technical.

Brimicombe, A. (2003). *GIS, Environmental Modeling and Engineering.* London: Taylor & Francis.

Burrough, P. A. and Frank, A. U. (eds.) (1996). *Geographical Objects with Indeterminate Boundaries.* London: Taylor & Francis.

Curtis, A., Li, B., Marx, B. D., Mills, J. W., and Pine, J. (2010). A multiple additive regression tree analysis of three exposure measures during Hurricane Katrina. *Disasters, 35*(1), 19–35.

Fekete, A. (2012). Spatial disaster vulnerability and risk assessments: challenges in their quality and acceptance. *Natural Hazards, 61*(3), 1161–1178.

Fischer, M. M. (1996). *Spatial Analytical Perspectives on GIS.* Boca Raton, FL: CRC Press.

Gilbert, E. W. (1958) Pioneer maps of health and disease in England. *Geographical Journal, 124*, 172–183.

Goodchild, M. F. (1993). Data models and data quality: problems and prospects. In M. F. Goodchild, B. O. Parks, and L. T. Steyaert (eds.), *Environmental Modeling with GIS.* New York: Oxford University Press.

Kosar, B., Gröchenig, S., Leitner, M., Paulus, G., and Ward, S. (2011). The Application of Geospatial Technology in Hazards and Disaster Research: Developing and Evaluating Spatial Recovery Indices to Assess Flood Hazard Zones and Community Resilience in Austrian Communities. In A. Car, G. Griesebner, and J. Strobl (eds.), Geospatial Crossroads @ GI_Forum '11.

Longley, P. A., Goodchild, M. F., Maguire, D. J., and Rhind, D. W. (2005). *Geographic Information Systems and Science*, Second Edition. Europe: John Wiley & Sons.

Openshaw, S. and Clarke, G. (1996). Developing spatial analysis functions relevant to GIS environments. *Spatial analytical perspectives on GIS*, 21–37.

Pedro, S. (2006). *Delineating Hurricane Vulnerable Populations in Orleans Parish, Louisiana*. Master's Thesis, Department of Environmental Sciences, Louisiana State University.

Pine, J. C., Brian D. M., and Lakshmanan, A. (2002). An examination of accidental release scenarios from chemical processing sites: the relation of race to distance. *Social Sciences Quarterly*, *83*(1), 317–331.

Rosenberg, M. S. and Anderson, C. D. (2011). PASSAGE: pattern analysis, spatial statistics and geographic exegesis. Version 2. *Methods in Ecology and Evolution*, *2*(3), 229–232.

Turner, B. L., Kasperson, R. E., Matson, P. A., McCarthy, J. J., Corell, R. W., Christensen, L., and Eckley, N., et al. (2003). A framework for vulnerability analysis in sustainability science. *Proceedings of the National Academy of Sciences*, *100*(14), 8074–8079.

Chapter 5

Risk Analysis: Assessing the Risks of Hazards

Kevin L. Shirley and John C. Pine

Objectives

1. Explain the process of risk analysis.
2. Explain what risk is.
3. Compare and contrast quantitative and qualitative approaches to risk analysis.
4. Identify and discuss related to using historical data in determining risk.
5. Explain the concept of uncertainty and how it impacts risk analysis.
6. Discuss the concept of acceptable risk and how we determine it.
7. Explain how we describe the likelihood and consequences of risks.

Key Terms

Acceptable risk
Flood flow frequency flood discharge uncertainty
Hazard models
Logic tree
Monte Carlo method
Qualitative analyses

Quantitative analyses
Risk
Voluntary risks and involuntary risks
Vulnerability

Introduction

Risk analysis is the determination of the likelihood of a disaster and possible consequences. To begin this analysis, the hazards affecting a community must be identified. After identification, data are collected to prepare a community hazards profile that will characterize the nature and extent of these hazards. Finally, we do an analysis of the risks present. This analysis includes identifying community vulnerability indicators and the probability of the hazard occurring. From this information, we explore the likely impact of the hazards on the community. This exploration may include the use of hazard models and software such as geographic information systems (GIS). This analysis should provide useful and accurate information for decision makers working in risk management or responsible for community hazard mitigation initiatives. Our goal is to provide decision makers with the right information, at the right level of complexity and detail at the right time.

Process of Risk Analysis

We are all vulnerable to some hazards. A community may be vulnerable to natural hazards because it is located on a coastline and subject to hurricane winds or storm surge while another community located in the mountains may be vulnerable to fire and floods. Some organizations may be constrained in the location of their business. For example, a business may need to be located near a major transportation route. As a result, it may be necessary to locate their operation in a floodplain or near a coast. By using risk analysis, the community or organization can make informed decisions about their exposure to the local hazards. The risk analysis process is used to assess this vulnerability.

The process of risk analysis examines the nature of the risk from a hazard, when and where it might occur, potential intensity, and the potential impact on people and property. The level of risk for a disaster of any scale is expressed as a likelihood of the occurrence or frequency times its consequences. Hence, risk analysis must begin with hazard identification. With each hazard identified, the probability or frequency of occurrence of the hazard event and the consequences if the event should occur is explored. The consequences could be loss of life; the sociocultural impact; or economic, recovery, and environmental costs of a disaster. Once the analysis is completed, the results of the risk analysis can be used in the problem-solving and decision-making process to adopt strategies to reduce organizational or community vulnerability.

In light of the inevitability of facing risk, individual families, organizations, and communities must make conscious choices about what is an "acceptable risk." Hazard reduction policies can be made with an understanding of what choices are possible and the consequences for any option. The level of risk may be very limited so that nothing needs to be done to address it. Other hazards may be more likely to occur and have the potential to cause extensive damage. The fact is that some organizations and communities may be willing to live with a specific risk or not willing to expend the resources necessary to reduce the adverse consequences that come with it. To assist in this decision-making process, relevant analysis is conducted. This analysis might include mapping the hazard to determine the spatial distribution of risk, such as the risk associated with a gas or chemical leak. In the case of an area vulnerable to landslide, this analysis could include the collection of data on the frequency and intensity of past landslides and the local areas most vulnerable to future landslides. In many cases, especially when data are lacking or need to be interpreted, judgments are made concerning specific risk factors (i.e., factors that may significantly increase or decrease the risk of disaster or the threat to life and property) and the vulnerability of the people and property within the risk area. The analysis stemming from the available data and expertise as well as the use of judgment are part of the hazards analysis process.

Risk managers consider the likelihood and consequence of all (identified) hazards faced by their jurisdiction, and they rank them according to priority. However, to understand the likelihood component of the risk analysis, one needs to understand probability. It is the probability of an occurrence that informs a risk manager whether or not they should expect a hazard to affect their community. Jardine and Hrudley (1997) suggest that a classical or frequency concept of probability be used and focus on discrete events, which examine all possible outcomes and the numerical relationships among the chances of these outcomes. In the real world, however, such complete information is seldom, if ever, available. Therefore, risk analyses must include subjective information along with detailed historical information.

What Is Risk?

Risk is the product of likelihood or probability of a hazard occurring and the adverse consequences from the event and viewed by many as our exposure to hazards. Figure 5.1 provides a model for defining risk.

Risk = (Likelihood or Probability) × Consequence

Figure 5.1 Defining risk.

This approach is based on the Royal Society Study Group (1992) defining risk as "the probability that a particular adverse event occurs during a stated period of time, or results from a particular challenge" (p. 2). The Society provides a basis for an analysis of risks associated with hazards by measuring the likelihood and consequence of hazards in the community. How one perceives the adverse impacts of risk either from an individual, organizational, or societal perspective certainly influences strategies to address the risk of natural hazards. Also, the process used in the analysis will help to shape the individual and institutional approaches in addressing risk. Although individuals, organizations, and public policy positions may be viewed differently, an open analysis of hazards is constructive in preparing a sound hazard risk management policy and community hazard mitigation plans.

Risks may be viewed as voluntary where we agree to participate in activities that increase our chances of harm or injury including driving fast or participating in high-risk sporting events. Other risks that we do not choose to participate in are classified as involuntary risks where we unknowingly or unwittingly are exposed to harm. For involuntary risks, one may be exposed simply because the nature of the risk has changed as in a potential for wildland fire or a hazardous material spill. Some communities may not appreciate the actual risk from some hazards because they have adjusted to the threat presented by the hazard and not examined alternatives that would reduce their vulnerability.

The Royal Society Study Group (1992) acknowledges that risk management as a concept involves making decisions concerning risks and that this concerns both hazard identification and risk analysis. Our use of the term risk analysis fits within this context and reflects our determination to understand the likelihood of a hazard event and the consequences of the disaster on a community, region, or an organization. This definition of risk analysis comprises the identification of the outcomes and estimations of the magnitude of the consequences and the probability of those outcomes. Finally, organizations use the outcomes of risk analysis to determine what is an acceptable level of risk and if anything can be done to reduce the adverse effects of the risk of a specific hazard. The determination of risk reduction measures at an organizational level is regarded as hazards risk management.

Quantitative Analysis of Risk

There are predominantly two categories of analysis: quantitative analyses and qualitative analyses. Quantitative analysis uses statistical measures to derive numerical references of risk. Qualitative analysis uses categorical variables in describing the likelihood and consequences of risk. Quantitative analysis uses specific measurable indicators (whether dollars, probability, frequency, or number of injuries/fatalities), whereas qualitative analysis uses qualifiers to represent a range of possibilities.

Quantitative Analysis of Likelihood

Quantitative analysis of the likelihood component of risk seeks to find the numerical statistical probability of the occurrence of a hazard causing a disaster. These analyses tend to be based on historical data. A standard for the numerical measurement for this likelihood of occurrence must be established. One of the most commonly used quantitative measures of likelihood is the number of times a particular hazard causes a disaster per year. For example, a measure indicating the frequency of a hazard occurrence 3 per year (3/year) would indicate an historical average of three hazard events occurring annually. Other time frames may also be used such as 1/decade or 10/week. An alternative technique for a quantitative measure of likelihood is to express the frequency per time frame as a probability that reflects the same data, but expresses the outcome as percentage between 0 and 100. For example, a 100-year flood has a 1/100 chance of occurring in any given year or expressed as a probability of 1% or 0.01, whereas a hazard that occurs 1/decade has a 10% or 0.1 chance of occurring in a given year. We interpret these probabilities based on how close they are to 0%, 50%, or 100%. For example, a 0% chance for occurrence indicates the hazard will not occur and a 100% chance for occurrence indicates the hazard is certain to occur in the specified time. The closer the percentage is to 100%, the more likely it is to occur, whereas a 50% chance indicates the hazard is equally likely to occur as to not occur.

Quantitative Analysis of Consequence

As was true with the likelihood component of risk, the consequences of risk can also be described according to quantitative or qualitative reporting methods. The quantitative representation of consequence can be represented by the number of deaths or injuries or by estimating actual damages from various events. For example, the final death toll for Hurricane Katrina was 1836 and caused $81 billion in property damage (Zimmermann 2012). These quantitative measures are sometimes used to rank and compare disaster events, such as the deadliest or most expensive. Figure 5.2 (National Oceanographic and Atmospheric Administration [NOAA],

Hurricane	Year	Category	Deaths
Great Galveston Hurricane	1900	4	8,000–12,000
Okeechobee Hurricane	1928	4	2,500–3,000
Hurricane Katrina	2005	3	1,500+
Louisiana Hurricane	1893	4	1,100–1,400
S. Carolina / Georgia	1893	3	1,000–2,000

Figure 5.2 Deadliest U.S. hurricanes.

NWS NHC [National Weather Service National Hurricane Center] 47) shows the deadliest hurricanes on record to hit the United States.

Qualitative Analysis of Risk

Qualitative Analysis of Likelihood

Qualitative representation of likelihood uses words to describe the chance of an occurrence. Each word, or phrase, will have a designated range of possibilities attached to it as illustrated in the categories in Figure 5.3.

Individuals determine the risk of a specific hazard by making a judgment among these alternatives. We base these judgments on many factors, which could include our recent experience, how hazards have affected others, information provided by the media, and/or community meetings that may have addressed potential hazards. We may also convert a calculated quantitative measure to a qualitative variable. For example, if it is calculated that there is a 1.5% chance of a wildfire in a specific area, then we would assign the categorical variable rare to this event according to Figure 5.3.

Qualitative Analysis of Consequence

As was true with the qualitative representation of likelihood, words or phrases that have associated meanings are used to describe the effects of a past disaster or the anticipated effects of a future one. These measurements can be assigned to deaths, injuries, or costs (oftentimes, the qualitative measurement of fatalities and injuries are combined). Figure 5.4 provides an illustration of the subjective ranges to help quantify the measurement indicator associated with injuries and fatalities.

Critical Thinking: We attempt to understand risk using both quantitative and qualitative tools that allow us to examine hazards and their impacts using both the physical and social sciences. Not only is an understanding of risk shaped on an individual basis by the individual's familiarity with local hazards (Slovic et al. 1979) but

Chance of occurring in a given year	
Certain	>99%
Likely	50%–99%
Possible	5%–49%
Unlikely	2%–5%
Rare	1%–2%
Extremely rare	<1%

Figure 5.3 Qualitative representation of likelihood.

	Injuries	**Fatalities**
Insignificant	None	None
Minor	Small number first aid treatment required	None
Moderate	Medical treatment required, some hospitalization required	None
Major	Extensive injuries, significant fatalities	Some
Catastrophic	Large number of severe injuries[a]	Some

[a] Emergency Management Australia (2001).

Figure 5.4 Qualitative consequence indicators.

also from elements of local culture that includes how hazards have been viewed locally over time. For example, residents in the northeast seemed relatively unprepared when hurricane Sandy struck the northeastern states in 2012. However, just 1 year after Sandy struck, 49% of the residents surveyed in New York and New Jersey believed that Sandy made them more urgent and thorough in their hurricane preparedness (Breslin 2013). For hazards that are possible in your area, what influences your view of risk?

Views of Risk

Rejeski (1993) notes that discussions of risk have included three primary groups including scientists who form their opinions through a rational process, policy makers who establish their perspectives based on multiple sources of information including quantitative and qualitative data, and finally the public whose perceptions and judgments of risk are formed from their own perspectives in some circumstances despite data provided by the other two major groups. He observes that environmental risks and risk are full of ambiguities that may not be resolved especially when interested groups have such different perspectives on the issues. He believes that a common view of risk can only be obtained when groups agree to share their perceptions and basis for their positions. He stresses that there is a great difference between uncertainty and ambiguity. For ambiguity, there are intrinsic elements of public policy that separate risk management strategies from the risk analysis process. One of the key elements in debates concerning risk and uncertainty is the relative level of trust that is established between the three groups, that is, scientists, policy makers, and the public. One possible option that may lead to a consensus is to encourage a more participatory process and open dialogue. Using tools to visualize the data and scientific results concerning a hazard can provide a point of access into this meaningful discussion. As an example, GIS provides a tool for examining both hazards and risk. It can be used to visualize the nature and extent of a risk zone, as shown in Figure 5.5. Three risk zones are displayed in this simulation of a hazardous spill release as well as an additional zone reflecting uncertainty. Unfortunately, this tool cannot solve the problem of disagreement but it may enable those interested in the risks of hazards with the means of building a consensus.

To examine how hazard models and spatial analysis tools may address some issues, let us scrutinize the GIS example more closely. Figure 5.5 provides an estimate of an accidental release of ammonia on a cool cloudy February day at 10 AM, and the wind is assumed from the east at 10 mph. The release is assumed to occur near a hospital when a 600-pound tank drops from a truck unloading a shipment of various cylinders. The model output provides three estimates of risk using alternative exposure limits of 25, 150, and 750 ppm. The model shows most vulnerable areas and provides an estimate of the exposure limits within these areas. This zoning helps to clarify the spatial uncertainty inherent in such a disaster. The goal then is to more fully understand the limitations of our hazard model and the data that are used in the spatial analysis indicating the area of vulnerability.

The three exposure limits for the scenario in Figure 5.5 were drawn from Emergency Response Planning Guidelines (ERPGs), which are used in the ALOHA (Areal Locations of Hazardous Atmospheres) chemical dispersion model

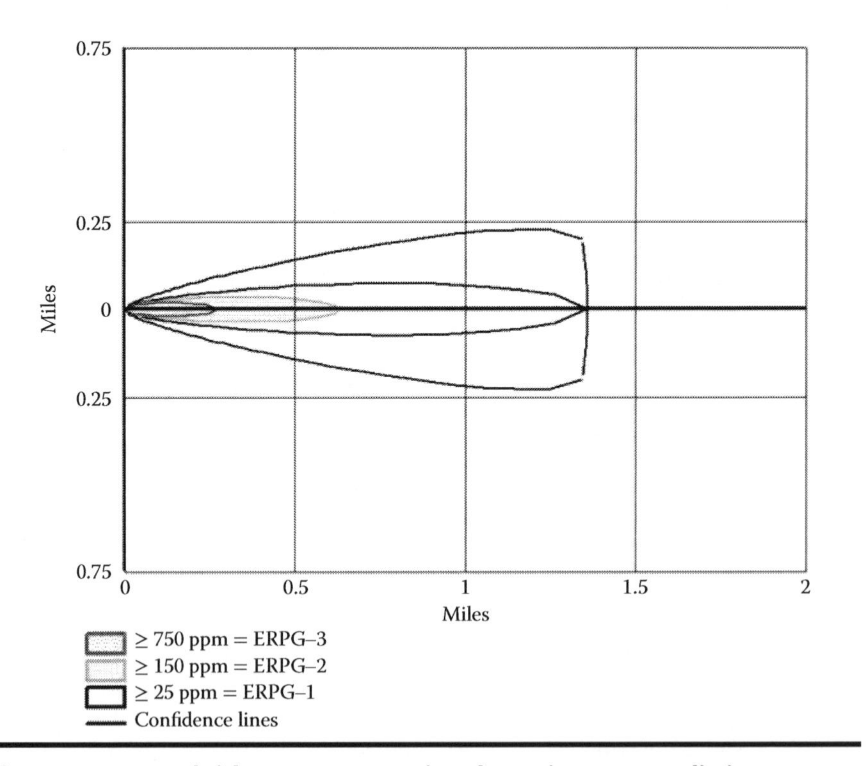

Figure 5.5 Hazard risk zones representing alternative exposure limit.

to predict the area where a toxic gas concentration might be high enough to harm people. A committee of the American Industrial Hygiene Association developed three sets of exposure limits to toxic chemicals for use as planning guidelines and to anticipate human adverse health effects caused by exposure. The three-tiered guideline, Figure 5.6, uses a 1-h direct exposure duration. Each guideline identifies the substance, its chemical and structural properties, animal toxicology data, human experience, existing exposure guidelines, the rationale behind the selected value, and a list of references.

Understanding the limitations inherent in the model and guidelines described above is useful in their application. First, the categories in these guidelines do not protect everyone. Very sensitive individuals, including younger children and the elderly, might suffer adverse reactions to concentrations far below those suggested in the guidelines. Further, these exposure limits are primarily based on animal studies and not human studies. In addition, the exposure limits are based on a 1-hour time period and do not account for any personal safety measures that might be taken to reduce our exposure. The fact is that we might experience exposure for a period longer than 1 hour but seek shelter at the initial signal of the release, thus subjecting ourselves to less toxicity or harm than assumed in the guidelines.

Critical Thinking: The question that the scenario of the accidental ammonia release presents centers on our risk of harm for a specific exposure limit in a chemical release. The question of risk in this case is not simple and depends on many factors such as where we are in the risk zone (are we close or further away from the actual release), if we are inside a building or are exposed in the outside environment, our individual health and if we suffer from asthma or other breathing

ERPGI
The maximum airborne concentration below which it is believed that nearly all individuals could be exposed for up to 1 h without experiencing other than mild transient adverse health effects or perceiving a clearly defined, objectionable odor.
ERPG2
The maximum airborne concentration below which it is believed that nearly all individuals could be exposed for up to 1 h without experiencing or developing irreversible or other serious health effects or symptoms, which could impair an individual's ability to take protective action.
ERPG3
The maximum airborne concentration below which it is believed that nearly all individuals could be exposed for up to 1 h without experiencing or developing life-threatening health effects.

Figure 5.6 Emergency Planning Guide Exposure Guidelines.

Figure 5.7 NFPA HAZMAT Diamond.

handicap, our age, and physical size. How aware are you of potential risks in your area? If you see the sign in Figure 5.7, that is, NFPA Hazmat Diamond on the side of the building next to you, how would you interpret it? Who is it meant to inform?

See http://www.compliancesigns.com/nfpadiamonds.shtml.

Using Historical Data in Determining Risk

Whether we are determining parameter values for a hazard model or just trying to get a fix on the vulnerability of a community to a specific hazard, risk experts turn to historical data to gain insight and understanding. Complete data are normally not available as methods and resources for collecting data have changed over time. However, even when it is known that a data set is accurate, care must be taken when it is used. For example, losses following a disaster are often measured in U.S. dollars. However, dollars in one country may have a different value than dollars in a different country making comparisons between the impacts of disasters in different countries problematic.

Over time, our ability and the methods used to collect information on disasters have changed. Scientific instruments are more sensitive and accurate than in the past. New technologies such as satellites provide opportunities for collecting data that previously did not exist. Indeed, our capacity to detect and accurately classify disasters since satellites have been in use means that the frequency data since the 1960s may be far more accurate than the frequency data sets of the early twentieth century. As an example, our ability to accurately detect and classify earthquakes or tropical cyclones today is far greater than ever before. Hence, we have observed a dramatic increase in disaster frequency in many data sets over the past 20 years.

Numerous data sets reflecting the frequency of disasters and their consequences worldwide are now available. These data sets may come from governing bodies, such as the United Nations, or private companies such as Munich Reinsurance Company. The Centre for Research on the Epidemiology of Disasters at the University of Louvain, Belgium, maintains one of the largest data sets relating to disasters, Emergency Events Database (EM-DAT). The EM-DAT data cover both natural- and human-caused disasters since 1900. These data sets may be of value in establishing a benchmark for a specific type of hazard that in turn may be adjusted for the same hazard in a specific part of the world. For example, parameters established from landslide data taken from communities in central Europe may need to be adjusted to specific soil layers and building codes when applied to analyzing landslides in Africa. Weather-related data obtained from domestic sources such as the NWS or NOAA may provide a more accurate determination of specific risks of hazards in a specific part of the United States.

Figure 5.8 shows centers for major natural hazard data sources. For the United States, the National Climatic Data Center (NCDC) serves as a national resource for climate information. As a climate resource, the NCDC works with scientists and researchers worldwide. They provide both national and global data sets for weather and climate information. In addition to the NDCD, the U.S. Geological Survey (USGS) Center for the Integration of Natural Disaster Information is a clearinghouse for disaster information and provides links to disaster data distributed by other agencies (Thomas 2001). The U.S. Environmental Protection Agency (EPA) and the U.S. Department of Transportation (DOT) provide information on accidental releases of hazardous chemicals. The DOT focuses its data collection on transportation accidental releases, whereas EPA focuses its attention on fixed site releases. Thomas (2001) notes that there has been some integration of hazard event data within a single agency such as the NWS, although he acknowledges that "a true systematic integration of multiple types of hazard data currently does not exist" (p. 64).

Hazard	Agency	Time Covered
Tornadoes	Storm Prediction Center Norman, IL	1959–present
Thunderstorm wind	Storm Prediction Center Norman, IL	1959–present
Hail	Storm Prediction Center Norman, IL	1959–present
Lightning	National Climatic Data Center Asheville, NC	1959–present
Storm data[a]	National Climatic Data Center Asheville, NC	1959–present
Hurricanes	National Hurricane Center Colorado State University	1886–1996
Floods	National Weather Service Council of National Seismic Systems	1903–present
Earthquakes	National Geophysical Data Center Earthquake Research Institute University of Tokyo, Japan	1970–present 2150 B.C.–1994 3000 B.C.–1994
Volcanoes	Global Volcanism Program Smithsonian Institution	8000 B.C.–present

[a] Meteorological events including wind, hail, lightning, water hazards, tornadoes, flooding, drought, landslides, hurricanes, wildfires, and thunderstorms

Figure 5.8 Natural hazard data sources with time covered.

Need for Complete Accurate Data for Decision Making

To reduce the adverse impacts of disasters, those involved in the hazards analysis process must have accurate and timely information to support effective decision making. Information that results from our hazard modeling exists to support this decision-making process. Transparency with regard to the information normally promotes confidence in the information and those who provide it by the user. Showing transparency includes revealing the sources of the data relied on, any errors found in the data, data that the expert chose to omit and why, and whether the data were complete or incomplete.

Since the data are used as input to hazard models or to find parameter values used in these models, understanding its quality and accuracy is important. Inaccurate data or data with a large amount of measurement uncertainty will result in an inaccurate model result or a large amount of uncertainty in the model result. The saying among experts is "trash in, trash out." Hence, it is important for the end user, the decision makers in this case, to know how much they should rely on the information given to them. In the end, the information coming from this rather technical report or complex hazard model must fit into a framework established for dealing with the hazard. The data requirements for supporting the emergency management process will vary both for the type of hazard and how the outputs will be used in supporting decision making (Cutter 2001). For example, suppose a village is seeking to mitigate and manage the effects of frequent landslides that plague the region. Accurate elevation, soil and water flow data are needed to show high-risk areas and identify past landslide areas. Community leaders may use this data to write regulations indicating where building is prohibited and building codes for areas where it is allowed. Inaccurate data could lead to regulations that do not provide sufficient protection for businesses or families or possibly too much regulation that becomes an economic hardship for expansion.

Using Technical Data in Decision Making

The description and categorization of hazard areas, critical infrastructure, and disaster zones is greatly facilitated by the use of geospatial technologies and hazard models. The use of scientific data from hazard models and risk analysis requires that decision makers fully understand the limitations of these tools and how to communicate information. An informed user of complex data is critical to minimizing legal challenges and law suits. Hazard models can provide different results with just small changes in data inputs. Clarifying the model sensitivity to the inputs and the limitations of the data used in the model will help to avoid challenges to the use of these models in emergency management.

Also, there may be a discrepancy between an objective assessment of risk by the hazards analysis team and the public (Kirkwood 1994). An objective view of risk by a knowledgeable professional who understands the nature and limitations of hazard modeling and how it is described may not be shared by the public. An objective evaluation of risk must be nonjudgmental and explained in a way that the public or other stakeholders can understand.

The discrepancy between risk analysts and the public in their view of risk has been changing for many hazards. For example, with satellite imagery, a storm can be tracked over long periods of time and distances providing ample warning to those in its path. The radar image of the storm and its motion through time provides a concrete way that experts can use to communicate the hazard information to the public. However, when hazards occur infrequently, such as in the case of volcanic eruptions, both experts and the public may be caught off its guard (USGS, Hawaiian Volcano Observatory 2012). In 1982, the eruption of El Chichón in Chiapas Mexico became North America's most deadly volcano, killing 2000 within a radius of 10 km. Its last eruption had been about 500 years earlier. Its peak appeared frosted and calm for dozens of generations. Even though there were seismic precursors, hazards analysis for volcanic eruptions was and is still in its infancy. According to Marzocchi and Woo (2009), since the 1982 eruption volcanic risk has been quantitatively defined but not effectively measured. Their paper proposes a framework for volcanic risk metrics in an attempt to provide rules to local authorities for managing this risk. As challenging as it will be to develop effective volcanic eruption risk management measurements, communicating this risk to authorities and local inhabitants in the face of a peak that has been frosted for generations will be even more difficult.

Analysts and decision makers can find ways to leverage the increased power in modern technology. As technology has increased and the speed and memory capacities of computers have grown, information can be stored and accessed more easily than in the past. Systems supporting decision-making activities from technical data have evolved. Today, decision support systems (DSS) allow the software and the user to interact in a way to solve problems and make decisions with the warehoused technical data. Indeed, Wallace and De Balogh (1985) stress the need for DSS for using technical data. They stress a DSS must address the following:

- Provide support to decision makers and their stakeholders
- Evolve as the users become more familiar with the technology
- Be interactive and controllable
- Recognize their nonroutine but consequential use
- Adapt to the idiosyncrasies that are inherent in human decision making

Indicators of Direct and Indirect Losses

We measure the consequences of disasters using indicators of disaster impacts. They could include social disruption, economic disruption, or environmental impacts. Social disruption measures include the number of people displaced or made homeless and incident rates of crime (murders, arrests for civil disorder, or fighting). Economic disruption measures include unemployment rates, days of work lost, production volume lost, and decrease in sales or tax income. Environmental impacts can be valued at total cleanup costs, costs of repair or restoration of water or sewerage systems, the number of days of unhealthy air, or the number of warnings involving fish consumption or restrictions on recreational use of a water feature. Direct tangible losses such as fatalities, injuries, cost of repair, loss of inventory, response costs by a business or community, or relocation costs are first-order consequences that occur immediately after an event (Smith 2004). Indirect losses associated with a disaster evolve after the event and include loss of income by displaced employees, sales that did not occur, increased costs for skilled employees, losses in productivity of employees, employee sickness, and increases in disruptive behavior (fights) at schools or crime in a neighborhood.

The indicators for social, economic, or environmental impacts may be based on historical data and collection of data after a disaster event or modeling techniques. To estimate the impact on the population in a disaster zone using historical data, one would determine from past disasters the number of injuries, fatalities, displaced persons, and those requiring shelter or left unemployed. To measure the relative impact of the disaster, the population size and economic data need to be known before and after the event. For example, a rate comparing the number of injuries for the total population would provide a means of comparing injuries at different disaster events allowing for population changes over time.

Allowing for population changes over time does not account for other related changes that could impact injuries, fatalities, and indicators for disaster impacts. Significant errors can result when projecting past disaster consequences forward based solely on projected population changes. The impacts from more recent disaster events may reflect legal changes (code requirements or floodplain management programs), changes in development patterns, or cultural and social changes causing movement in populated areas.

The use of measurable indicators to help understand risks could be enhanced if all of the indicators used the same units of measure or the same reference points. An example would be to quantify deaths, injuries, and damages in a common measure such as U.S. dollars. Unfortunately, it may be impossible to associate a dollar amount to some indicators. The alternative is to use measurable indicators that may be compared over time. As an example, consider the indicators used when assessing population vulnerability to disasters. It is often the case that countries with high poverty levels show increased vulnerability to many natural hazards. This increased vulnerability can be attributed to the lack of resources for planning and reduced government enforcement of codes and restrictions. The World Bank classifies each national economy by its gross national income per capita to reveal low-income, middle-income, and high-income countries (ISDR Secretariat 2003). A more complex measure of population vulnerability is provided by the United Nations in its

Human Development Index that uses life expectancy, educational attainment, and income as indicators of sustainability.

Intangible losses are those that cannot be expressed in universally accepted financial terms and are not generally included in damage assessments or predictions. Despite the difficulty in associating some intangible losses to specific indicators, we may want to identify some type of indicator that reflects the losses associated with cultural changes, individual and family stress, mental illness, sentimental value, and environmental losses. We need to identify appropriate measures of both tangible and intangible losses associated with disasters. It is not uncommon for the intangible impacts to exceed the tangible ones in terms of the overall effect they have on a community (United Nations 2006).

As we examine potential losses from disasters, we may find that the community or business organization actually has gains. Though it is extremely rare for gains to be included in the assessment of past disasters or the prediction of future ones, it is undeniable that benefits can exist in the aftermath of disaster events. Gains could be observed in increases in employment, business volume, tax collections, or the number of residents, or decreases in the volume of traffic or crime rates. Post-Hurricane Katrina data show that many cities within a 100-mile distance of the City of New Orleans had positive gains from the displacement of the metropolitan New Orleans population. Although the impacts were temporary, some gains remained even years after the disaster.

Critical Thinking: How might you measure the intangible losses related to the impact of a disaster in an education system that has to accommodate an increase of 25% more students or increases in traffic in a community that absorbed 40% more residents who have been displaced?

Issues in Risk Analysis

Changes in Disaster Frequency

Changes in disaster frequency may be the natural result of climatic variations that occur over a long-time interval or from changes in factors that impact the frequency or severity of an environmental change such as an increase in human activity where the hazard already exists. The number of hurricanes that enter the Gulf of Mexico varies over a long period, especially with the rise or fall of sea surface temperatures or wind patterns from an El Niño. Flooding or hurricane storm surge might cause more physical damage because of increased development in coastal areas (Smith 2004). The trend in population shifts to high-hazard coastal zones will likely result in higher losses from tropical cyclones. Environmental changes resulting in natural system degradation may also increase the severity of hazards. As infrastructure is added and more buildings are constructed, the potential for hazard impacts increases. With changes in technology, people expect to have access to a certain level of services, including availability of water, electricity, and easy long-distance transportation. As these systems expand and develop, they become more vulnerable to hazards. Major blackouts, the spread of computer viruses, or communication of terrorist threats have occurred worldwide in the past and will likely occur in the future. The interdependence of our societies globally contributes to our increasing vulnerability to epidemics and disease. This interdependence also increases our economic vulnerability. For example, when Greece defaulted on its government debt in 2012, there were major political and economic ramifications within and outside of the European Union. Natural resources such as oil, water, and air quality are increasingly being recognized as threatened from human activity. To understand these changes in disaster severity and frequency and its causes, trends may need to be examined over longer periods of time than the current data reflect. Hence, the continued measurement and collection of data is needed to help risk experts understand the natural variation of specific hazards and the impacts they will have in an increasingly complex and interdependent society.

Availability of Essential Data

The availability of essential data for modeling hazards and determining the frequency of occurrence is critical in a valid risk analysis. This essential data come from both historical data, as previously discussed, as well as measurements and data taken in real time. Examining the management of flood hazards illustrates how the availability of historical and real-time data is used when attempting to understand the nature of the risk presented by a natural hazard.

The National Flood Insurance Program (NFIP) was established in 1968 and made affordable flood insurance rates to individuals through participating local communities such as towns, cities, counties, or parishes. In 1983, a common standard for risk assessment and management of flood hazards was adopted by federal agencies and known as the 100-year event or 1% annual chance of flood as the standard for floodplain management. This standard was considered to represent a degree of risk and damage worth protecting against, but was not considered to impose excessive burdens or cost to property owners. The 100-year event standard represents a compromise between minor floods, and the greatest flood likely to occur in a given area. In many cases, the 100-year flood level is less than the highest recorded flood.

As part of its role in floodplain management, this 1% annual chance of flood is used to determine the need for flood insurance. Further, the development of flood models and flood maps was considered by the NFIP as a primary means of reducing flood hazards. The flood maps would provide a basis for managing the development and use of flood-prone areas and lead to a better understanding of the magnitude and likelihood of large flows. As federal agencies enhanced their efforts to assist in flood mapping for communities joining the NFIP, information on water feature flow frequencies grew in importance. In 2002, The U.S. Water Resources Council

(USWRC) published a report describing flood regionalization techniques used in the National Flood Frequency (NFF) Program (Benson 1967; Ries and Crouse 2002). These techniques were adopted by USWRC for use in all federal planning involving water and related land resources.

All flood modeling programs that are used to create flood maps for local communities need a discharge value for a water feature, such as a stream, canal, lake, or reservoir. The discharge value is determined by measuring flow rates directly by the USGS through a river gage or indirectly by statistical methods. A USGS River Gage Station measures a water feature's discharge at particular site. Figure 5.9 shows the locations of the gage stations. For these stations, data are collected on a real-time basis by automated instrumentation and analyzed quickly enough to influence a decision that affects the monitored system. The discharge measured at the gaged sites characterizes the volume of water passing a point of the river gage station and is commonly expressed in a hydrologic unit per unit of time, such as cubic feet per second, million gallons per day, gallons per minute, or seconds per minute per day.

The table or graphical representation of the discharge data over a specific time period is called a hydrograph. It provides real-time and historical values. Figure 5.10 provides an example of a hydrograph at a site that has a USGS river gage. For examples of real-time hydrographs, see the following USGS Internet site: http://cfpub.epa.gov/surf/locate/index.cfm. For ungaged areas, the NFF program produces estimates of the magnitude and frequency of flood-peak discharges as well as the corresponding flood runoff hydrographs. The estimates for ungaged areas are based on statistical methods using regression equations. The estimates are used in

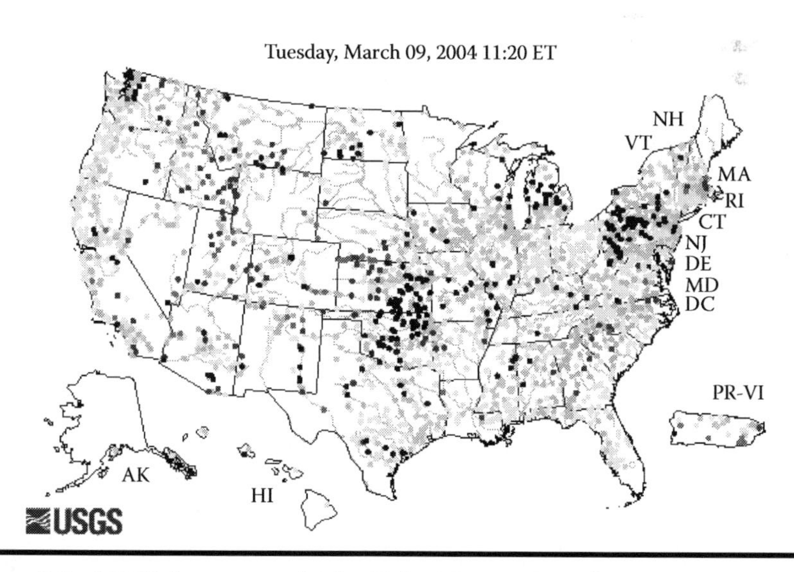

Figure 5.9 USGS river gauges in the United States. Go to http://water.usgs.gov/waterwatch to review active state stations.

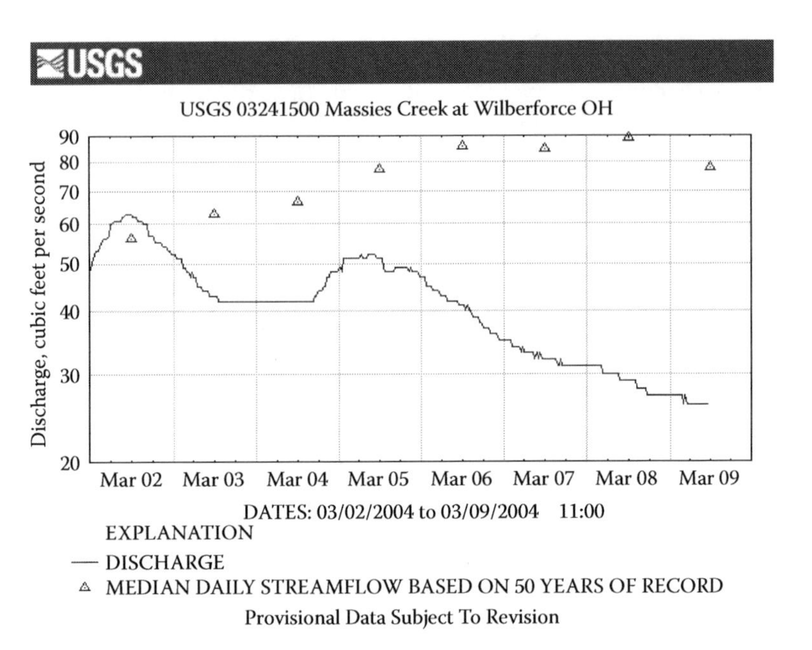

Figure 5.10 USGS hydrograph for a water feature.

floodplain management, flood control, and the design of different structures used in these areas (Ries and Crouse 2002).

Depth of Analysis

Each hazard that is analyzed must be considered according to the range of possible intensities that could be exhibited by the particular hazard. Depending on the hazard, we may need to examine it based on its intensity since the frequency of occurrence is so rare.

We generally see that lower-intensity hazard events occur more often than more severe ones as in the case of a hurricane or an earthquake. More hazard events provide more data that can be broken down into more classes. This increased granularity allows for a more comprehensive assessment. We can, thus, determine a broad-based frequency by calculating the likelihood of each identified hazard broken down by magnitude or intensity, if appropriate. Likewise, the consequences that are expected to occur for each hazard can be calculated and broken down by magnitude or intensity, if appropriate. Finally, a locally tailored qualitative system for each hazard identified as threatening to the community can be produced. The qualitative measures may be determined from the quantitative calculations as described above or at least should reflect them.

In calculating the consequences as described above, damage resulting from past major disaster events may form the basis for examining the impacts of future disasters. The massive floods of 1993 or hurricanes such as Andrew (1992) or Katrina (2005) could provide a basis for estimating damages in similar future floods or storms. However, estimates must be adjusted for local characteristics. For example, levy failure caused much of the flooding in New Orleans due to Katrina. Hence, with improved levy conditions and changes made in the location of future structures, the damage estimates from past data would need to be adjusted to reflect these changes for future estimates. Also, estimates for the cost of damages from future landslides in one region may differ greatly from a similar site in a different region due to the specific location of local structures relative to the landslide risk, local building code differences, and differences in local planning. Note that inflation factors may need to be used to help us project damages from one time period to another.

For major weather-related events, granularity in the data may be difficult to achieve for small regions. The bulk of the data may be available for large areas rather than smaller ones. Unfortunately for most hazards, sufficient information does not exist to accurately quantify the likelihood of a future occurrence of the disaster to a high degree of confidence. This is especially true for those occurring infrequently or those occurring in no apparent pattern such as earthquakes, droughts, terrorism, or nuclear accidents.

Ranking of Risks

Quantitative Data

For quantitative data, the relative ranking of risk can be obtained by numerical calculation. We have seen that certain risks may be quantified by the numerical formula Probability × Consequence. Hence, for a list of local hazards that are quantifiable numerically, their relative risks can be ranked and compared. The EPA uses this type of relative ranking of risk in their assessment of the inland waterways oil spill hazard. In Etkin (2006), this risk is assessed in aggregate and relative to oil type, EPA region, and transportation mode. For example, consider the assessment of risk of inland waterways spills by oil type, as shown in Figure 5.11.

Oil Type	Number of Spills	Probability of Spill	Avg. Spill Size (gal)	Approx Cost per Gallon	Relative Risk
Crude	11,809	0.2581	11,445	384	1,133,412
Light fuel	21,220	0.3754	2,152	533	430,569
Volatile distillate	7,417	0.1256	7,661	423	406,724

Figure 5.11 Risk of inland waterway oil spills by oil type.

From this assessment, it is clear that the greatest risk across oil types and EPA regions is from crude oil. Crude oil spills cost less per gallon, but the relative average size of the spills is large. Even though there are many more light fuel spills and their average cost per gallon is higher, the spill volume is about five times less than the average crude oil spill volume, lessening the overall measure of consequence for light fuel spills relative to crude oil spills. Hence, in using this analysis, one may be led to focus on reducing the number and size of crude oil spills to reduce the overall risk. However, Etkin also performs a trend analysis that shows that the proportion of light fuel spills are increasing sharply relative to other types of oil spills. This indicates the need to also focus on reducing the number of light oil spills to mitigate future overall risk increases.

Likelihood–Consequence Matrix

Risk evaluation involves the determination of the relative seriousness of the risk of a hazard as they could affect an organization or a local community. Organizations and communities face a range of natural and technological hazards, each of which requires a different strategy to reduce the risk factors of likelihood or consequence. To facilitate the relative ranking of risks, organizations should determine if a risk may be addressed by another agency, identify which risks require immediate attention, and clarify if the risk associated with a hazard requires further evaluation (Cameron 2002).

We can determine the relative ranking of risks associated with hazards facing our organization or community by considering the following factors:

The likelihood and consequences of the hazard.
The voluntary or involuntary nature of the risk (Smith 2004).
Is there a benefit to cost ratios of mitigating different risks?
Are there political and social ramifications of certain mitigation decisions?

The final output of risk evaluation is a prioritized list of risks, which will be used to decide treatment (mitigation) options.

In assessing risk, the first step is to identify the hazards of interest. We next assess the hazard for its level of likelihood and the impact or intensity of its consequence. We use quantitative values when possible. However, to apply the risk assessment matrix method for ranking risks, the likelihood and consequence variables must be categorical. Hence, hazards that are known to exhibit a numerical range of likelihood and intensity values are assigned categorical values across the range of possibilities. Assigning these levels to likelihood and consequence allows for a direct comparison of the risks faced by a community.

It is common to use four or five categorical values for the probability of occurrence. A summary of five values and their description as given in the Army Reserve Officers' Training Corps (ROTC) risk management worksheet is shown below

(Army ROTC 1997). In these descriptions, it is assumed that a time horizon is specified. The parenthetical values shown are also used.

Frequent—Occurs often, continuously experienced
Likely—Occurs several times
Occasional (possible)—Occurs sporadically
Seldom (rare)—Unlikely, but could occur at some time
Unlikely—Can assume it will not occur

Values are also assigned to the severity describing the expected consequence of the event in terms of degree of injury, property damage, or other impairments to the organization or community doing the assessment. The summary below uses the terms as specified in the Army ROTC worksheet with alternative values defined earlier for injury and death shown parenthetically.

Catastrophic—death or permanent and total disability, complete system loss, major damage or significant property damage, and mission (organization) failure (or complete community disruption)
Critical (major)—Permanent partial disability, temporary disability in excess of 3 months, major system damage or significant property damage, and significant mission (organization or community) disruption
Marginal (moderate)—Minor injuries, lost workday accident, minor system damage, minor property damage, and some mission (organization or community) disruption
Negligible (minor)—First aid or minor medical treatment, minor system impairment, and little or no impact to the mission accomplishment (organization or community)

As seen in Figure 5.12, once the values have been assigned for the identified hazards, we can then summarize the likelihood and consequence for the risks associated with each hazard in the first column using the risk description category.

To compare hazards, the values in the fourth column can be determined through the use of a risk matrix. A risk matrix plots the likelihood and consequence of hazards together in various combinations, with one risk component falling on the *x* axis and the other on the *y* axis, similar to how a multiplication table is laid out. By plotting these values on the matrix, individual boxes representing unique combinations of likelihood and consequence can be determined. Each hazard listed in the likelihood–consequence matrix can then be placed in the box of the risk matrix that best reflects its risk.

Figure 5.13 provides a risk matrix for assessing the likelihood and consequences of risks presented by natural hazards. The labeling of the boxes with the risk categories, that is, extreme, high, moderate, and low may vary with the organization. In this figure, the classifications as defined by the Emergency Management Australia

Hazard	Likelihood	Consequence	Risk
Flood	Possible	Major	Extreme
Drought	Likely	Minor	Low
Extreme heat	Possible	Moderate	High
Extreme cold	Possible	Moderate	High
Thunderstorm/lightning	Almost certain	Minor	High
Tornadoes	Likely	Major	Extreme
Severe snowstorms	Likely	Moderate	High
Ice storms	Unlikely	Major	High
Land subsidence	Rare	Minor	Low
Earthquake	Rare	Major	High
Transportation accidents	Possible	Catastrophic	Extreme
Hazmat transportation accidents	Unlikely	Moderate	Moderate
Closure of critical transportation routes	Unlikely	Moderate	Moderate
Power failures	Possible	Moderate	High
Water/sewer line failure	Unlikely	Moderate	Moderate
Telecommunications failure	Unlikely	Minor	Low
Computer systems failure	Possible	Minor	Moderate
Gas line break	Unlikely	Minor	Low
Stored chemical leak/accident	Unlikely	Moderate	Moderate
Sabotage/intentional destruction	Possible	Major	Extreme
Biological communicable disease (plague)	Possible	Major	High
Laboratory accidents	Possible	Major	Extreme
Building collapse	Rare	Catastrophic	High
Building fire	Unlikely	Catastrophic	Extreme
Epidemic	Unlikely	Major	High
Widespread poisoning	Unlikely	Major	High
Water/air contamination	Rare	Major	High
Contaminated medical facilities	Possible	Major	Extreme
Terrorism	Unlikely	Major	High
Terrorism-federal and international property-destruction	Possible	Major	Extreme
Protest	Possible	Minor	Moderate
Riots	Possible	Moderate	High
Strikes	Possible	Minor	Moderate
Crime	Almost certain	Moderate	Extreme
War	Rare	Major	High

Figure 5.12 Likelihood–consequence matrix.

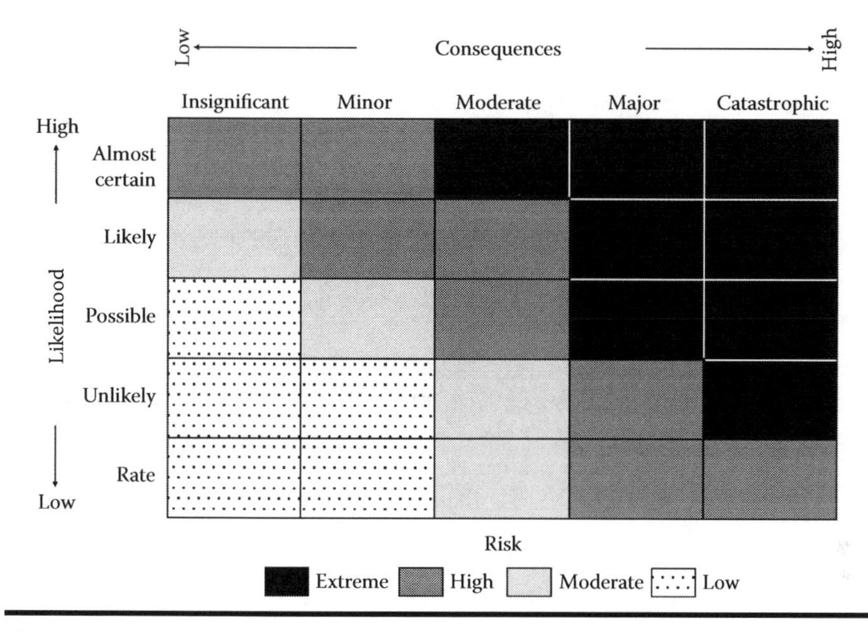

Figure 5.13 Risk matrix.

are used. The cells down and just above the diagonal are labeled high risk with the cells just below the diagonal labeled moderate risk. The cells in the upper right-hand corner with high likelihood and catastrophic severity are labeled extreme risk, and the cells in the lower left-hand corner of the matrix are labeled low risk. The risk categories are then assigned risk mitigation priorities, as shown in the figure.

The following definitions are used for the risk description categories:

Extreme: High-risk condition with highest priority for mitigation and contingency planning (immediate action).
High: Moderate-to high-risk condition with risk addressed by mitigation and contingency planning (prompt action).
Moderate: Risk condition sufficiently high to give consideration for further mitigation and planning (planned action).
Low: Low-risk condition with additional mitigation contingency planning (advisory in nature).

Federal Emergency Management Agency (FEMA) classifies risks in a similar way in their "MultiHazard Identification and Risk Assessment" publication using the following categories:

Class A: High-risk condition with highest priority for mitigation and contingency planning (immediate action). High likelihood and high consequence block.

Class B: The likelihood of a risk is high, but the consequence is low. Risk addressed immediately by mitigation and emergency preparedness and contingency planning (prompt action).
Class C: Risk likelihood is low, but consequences are high. Consideration for mitigation and preparedness is critical.
Class D: Low-risk condition with additional mitigation contingency planning (advisory in nature). Low likelihood and low consequence block.

Because a "risk level" may be assigned to more than one matrix box, an ordered list of risk priorities is not created, but rather several categories of risk with several hazards fall within each category group. For instance, if a 50-year flood was determined to be a Class C risk, and an accident involving a truck carrying hazardous materials was determined to be a Class C risk, then these two would be considered as equal risks according to the risk matrix.

The evaluation can then combine these categories into a spreadsheet to reflect

Likelihood: Is the hazard likely to occur
Consequences: What is the seriousness of the impacts of the hazard
Level of risk as determined by evaluation on the risk matrix (extreme risk, high risk, moderate risk, and low risk)
Additional considerations include
Other organizations or entities that are impacted by the hazard (potential partnerships, resources, or interdependence of risk management or hazard mitigation strategies)
Manageability: Adequacy of existing or potential risk management or hazard mitigation measures or controls
Acceptability: Is the risk acceptable from social, political, economic, or environmental impacts
Change in the risk from the hazard: Will the risk remain the same priority rating (Lunn 2003).

Risk Strategies

The strategies used to mitigate or eliminate risks involve decisions about what risks to treat, what risks to prevent at all costs, and what risks can be disregarded because of either low consequence, low frequency, or both. The risk analysis process is not working in a vacuum. There are many factors such as political, social, or economic systems that could affect the determination of what risks are acceptable, and what risks are not.

Once hazards have been identified, analyzed and evaluated, a priority list of risks that must be considered for treatment is generated. Ideally, communities would treat all risks in a way such that nobody would have to worry about them ever again, but that risk-free world scenario is inconceivable despite modern technology and engineering. While most risks can be reduced by some amount, few can be completely eliminated, and rarely do the funds exist to reduce all of the risks by an amount that is acceptable to all people in the community.

Another factor in the problem of risk acceptability and mitigation relates to the benefits associated with almost every risk. It is almost universally true that a benefit enjoyed by a community or organization requires some acceptance or tolerance of an associated risk. Locating a business near a water feature used as a major transportation route may expose it to flooding. Locating a subdivision in or near a forest may be desirable to future homeowners but could expose the residents to a wildland fire hazard. To completely eliminate the risk will in many cases eliminate associated benefits as well.

Eliminating certain risks may directly or indirectly create new ones. That is, one problem may be solved only to create another. For instance, to completely eliminate the risk from nuclear power generation plants, those plants would need to be dismantled and taken out of service. The resulting shortage of power would require that fossil fuel–burning plants increase their production, which in turn would create increased carbon-based pollution likely resulting in increased health and environmental risks.

With these concerns in mind, a thoughtful response to the assessed risks must be determined. This response becomes the organization's risk strategy. The risk strategy for a particular risk depends on the risk level for that risk. A mandatory risk level indicates a risk requiring immediate attention, whereas, a de minimis risk level may only require continued observation and data collection. It may also be determined that the risk is acceptable.

Mandatory Risk Level

This type of risk is considered as one that is so great because it is mandatory that action be taken to deal with it. It is viewed as an "obnoxious risk," which cannot be ignored, and strategies to reduce vulnerability to it are mandatory. In practice, this level is generally set at 1 in 10,000 risks per vulnerable individual. This practice is often cited in regard to secondhand smoke exposure or accidents in the workplace, where safety measures or procedures are required.

Extremely Low Likelihood of Risk

A "de minimis" risk level suggests that the statistical probability of a specific risk from a hazard is so low that concern is not merited. This level is often set at either 1 in 100,000 or 1 in 1,000,000 and is set either for a 1-year period or for a lifetime (70 years). The term de minimis is a shortened version of the Latin phrase "de minimis non curat lex," which means "the law does not care about very small matters." This concept is widely used to set guidelines for levels of risk exposure to the general population such as the chance of personal injury in an airline crash or train derailment or a reaction from an over-the-counter medication. For instance, the EPA has set de minimis risk levels for human lifetime risk from pesticides at 1 in 1,000,000 over a 70-year lifetime. The Food and Drug Administration and the U.S. Department of Agriculture are working on similar regulations of risk for food safety.

Accept the Risk

One option is to simply accept the risk given the present situation and resources of the community. A specific hazard event may have a very low probability of occurrence and as a result spending any amount of money to mitigate it would be counterproductive considering some greater risk reduction that could be achieved by using the money to treat another more probable or severe hazard. The risks that fall within the lowest category of both consequence and likelihood are generally the risks that are considered as acceptable. Members of a community may also believe that the level of a risk can be mitigated so as to reduce the most adverse consequences. Homeowners who have invested in reducing the vulnerability of their homes and businesses that have spent funds in reducing the vulnerability of their business property may believe that they can withstand a disaster event with limited property damage. Therefore, they accept the current level of risk to their property.

Critical Thinking: What steps could a homeowner or business owner do in your area to reduce the vulnerability of their property to the risk of a local hazard? In general, how do we transfer our financial exposure to risk in a modern economy to a willing party?

Determining Risk Acceptability

Personal, political/social, and economic factors influence the determination of risk acceptability. While the three are interrelated, different processes drive each of them.

Personal

Differences in individual acceptance between risks that are voluntary or involuntary in nature are greatly determined by what we see as the benefit from the risk. An individual is more likely to accept a voluntary risk if he or she perceives that the benefit is great. Many recreational activities and sports involve considerable levels of personal risk entered voluntarily. Indeed, the thrill of the risk may be part of the enjoyment of the recreation. When the benefits of a risk outweigh the costs, then the perception of the risk is reduced. In this case, the threat level may be considered acceptable. In some cases, a high risk may be accepted voluntarily and a lower risk imposed from outside may not be acceptable. Skydivers are normally well acquainted with the risks in their activity but find the experience extremely rewarding. In 2007, the U.S. parachute association data suggest about 4 out of 10,000 jumps resulted in injury and 1 out of 100,000 jumps resulted in death (Hsu 2009). Hsu says that this death rate is roughly equivalent to the death rate of women in childbirth. However, a woman will only give birth to one child in a given year while a skydiver may take as many as 10 jumps in a single day, thus greatly increasing the chance of their involvement in an accident and hence increasing the risk. In this case, the skydiver is clearly willing to tolerate the voluntary risks associated with their activity.

On the other hand, our personal experience and knowledge of a risk or hazard may lead us to reject voluntarily accepting the risk. We may have seen firsthand the potential outcomes from accepting a voluntary risk and believe that the likelihood of harm is so great or the consequence is so severe that we avoid the risk all together. Consider an individual who was once at ease swimming in the ocean who then witnesses a shark attack. Their perception of the risk of a shark attack may be altered by the experience if not their voluntary acceptance of it.

In addition to our willingness to accept risk voluntarily, risk may be associated with our individual values, educational experiences, exposure to the media coverage of risk, and our individual tolerance of risk.

Political/Social

Because of the differences in the makeup of different communities and populations, risk acceptance will not be universal in all communities and cultures. Risk acceptance is likely to change from place to place, from time to time, and from hazard to hazard (Alesch et al. 2001). Acceptability is likely to change even within individual communities over time as the makeup of that community changes. It is these differences that make the wide public participation in the hazards analysis process so important. Communities that have recently experienced the impacts from a disaster will likely be more willing to learn more about the hazard that caused the disaster and take some type of action, including risk assessment and risk mitigation.

Economic

Economic considerations of risk are viewed by federal agencies such as the U.S. Army Corps of Engineers from a cost–benefit perspective. The costs of reducing a risk will need to be compared to the benefits (actual risk reduction) that would result. Regulatory agencies such as the U.S. Department of Energy, Transportation, or Environmental Protection assess risk for private enterprises, which directly deal with the hazardous substances. Their consideration includes cost factors, but the overall public health and environmental sustainability is a higher priority. Local governmental agencies that have building departments issue permits and conduct inspections to enforce building codes and promote safety. Cost–benefit may be a consideration in the initial adoption of the regulations, but extreme events such as hurricanes can motivate public officials to strengthen the codes to provide more protection for people and property. Hence, cost considerations may not be the primary driver of the new standards.

Critical Thinking: Cost–benefit analysis is a tool that can be helpful in understanding the implications of risk, where alternative risk strategies and their costs are examined. The benefits gained by the funds expended are examined. Associated with a cost–benefit analysis is the cost-effectiveness assessment that examines the minimum unit cost to reduce a maximum level of risk. Consider the risks associated with flooding in a flood zone for existing homes. Suppose, a house is constructed using a slab on grad foundation. How might the unit cost of raising the house using piers be reduced? What modifications to oil-carrying tankers have been introduced to reduce the cost associated with oil spill impact? Do these modifications make sense relative to a cost–benefit analysis?

Hazard Models

In 1994, the FEMA report titled, "Assessment of the State of the Art Earthquake Loss Estimation Methodologies (FEMA 249)" summarized the current state of methodologies used in the estimation of earthquake losses. This report led to the development of a catastrophe model, Hazard U.S. or HAZUS. This catastrophe model helped to standardize how hazard losses are estimated. It has since been extended to wind and flood hazards (Grossi et al. 2005).

The end result of the catastrophe model can either be a GIS map of potential losses or an exceedance probability (EP) graph. The EP graph gives the probability that a given level of loss will be exceeded. For example, for a certain inventory of covered buildings by an insurer, the EP graph may tell the insurer that there is a 2% chance that losses will exceed $5 million. The insurer can then use this information to help judge if insuring this inventory poses an unreasonable risk to the company.

Grossi et al. (2005) identify the following four components that must be quantified for a catastrophe model: the hazard, the inventory, vulnerability, and finally the loss. The hazard component includes the probability or frequency of occurrence and different characteristics of the hazard. The inventory includes the physical structures and property in the geographic area being assessed. The vulnerability component then takes the hazard component and inventory component as inputs to quantify the impact of the hazard on the inventory elements. This may include damage curves for buildings or other structures, property damage, contents damage, or business interruption expenses. From the vulnerability output, potential losses are then assessed and a risk category can be assigned to the inventory element.

Uncertainty

No matter the methodology used to assess the risk of a hazard, there remains the uncertainty to consider. Uncertainty can exist in two forms. There is uncertainty in the natural hazard itself. This type of uncertainty reveals itself in the randomness of occurrence and the randomness in the severity of the event when it does occur. This uncertainty is normally captured through constructing probability distributions, which are then used in the risk assessment. This type of uncertainty cannot be reduced, but it can be quantified and better understood. A second type of uncertainty is due to our limited or incomplete data related to the hazard and limited or incomplete understanding related to the science describing the hazard. This type of uncertainty can be reduced or mitigated through more thorough data collection, data collected over a longer period of time, and advancement in the areas of science related to understanding the events underlying the hazard. We will describe three ways that the uncertainty can be incorporated into the risk assessment, that is, logic trees, simulation, and use of the probability distributions.

Logic Tree

The logic tree analysis method as used in hazard risk assessment is a special case of event tree analysis (ETA). In ETA, engineers studying the risk of a system's failure will break the system into component events. A particular component in the system may fail or not fail or possibly be in one of several states with each state having a certain probability of occurring (Figure 5.14). A logic tree is then constructed showing all possible outcomes (Figure 5.15). The probability of a path occurring is then found by taking the product of the probabilities of the individual events in the path. For example, suppose an engineer wishes to study the failure of system with components A and B. The components have the following failure characteristics.

To construct the logic tree diagram, the first stage after initiation represents the state of component A and the second stage represents the state of component B. In the above logic tree diagram, the system only works if both components are successful in performing their function. Hence, the system works only if the top branch illustrates the system's functionality. The probability of the system working is $0.97 \times 0.95 = 0.9215$ or 92.15%. The other three branches, that is, SF, FS, FF show the three ways in which the system can fail.

Component	Failure (F)	Success (S)
A	0.03	0.97
B	0.05	0.95

Figure 5.14 Component failure.

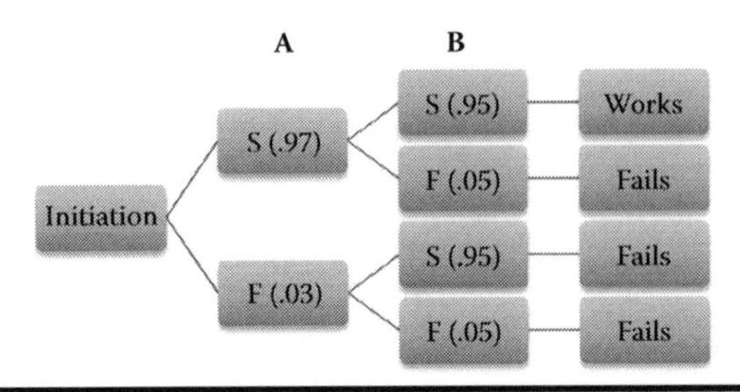

Figure 5.15 Logic tree for system success.

To apply the logic tree to risk assessment, we define the component events as either intermediate events or procedures leading to the final outcome of a hazard event or as model parameters leading to the final model result. For example, suppose the risk being assessed is a nuclear terrorism event. Event A may represent the type of attack (a nuclear explosion or a nuclear dispersion), event B may represent the type of material (high half-life, low half-life), event C may represent density of population in affected area, and so on. Or suppose the event being assessed is the risk of fatalities from a Category 3 or higher hurricane striking a particular land area in the Gulf of Mexico. To illustrate the logic tree model, a simplified version of an example given in Chapter 3 of the book *Quantifying and Controlling Catastrophic Risks* by B. John Garrick is constructed (2008). The reader may wish to consult Garrick (2008) for a more detailed and comprehensive discussion of the hurricane impact model. Also, Grossi et al. (2005) apply the logic tree method to demonstrate the incorporation of uncertainty into landslide risk (pp. 74–79). The parameters in the logic tree diagram are defined as follows:

A—The time the hurricane spends in the Gulf before landfall [<48 h, 48–72 h, >72 h].
B—Whether the hurricane impacts the area being studied [Yes, No].
C—The category of the hurricane [3, 4, 5].
D—The type of evacuation prior to landfall [minimal, medium, full].

To fully use this method in obtaining a numerical value for risk, a consequence with damage estimation for each branch of the logic tree needs to be determined. Garrick (2008) uses six damage states for the final stage in his logic tree assessment for hurricane fatalities affecting New Orleans. He labels his damage states 1 through 6 with 1 being the most severe. Therefore, damage state 1 would occur if there was a Category 5 hurricane in the Gulf for less than 48 h with minimal evacuation that affected New Orleans. Whereas damage state 6 would occur if there was a Category 3 hurricane in the Gulf for more than 72 h before landfall with full evacuation that affected New Orleans. In our illustration, we will refer to the damage states as impact states (ISs) and reverse the meaning of the numbers and use lower ISs for less consequence.

To provide a numerical example, we assume some hypothetical values:

A—The time the hurricane spends in the Gulf before landfall [<48 h (0.2), >48 h (0.8)].
B—Whether the hurricane Categories 3–5 impacts New Orleans [Yes (0.05), No (0.95)].
C—The category of the hurricane [3 (0.6), 4 (0.4), 5 (0)].
D—The type of evacuation prior to landfall [minimal, medium, full].
E—Impact state [0, 0 lives; 1, 50 lives; 2, 750 lives; 3, 1,500 lives; 4, 10,000 lives].

The probabilities appear in parenthesis. Since we are simplifying the example, we have assigned a probability of 0 to a Category 5 storm. Hence, it will not appear in the logic diagram. For simplicity, the probability for type of evacuation will only be a function of the time the storm spends in the Gulf before landfall. Of course, in reality, this probability will depend on many events including the size of the storm reported and the action of state and local officials. For our example, we use the following values in Figure 5.16.

IS (expected number of fatalities) will be function of the size of the storm and type of evacuation according to the table in Figure 5.17.

Finally, the initiation event is the moment a major hurricane is reported to appear in the Gulf of Mexico. The logic tree describing this model is shown in Figure 5.18. The stages in the logic tree model beyond initiation illustrated from left to right are the time spent in the Gulf (TIME), whether it will impact New Orleans

Time Spent in Gulf	Minimal	Medium	Full
<48h	0.8	0.2	0
>48h	0.2	0.8	0

Figure 5.16 Evacuation success probabilities for time the hurricane spends in the Gulf.

Category	Minimal	Medium	Full
3	3	2	1
4	4	3	2

Figure 5.17 Impact states by type of evacuation.

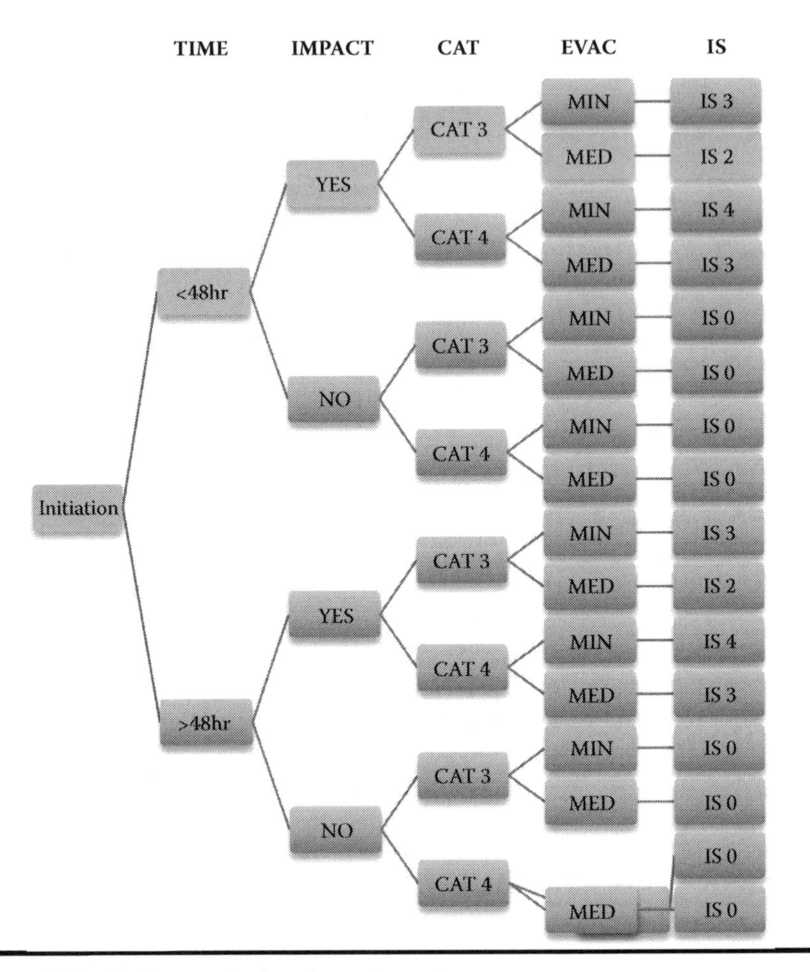

Figure 5.18 Logic tree for hurricane impact.

(IMPACT), the category (CAT), the evacuation success (EVAC), and the storm impact status (IS).

We can now use the logic diagram to calculate probabilities and to quantify the risk. For example, following the branch in the diagram in a lighter shade of orange, after a major hurricane appears in the Gulf, the probability of the storm appearing in the Gulf less than 48 h before landfall (0.2), impacting New Orleans (0.5), being Category 3 (0.6), with medium evacuation success (0.2) is $0.2 \times 0.05 \times 0.6 \times 0.2 = 0.0012$.

Since the scenario of a Category 3 storm with medium evacuation success yields IS 2, the fatality risk for the stated scenario is $0.0012 \times 750 = 0.9$. Notice that we have calculated the risk for 1 of 16 possible scenarios. To calculate the total risk, we add the risk from all 16 paths.

Distribution for Impact States	
0	0.950
1	0.000
2	0.024
3	0.023
4	0.006

Figure 5.19 Distribution for hurricane model impact states.

We can also calculate the distribution of probabilities for the ISs from the logic tree. For example, there are two paths in the tree leading to IS 2:

1. <48 h, YES, CAT3, MED: 0.2 × 0.05 × 0.6 × 0.2 = 0.0012.
2. >48 h, YES, CAT3, MED: 0.8 × 0.05 × 0.6 × 0.8 = 0.0192.

Therefore, the probability that after a storm enters the Gulf that it will lead to IS 2 in New Orleans is 0.0012 + 0.0192 = 0.0204. This can be done for each IS yielding the distribution for the ISs shown in Figure 5.19.

Critical Thinking: Describe the four paths in the logic tree leading to IS 3. Calculate the probability that after the storm enters the Gulf that it will lead to IS 3 and compare your solution to the probability in Figure 5.19.

Monte Carlo Method or Simulation

After constructing a model for all of the inputs to our hazard analysis, which may consist of probability distributions, calculus type models, or probabilities for categorical descriptions, we can in theory determine the risk of the hazard. We could simplify the uncertainty involved by making enough simplifying assumptions to construct a logic diagram as previously shown. However, modern technology also allows us to simulate the hazard event. Using random values to simulate the value for a model or a model parameter is called a Monte Carlo method.

For example, in the hurricane model above, either before or at the time the storm enters the Gulf, we may use historical data to arrive at the probability that the storm will impact New Orleans. However, modern hurricane models make projections as to where the storm will make landfall using probability distributions. The probability that the storm will affect New Orleans will not remain fixed. It will change as the storm progresses through the Gulf of Mexico. At a point in time, computer models may indicate a probability distribution for the storm to impact New Orleans. For a numerical example, suppose that distribution is given by Figure 5.20.

Probability	Cumulative Probability	Chance of N.O. Landfall
0.75	0.75	0%
0.13	0.88	5%
0.07	0.95	30%
0.05	1.0	50%

Figure 5.20 Distribution for landfall.

Reading the table given in Figure 5.20, there is a 0.75 probability of no landfall in New Orleans; there is a 0.13 probability of a 5% chance of landfall, and so on. The second column is the cumulative distribution. We use the cumulative distribution to perform the simulation. In this case, the computer program will choose a random integer from 1 to 100. If the random integer is between 1 and 75 inclusive, then a 0% chance would be used for the hurricane making landfall in New Orleans. If the integer is between 76 and 88 inclusive, then a 5% chance would be used for the hurricane making landfall in New Orleans. If the integer is between 89 and 95 inclusive, then we would use 30% and if it is between 96 and 100 inclusive, then we would use 50%. Each choice of a random number will provide a probability of landfall to use in determining the risk from the storm. For example, suppose the random integer 23 is generated. Then a probability of 0 impacting New Orleans is used rather than 0.05 (5%) in the logic tree above. In this case, the only relevant IS would be 0. However, if the integer 90 is generated, then a probability of 0.3 (30%) impacting New Orleans would be used rather than 0.05. A large number of scenarios can then be used, such as 10,000, with the risk calculated for each scenario. The final result will be a distribution of values for either the damage state or the risk.

The previous example shows how to perform the simulation in the case of one input having uncertainty. This method can be further extended in case simulation is needed for more than one input or model parameter.

Uncertainty Expressed in Interval Estimates

Suppose the damage states for the hurricane model represent the consequence in terms of structural loss, recovery expense, and other economic losses rather than loss of lives. Using damage states 1—\$0 through 9—\$10M with each damage state representing \$1.25M more damage than the previous damage state, we may obtain a distribution of losses due to a Category 3 hurricane striking a specified city, as shown in the table in Figure 5.21.

We can also view this distribution graphically in the form of a histogram (Figure 5.22). In this graph, the damage states are on the horizontal axis and the probabilities are on the vertical axis. The height of the bar above the damage state shows the probability the damage state will occur. For example, the height of the bar above damage state 7 is 0.25 indicating a probability of 0.25 that there will be \$7.5M in damage if a Category 3 storm should strike this particular city. Since each bar in the histogram has width 1, it is also convenient to view the area of each bar as the probability of the occurrence of the damage state.

The expected damage can be calculated by multiplying each damage value by the probability of its occurrence and summing over all of the damage states. The expected damage from the distribution above is Cost times Probability for each Damage State or $1.25 \times 0.04 + 2.50 \times 0.06 + 3.75 \times 0.07 + 5.00 \times 0.11 + 6.25 \times 0.18 + 7.50 \times 0.25 + 8.75 \times 0.19 + 10.00 \times 0.10 =$ \$6.675M.

\$6.675M can then be used as the consequence in calculating the risk; however, the distribution of damages provides more information than just the expected damage. It may also be used to express the uncertainty in the estimates. For example, there is a 0.54 probability that the Category 3 hurricane will cause damage in the interval [\$5M, \$7.5M]. We obtain this probability by adding the probabilities for damage states 5, 6, and 7. We can view this and its probability graphically by observing the area of the bars over damage states 5, 6, and 7 in the histogram (area between the vertical bars) as shown in Figure 5.23. Using this distribution, we can

DS	**1**	**2**	**3**	**4**	**5**	**6**	**7**	**8**	**9**
Probability	0	0.04	0.06	0.07	0.11	0.18	0.25	0.19	0.1
Cost	\$0	S1.25M	\$2.5M	\$3.75M	\$5M	\$6.25M	\$7.5M	\$8.75M	\$10M

Figure 5.21 Distribution for hurricane model damages.

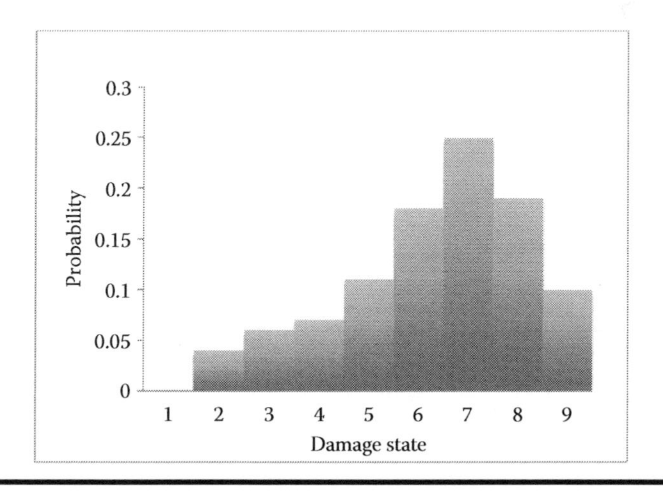

Figure 5.22 Histogram for the distribution for hurricane model damages.

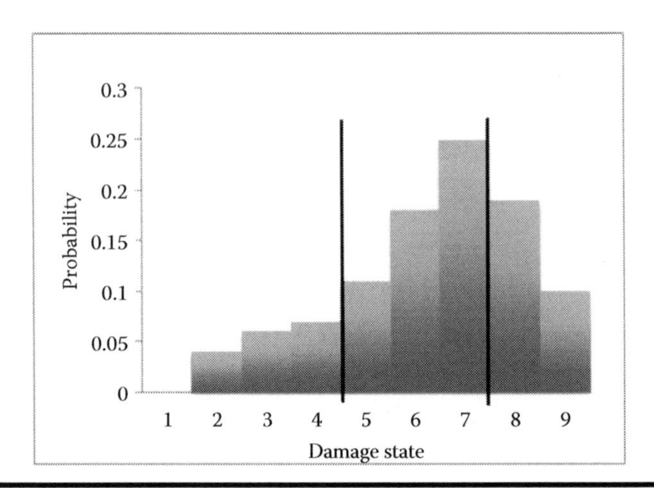

Figure 5.23 Histogram view for the interval of estimation (\$5M, \$7.5M).

also conclude that there is a 10% chance of damage being \$10M. Therefore, in this example, \$8.75M (damage state 8) is called the 90th percentile since the probability that the damage will be less than or equal to \$8.75M is 0.9.

When calculating parameter values from historical data, it is common to use the sample mean for the parameter as the point estimate for the true mean of the parameter value and use the 95% confidence interval for the interval estimate. In this case, if the data set is large, then the sample mean has an approximate normal distribution that can be used to identify the interval estimate. In general, if $\overline{x}$ denotes the sample mean, σ denotes the sample standard deviation, and these statistics are calculated from n data values, then the 95% confidence interval for the parameter as measured in the data set is given by

$$\left[\overline{x}-1.96\frac{\sigma}{\sqrt{n}},\ \overline{x}+1.96\frac{\sigma}{\sqrt{n}}\right]$$

For example, in an earthquake hazard model, suppose a parameter in the model is the average time a 3.0 earthquake lasts at a particular fault. If historically there are 60 such earthquakes recorded for this fault and the average duration is 45 s with a sample standard deviation of 12.4 s, then the 95% confidence interval estimation for a 3.0 earthquake average duration is

$$\left[45-1.96\frac{12.4}{\sqrt{60}},\ 45+1.96\frac{12.4}{\sqrt{60}}\right]=[41.9,\ 48.1]$$

We interpret this interval of estimation by stating that we are 95% confident the true mean duration of 3.0 earthquakes at this particular fault is between 41.9 and 48.1 s. Notice, if we base our estimate on fewer data values, this causes a wider interval estimation reflecting the fact that we should view our point estimate for the parameter as having less precision. Hence, if the statistics above were the same, but based on 25 data values, our interval becomes

$$\left[45-1.96\frac{12.4}{\sqrt{25}},\ 45+1.96\frac{12.4}{\sqrt{25}}\right]=[40.1,\ 49.9]$$

indicating more uncertainty about the true parameter value.

Landslide hazards present a complex scenario to analyze. Among the complexities identified by Glade (2003) and Uzielli et al. (2008) is the site-specific nature of the phenomena and the difficulty in quantifying the spatial aspect of the hazard. To help overcome these difficulties, in landslide hazard analysis, it has become accepted practice to divide consequence into two components: vulnerability (V) and the cost of items at risk (C_I). Hence, the risk equation becomes $P \cdot C = P \cdot V \cdot C_I$. The vulnerability is then defined both in terms of landslide intensity and the susceptibility of the inventory at risk $V = I \cdot S$ (Uzielli et al. 2008). This enables the authors to develop separate and independent models for the landslide intensity and the susceptibility of the local structures, which are of varying ages and subject to local building codes and construction techniques. Further in a follow-up paper, the authors explore the uncertainty in the vulnerability component by providing upper and lower thresholds resulting in an interval of estimation for vulnerability (Kanyia et al. 2008).

Summary

We have explored how experts analyze risk in human- and natural-caused hazards. Experts identify hazards that are probable to impact a location. The structures and inhabitants in that location are assessed for their vulnerability to each risk. This assessment relies on historical data, current data, socioeconomic data, local demographics, and judgment. In some cases, hazard models are constructed to incorporate scientific and statistical information into the assessment. Software packages such as GIS can provide insight into the spatial distribution of the risk being considered. The amount, quality, and the depth of the data are important factors in establishing parameter values, inputs for hazard models, and providing a basis for decisions made about risk strategies. In the risk analysis process, both quantitative and qualitative analyses are used. If the probability and consequence of a hazard event occurrence can be numerically established, then the risk for that hazard event can be calculated as the probability times the consequence. Tools such as the risk matrix help to place each risk into categories such as low, medium, high, or extreme. Using this relative ranking of risks, risk managers can then formulate risk strategies appropriate to each hazard. These strategies may include simple risk mitigation measures for low and medium risks; risk acceptance for some risks; and planning or immediate action, such as recommendations for regulation, for more extreme risks. Further, the risk managers and experts are faced with the challenge of communicating this technical information and its interpretation to the public and policy makers. As was said in the beginning, the goal is to provide decision makers with the right information at the right level of complexity at the right time.

Advanced instrumentation for collecting the data and the increase in scientific knowledge and technology have most certainly contributed to a more accurate assessment of risks and its communication to decision makers. However, hazard events seem to regularly catch us off guard and cause billions of dollars in damage as well as cause human suffering and loss of life. This reminds us that knowing the chance of something happening does not tell us when or where it will happen or even whether it will happen. Much of our expectations about the present and future are based on past trends and occurrences. However, we must explore more thoroughly whether past trends will continue or will change into new trends. Many hazards affecting large portions of world populations are climate related, such as droughts, floods, destructive storms, a rising sea level, long-term temperature changes, insect infestation, and potable water availability. The earth's climate is a dynamical system, difficult to model, and changes in climate are difficult to predict. It changes through natural cycles and due to natural causes as well as the impact of human activity. As human impact on the climate increases, some changes may be accelerated or even be different than expected. The emission of greenhouse gases and its environmental impact has been receiving increased attention as our consumption of carbon-based fuels continues to rise. As an example of how climate change affects our knowledge of a hazard, consider the coastal flooding hazard associated with a change in sea level. According to NOAA's ocean facts (NOAA 2013), in 2010, 39% of the U.S. population lived in coastal counties with a population density six times than that of inland counties. This is consistent with the estimates that 40% of the world's population lives within 100 km of the coast. For these residents, rising sea levels pose an alarming problem. The source of this problem lies in the thermal expansion of warming ocean waters and the melting ice sheets. Both of these factors are related to a warming climate. The IPCC (Intergovernmental Panel on Climate Change) regards ice sheet melting as the major unknown factor to predicting future sea level rises (Quaile 2013). Just a generation ago, it was difficult to imagine an ice-free artic summer. With satellite data, scientists are now predicting an ice-free artic summer within decades. Erik Ivins, who coordinated a new study for National Aeronautics and Space Administration on ice sheet melting, says that the rate of ice loss from the Greenland ice sheet has increased fivefold since the mid-1990s, and the melting ice from both poles is responsible for one-fifth of the global rise in sea level (Quaile 2013). Clearly, historical trends in coastal flooding and its causes will not be as useful in assessing this risk in the future. Hence, the challenges facing risk assessors and risk managers include the past challenges of data collection and mining, building or using models and new technology, and effectively communicating their results to decision makers and the public, as well as the new challenge of how to use our knowledge of a changing climate.

Discussion Questions

Would you classify a risk as voluntary or involuntary where changes in factors influencing the frequency and severity of local flooding have changed without the knowledge of the community?

What risks would be included as obnoxious where mandatory action is required?

What risks do you consider to be unacceptable where others have determined that they can live with the hazard? What personal views, values, beliefs, or your personality contribute to this conclusion associated with a risk?

Applications

Using the risk matrix categories included in Figures 5.12 and 5.13, examine the hazards in your community and categorize their likelihood and damage consequences from low to extreme.

Websites

The ERPG guidelines are clearly defined and are based on extensive, current data. The rationale for selecting each value is explained, and other pertinent information is also provided. http://www.aiha.org

National Climatic Data Center (NCDC): http://www.ncdc.noaa.gov/oa/about/ncdcwelcome.html

Storm Prediction Center, Norman, OK: http://www.spc.noaa.gov/archive/index.html

National Hurricane Center, Colorado State University: http://www.nhc.noaa.gov/pastall.html

NOAA information on coastal populations: http://oceanservice.noaa.gov/facts/population.html

Earthquakes Epicenter: Council of National Seismic Systems: http://quake.geo.berkeley.edu/cnss/

Catalog of Significant Earthquakes. National Geophysical Data Center: http://www.ngdc.noaa.gov/seg/hazard/eqint.html

Earthquake Research Institute, the University of Tokyo, Japan: http://www.eri.u-tokyo.ac.jp/eng/

Watershed Flood Monitoring Program: http://cfpub.epa.gov/surf/locate/index.cfm

Environmental Research Consulting: http://www.environmental-research.com/publications.php

USGS River Gauges in the United States for active stations: http://water.usgs.gov/waterwatch

Damage Estimation: http://www.westegg.com/inflation/

References

Alesch, D. J., Holly, J., Mittler, E., and Nagy, R. (2001). *Organizations at Risk: What Happens When Small Businesses and Not-for-Profits Encounter Natural Disasters?* Fairfax, VA: Public Entity Risk Institute.

Army ROTC (1997). Sample Risk Management Worksheet. http://www.google.com/#q=rotc+risk+management+worksheet. Accessed November 4, 2013.

Benson, M. A. (1967). Uniform flood-frequency estimating methods for federal agencies. *Water Resources Research, 4*(5), 891–908.

Breslin, S. (2013). Many Coastal Residents Still Unprepared for Hurricanes. The Weather Channel, Hurricane Central. http://www.weather.com/news/weather-hurricanes/few-coastal-residents-prepared-hurricane-season-study-finds-20130618. Accessed November 4, 2013.

Cameron, G. (2002). *Emergency Risk Management: What Does It Mean?* ATEM-AAPPA 2002 Conference, Brisbane, Australia.

Cutter, S. L. (2001). Charting a course for the next two decades. In S. L. Cutter (ed.), *American Hazardscapes: The Regionalization of Hazards and Disasters*. Washington, DC: Joseph Henry Press.

De, M. B. and Ravetz, J. R. (1999). Risk management and governance: a post-normal science approach. *Futures, 31*(7), 743–757.

Derby, S. L. and Keeney, R. L. (1981). Risk analysis: understanding "How Safe Is Safe Enough?" *Risk Analysis, 1*(3), 217–224.

Emergency Management Australia (2001). *Decision Making Under Uncertainty in the Emergency Management Context*. Mount Macedon Paper 2001. Workshop held at the Australian Emergency Management Institute, August 27–28.

Etkin, D. S. (2006). Risk Assessment of Oil Spills to US Inland Waterways. Proceedings of 2006 Freshwater Spills Symposium. http://www.environmental-research.com/erc_papers/ERC_paper_16.pdf. Accessed September 15, 2013.

Federal Emergency Management Agency (1994). Assessment of the State of the Art Earthquake Loss Estimation Methodologies. FEMA 249: Washington, DC: FEMA.

Garrick, J. B. (2008). *Quantifying and Controlling Catastrophic Risks*. Burlington, MA: Academic Press.

Glade, T. (2003). Vulnerability assessment in landslide risk analysis. *Die Erde, 134*(2), 123–146.

Grossi, P., Kunreuther, H., and Windeler, D. (2005). An introduction to catastrophe models and insurance. In P. Grossi, P. Grossi, and H. Kunreuther (eds.), *Catastrophe Modeling: A New Approach to Managing Risk*. New York: Springer.

Hsu, J. (2009). *The Truth about Skydiving Risks*. http://www.livescience.com/5350-truth-skydiving-risks.html. Accessed March 26, 2009.

ISDR Secretariat (2003). *Living with Risk: Turning the Tide on Disasters towards Sustainable Development*. Geneva: United Nations.

Jardine, C. G. and Hrudley, S. E. (1997). Mixed messages in risk communication. *Risk Analysis*, *17*(4), 489–498.

Kanyia, A. M., Papathoma-Kohle, M., Neuhauser, B., Ratzinger, K., Wenzel, H., and Medina-Cetina, Z. (2008). Probabilistic assessment to vulnerability to landslide: application to the Village of Lichtenstein, Baden-Wurttemberg, Germany. *Engineering Geology*, *101*, 33–48.

Kirkwood, A. S. (1994). Why do we worry when scientists say there is no risk? *Disaster Prevention and Management*, *3*(2), 15–22.

Lunn, J. (2003). Community consultation: the foundation of effective risk management. *Journal of Emergency Management*, *1*(1), 39–48.

Marzocchi, W. and Woo, G. (2009). Principles of volcanic risk metrics: Theory and the case study of Mount Vesuvius and Campi Flegrei, Italy. *Journal of Geophysical Research*, *114* B03213, doi:10.1029/2008JB005908.

NOAA (2013). National Coastal Population Report: Population Trends from 1970 to 2020. http://stateofthecoast.noaa.gov/features/coastal-population-report.pdf. Accessed November 4, 2013.

Quaile, I. (2013). Polar Ice Sheets Melting Faster Than Ever. *Deutsche Welle*. N. Conrad (ed.). http://www.dw.de/polar-ice-sheets-melting-faster-than-ever/a-16432199. Accessed November 4, 2013.

Rejeski, D. (1993). GIS and risk: a three-culture problem. In M. F. Goodchild, B. O. Parks, and L. T. Steyaert (eds.), *Environmental Modeling with GIS*. New York: Oxford University Press.

Ries, K. G., III and Crouse, M. Y. (2002). The National Flood Frequency Program, Version 3: A Computer Program for Estimating Magnitude and Frequency of Floods for Ungauged Sites, 2002: U. S. Geological Survey Water-Resources Investigations Report 02-4168.

Royal Society Study Group (1992). *Risk Assessment, Report of a Royal Society Study Group*. London, United Kingdom: The Royal Society.

Slovic, P., Fischhoff, B., and Lichtenstein, S. (1979). Rating the risks. *Environment*, *21*(3), 14–20, 36–39.

Smith, K. (2004). *Environmental Hazards: Assessing Risk and Reducing Vulnerability*. New York: Routledge.

Thomas, D. S. K. (2001). Data, data everywhere, but can we really use them? In S. L. Cutter (ed.), *American Hazardscapes: The Regionalization of Hazards and Disaste*rs. Washington, DC: Joseph Henry Press.

United Nations (2006). *The Millennium Development Goals Report*. New York, NY: Department of Economic and Social Affairs, United Nations. http://mdgs.un.org /unsd/mdg/Default.aspx. Accessed November 4, 2013.

USGS, Hawaiian Volcano Observatory (2012). *Volcano Watch*. http://hvo.wr.usgs.gov /volcanowatch/view.php?id=155. Accessed December 20, 2012.

Uzielli, M., Farrokh, N., Suzanne, L., and Kanyia, A. (2008). A conceptual framework for quantitative estimation of physical vulnerability to landslides. *Engineering Geology*, *102*, 251–256.

Wallace, W. A. and De Balogh, F. (1985). Decision support systems for disaster management. *Public Administration Review*, 134–146.

Chapter 6

Social, Economic, and Ecological Vulnerability

John C. Pine

Objectives

1. Define the concept of vulnerability and its relationship to extreme events.
2. Clarify alternative approaches to understanding social/cultural, economic, and ecological vulnerability.
3. Identify and explain the dimensions of vulnerability.
4. Clarify how we can measure the dimensions of vulnerability.
5. Explain how sociocultural, environmental, and economic systems are interdependent?

Key Terms

Coastal sprawl
Economic systems
Extreme events
Hazards
Infrastructural systems
Natural or ecological systems
Resiliency
Smart growth
Sociocultural systems
Vulnerability

Critical Thinking: Some communities bounce back and even prosper from disasters while others take much longer to recover or experience delays in restoring their ecological, social, or economic systems. How can we measure the potential impacts from natural hazards on sociocultural, economic, and natural systems? How can we better understand how these systems are interconnected and interdependent?

Introduction

Vulnerability refers to the susceptibility or potential for harm associated with disasters and includes impacts to social, economic, and ecological systems (Turner II et al. 2003). Vulnerability results from conditions and processes that influence how social, economic, and ecological systems are impacted by disasters and extreme events. Vulnerability has its roots in geography and natural hazards research but is now used in risk management, ecology, public health, developments, sustainability science, and planning (Kumpulainen 2006). Vulnerability is closely associated with resilience and involves the capacity of social, economic, ecological, and constructed systems to bounce back from disasters or their capacity to respond, adapt, or cope with extreme events (Füssel 2007). O'Brien et al. (2004) define vulnerability in a similar manner and used the Intergovernmental Panel on Climate Change (IPCC) definition of vulnerability, which includes exposure, sensitivity, and adaptive capacity.

It was earlier noted that risk was the result of hazard potential, time, and vulnerability. Thus, vulnerability becomes central in understanding how communities deal with risks associated with disasters. Expressed in a different way, vulnerability is the result of an exposure to hazards and a capacity to cope and recover in a sustainable manner.

Dwyer et al. (2004) view vulnerability from a different perspective stating that how one fares in a disaster is influenced not only from exposure but also by community support, access to resources, and governmental management. Individual attributes are acknowledged but the community capacity to deal with a disaster is a key factor. This is a complex environment that involves community structure, capital, economy, the hazard condition, and social capital. Dwyer notes four levels of vulnerability including:

1. Individual within household (relating to personal attributes)
2. Community (relating to the interactions between the individual and the community)
3. Regional/geographical (relating to how far we are from services)
4. Administrative/institutional (relating to disaster funding and mitigation studies)

Critical Thinking: What factors contribute to the vulnerability of individuals, households, or a region? Are individual physical attributes, such as age or disability, more important than individual financial attributes (income, house ownership, or debt) in determining vulnerability? To what extent is vulnerability influenced by location or geographic factors? Can a single personal attribute determine vulnerability, or is vulnerability dependent on a combination of factors (Dwyer et al. 2004)?

No single investigation into vulnerability indicators will provide a holistic and comprehensive answer; however, there are aspects of vulnerability that can be explored and represented through the development and application of quantitative vulnerability indicators (Dwyer et al. 2004).

Approaches to Vulnerability

Recent literature suggests that vulnerability takes many forms, and different approaches have been developed that help us in understanding this phenomenon. An exposure model emphasizes the identification of conditions that make people and places vulnerable to disastrous conditions and is related to the relative frequency and intensity of the hazard, risk, or threat (Bruneau et al. 2003; Füssel 2007). Hufschmidt (2011) stresses the use of risk as a basis for understanding vulnerability for "risk offers a holistic perspective on the interplay of processes operating in social and environmental systems, and it is this interplay that produces the precondition for disaster."

Exposure models allow one to test the vulnerability of critical infrastructure, facilities, and other constructed or structural systems to impacts of hazardous events. Quantitative approaches in the engineering sciences attempt to assess infrastructure resilience with the goal of reducing losses through research and the application of advanced technologies that improve engineering, pre-event planning, and postevent recovery strategies (Bruneau et al. 2003). Vulnerability as a hazard exposure includes the distribution of people, economies, and the environment to hazardous conditions. Under this view, vulnerability is a result of a physical condition that is associated with place (Cutter 1996).

A second approach views vulnerability as a social condition that examines societal resistance or resilience to hazards (Hewitt 1997; Wisner et al. 2004). Under this view, vulnerability is an outcome of the relation between a hazard and social conditions that define the capacity of a community to respond and cope with a disaster event. Hufschmidt (2011) provides a related perspective stressing individual coping mechanisms including adjustment and adaptation. Adjustments are actions such as building a levee or designing a house that resists a hazard. Adaptation on the other hand is a long-term action toward reducing potential loss such as improving the overall quality of housing. Making adjustments is inherent within the exposure of families, groups, and communities within a geographic area.

The third is an integrated approach that examines potential exposures and social resilience (Cutter, 1996; Cutter et al. 2000; Kasperson et al. 1995). This integrated view combines vulnerability associated with risk and exposure with social response and geographic characteristics. Cutter (1996) notes problematic issues even in this integrated approach, suggesting that there may be a lack of consideration of the underlying causes of social vulnerability and a failure to consider distinct spatial outcomes that may vary over time. Variability of risk over a geographic area is central to Cutter's hazards of place model (Cutter et al. 2008). Social and biophysical conditions thus interact to produce an overall place vulnerability. Khazai et al. (2013) also suggest an integrated approach in understanding vulnerability but stress that human-caused industrial risks including socioeconomic and sociopolitical factors are relevant in aggravating and amplifying direct and indirect risks from industrial sites.

Birkmann (2007) contends that disaster risk is a product of exposure to hazards, the frequency or severity of the hazard, and the vulnerability. A key element of Birkmann's view of measuring vulnerability is the scale of analysis. Indicators reflecting conditions at the local level may not be appropriate at a higher scale for regional, national, or international scales. He also sees that the type of risk is important for some hazards are characterized as creeping including sea level rise or drought, and indicators of vulnerability may not be as obvious in many data sets. Further, Birkmann suggests that we view vulnerability as a process where we measure past, current, and potential future areas and people at risk. He notes that we need to strengthen linkages between global and local approaches in understanding vulnerability.

The integrated view of vulnerability fits with Adger's (2006) views on vulnerability. He stresses that social and ecological systems have many interactions and that an analysis of these relationships is needed to understand how external stressors impact these systems. This view of the connections between complex systems is also stressed by Berkes and Folke (1998). They examine vulnerability focusing on the interdependence of social and ecological systems. This view stresses the interactions between human action and social structures with natural systems. Natural systems refer to biological and biophysical processes, whereas social systems are made up of rules and institutions that mediate human use of resources as well as systems of knowledge and ethics that interpret natural systems from a human perspective. Resilience is viewed within the relationship between social and ecological systems and is examined by understanding the magnitude of disturbance that can be absorbed before a system changes to a radically different state or the capacity for adaptation to emerging circumstances (e.g., Adger 2006; Berkes et al. 2003; Carpenter et al. 2001; Folke 2006).

Birkmann (2007) stresses the integrated approach in understanding vulnerability that combines exposure, coping capacity as well as physical, social, economic, environmental, and institutional capacity. This view of vulnerability is reflected in Figure 6.1. Coping strategies should be included in an integrated vulnerability approach. Vulnerability is, thus, integrated into the development of action or coping strategies that can be implemented. These strategies reflect choices or public policies that are made by individuals, families, businesses, and public agencies and models

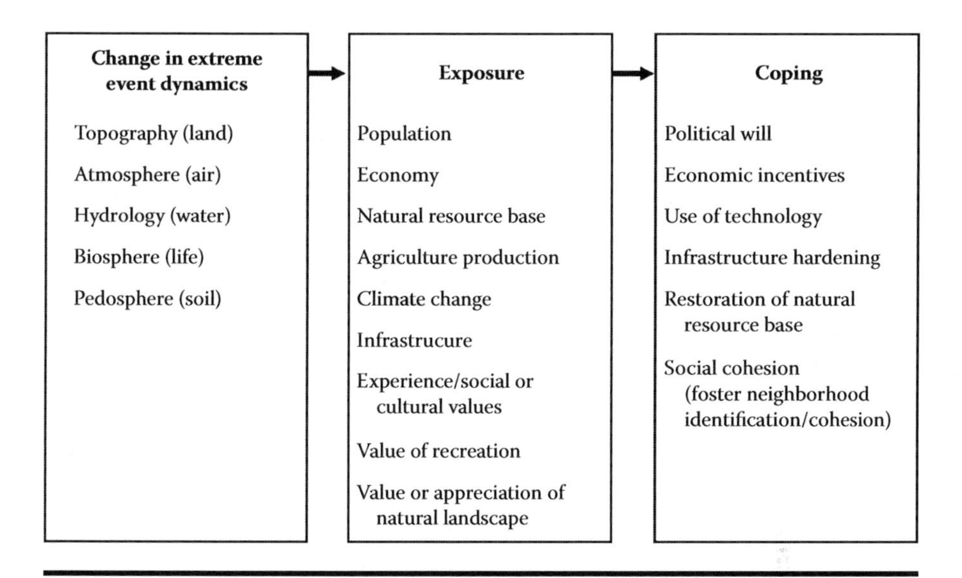

Figure 6.1 Conceptual view of vulnerability.

that allow testing them. We are interested in who lives in the community and where they reside, but it is the decision that people make on an individual and collective basis that really drive vulnerability. This is consistent with the approach of Cutter et al. (2009), which suggests that disasters are not caused by external events (such as flood or tornado) but by social systems that make people vulnerable. Vulnerability is reflected in a combination of factors that determine the degree to which someone's life, livelihood, property, and other assets are put at risk by a discrete and identifiable event (or series or "cascade" of such events) in nature or in society (Wisner et al. 2004, p. 11). The factors that influence vulnerability range from individual characteristics to attributes of whole communities or regions including the economy, the built environment, and the natural landscape (Holand et al. 2011).

The integrated view of vulnerability also stresses the contribution of social networks in a community, the strength of critical infrastructure to hazards, an area's risk of a hazardous event, and efforts by the community to reduce potential losses or to mitigate exposure. A community's vulnerability is thus filtered through its social fabric, efforts to strengthen infrastructure, business enterprises initiatives to reduce their exposure, and capacity to deal with disasters (risk management). Both community mitigation activities and organizational risk management initiatives thus impact the social, economic, and ecological exposure to hazards. Vulnerability is thus more complex than just the exposure of people to hazards for their efforts to prepare and cope along with community and organization initiative contributes to the community's capacity to respond and cope with disaster events. Natural, economic, and social systems are deeply integrated and interdependent in ways that must be considered in understanding some communities, people, and natural environments, which are better able to cope and recovery from disasters than others. Birkmann (2007) stresses that disaster risk is "a product of exposure to hazards, the frequency and severity of the hazard and vulnerability (p. 21)."

Critical Thinking: The scale at which vulnerability of place is examined may vary from large regions such as metropolitan areas to the neighborhood level. The analysis of vulnerability at the neighborhood level is present in isolated disaster case studies but rarely included in assessments of large-scale disasters in the United States or internationally. Tornadoes may impact a small community or neighborhood where a hurricane or flood could affect a much larger area. What other examples might illustrate the scale of disasters?

Dimensions of Vulnerability

Vulnerability is reflected within the social, economic, and ecological dimensions of communities. To understand a community's vulnerability, one must identify indicators in each of these dimensions. These indicators are measures that reflect how a community might be impacted in a disaster.

Social, economic, and ecologic indicators emerged independently during the 1960s and 1970s specifically designed to provide indices of exposure and environmental health (Cutter et al. 2003). Pelling et al. (2004) have used socioeconomic indicators to examine social and economic implications of regional partnerships. The Coastal Risk Atlas is one of the few attempts to link physical hazards and social vulnerabilities (Boyd et al. 2005). Richmond et al. (2001) concluded that "there exists no established methodology for determining the hazardous nature of a coastline," and Cutter et al. (2003) reconfirmed that metric standards do not exist to assess the vulnerability to environmental hazards.

Although there is little agreement in determining hazardous conditions, Nakagawa and Shaw (2004) believe that there are common features that suggest why some communities are more resilient than others. They see that there is a complex mixture of social, economic, religious, and political factors present that influence community resilience to disasters.

Environmental degradation can result in health and economic losses, and create exposures to extreme events. It also might be related to the root causes of a hazard outcome such as disease. As an example, water supply, air pollution (indoor), and sanitation are all related to the highest level of risk from disease. This would suggest that indicators are thus related to specific hazards but not to others.

Critical Thinking: It is widely agreed upon that social vulnerability is influenced by a lack of information, political representation, richer social networks, culture, infrastructure, age, gender, race, and socioeconomic status, non-English speaking, and disabilities (Cutter et al. 2003). Hazard potential, geography, and infrastructure conditions interface with the social and economic fabric of a region to influence risk (Cutter et al. 2009).

Social and Human Vulnerability

The social dimension of vulnerability arises from the exposure of people, neighborhoods, cities, and rural populations and their capacity to recover from hazard events. The hazards literature has noted that the poor, unemployed, single head of a household, elderly, handicapped, or carless households (Wisner et al. 2004; Yohe and Tol 2001) are much more likely to suffer the hardest from a disaster and have more difficulty in restructuring their lives than households who have more resources. The more vulnerable populations take more time than their counterparts to recover following a disaster and as a result suffer to a greater extent. Cannon et al. (2003) suggest that vulnerability also involves self-protection actions and access to political networks and institutions. Cutter et al. (2003) acknowledge these factors but stresses the geographic dimensions of vulnerability noting that place matters. Too often, the poor and most vulnerable populations reside in the most hazardous zones in a community.

Social vulnerability suggests a differential capacity of groups and individuals in dealing with the adverse effects of hazards based on their positions within the physical and social world (Dow 1992). Historical, cultural, social, and economic processes shape an individual's or social group's coping capacity (Wisner et al. 2004). Research studies suggest that specific populations are far more vulnerable to the risks from natural- and human-caused disasters (Cutter et al. 2003). These studies also indicate that there is a strong relationship between socioeconomic vulnerability and disasters and that social and economic costs of disasters fall unevenly on these population groups (e.g., Bolin and Stanford 1991; Cutter et al. 1997; Heinz III Center for Science, Economics and the Environment 2000; Mileti 1999; Cutter et al. 2000; Morrow 1999; Wisner et al. 2004; Pelling et al. 2004).

Critical Thinking: The literature suggests that indicators reflecting social capital are keys to understanding vulnerability. These indicators suggest that those who are poor, unemployed, single head of household, the elderly, and handicapped are appropriate measures for social vulnerability.

It is widely agreed upon that social vulnerability is influenced by a lack of information, political representation, richer social networks, culture, infrastructure, age, gender, race, and socioeconomic status, non-English speaking, and disabilities (Cutter et al. 2003). Social vulnerability is considered highly linked to inequality but to place/location inequalities as well. More valuable homes and higher income increase resilience to hazards and reduce risks (Cutter et al.). Place and socioeconomic characteristics interact to influence hazard potential (Cutter). Hazard potential, geography, and infrastructure conditions interface with the social and economic fabric of a region to influence risk (Cutter). The key question raised by these studies centers on the suggestion that some groups are at greater risk than others.

Adato et al. (2006) takes a different perspective on social vulnerability observing that for some, droughts, hurricanes, and other environmental disasters deal a blow to the poor and vulnerable populations in many parts of the world so as to trap them in poverty, despair, and dependency. They view patterns around the world to suggest that the poorest households struggle to overcome the desperate situation that disaster or shocks deal them. Their short- and long-term well-being and sustainability make it impossible to ever catch up with wealthier households. This dynamic is seen not only at the community level but also at a national scale globally (UNCTAD 1997).

A hurricane hazard vulnerability assessment conducted during 2005 for the Mississippi Gulf Coast combined a geographic information systems-based risk atlas and hurricane simulations (Boyd 2005). Risks were ranked such as flood zones and vulnerability examined using income, age, single parents, education, non-English, vehicle ownership, home ownership, and type of home to identify populations at risk and hurricane hazards.

Nakagawa and Shaw (2004) note that there are common features that suggest why some communities are more resilient than others. They see that there is a complex mixture of social, economic, religious, and political factors present that influence community resilience to disasters. They found that the resilience of communities to recover following a disaster is based on both social and economic activities that are heavily influenced by social capital or the level of trust present in the community, social norms, degree of community participation, and finally the presence of strong community networks.

Critical Thinking: Hoffman (2003) examined who might be hidden victims of disaster and suggests that some very vulnerable people fall through the cracks in disaster recovery, not getting the type of relief needed and endure ongoing suffering as a result of their situation. She explains that those less able to prepare or cope with disasters are poor or working classes and are some of the most unprotected people in a disaster. As a result of catastrophes, some people slip into a state of perpetual misery. These hidden victims could include undocumented workers; people who lost rental housing (owners or renters) and who did not have insurance; the mentally ill or those with chronic illnesses; severely incapacitated; or people who are viewed as "social parasites," such as beggars, trash scavengers, hustlers, or just the homeless. She raises the question of what happens when those hidden victims who are at the bottom of our society or bottom of the heap are not helped. What happens to the rest of society?

A broader discussion of hazards that include climate change suggests vulnerability from a variety of additional perspectives including:

1. Climate change—CO_2 and other greenhouse gas emissions associated with increasing global temperatures (IPCC 2007).
2. Ocean acidification—Bleaching of coral reefs and negative impacts on reef ecology relevant to sustainability of human life on low-lying coral atolls (Veron et al. 2009).
3. Stratospheric ozone—O_3 depletion is related to an increased incidence of cancers (Ni-Bin et al. 2010).

4. The biogeochemical nitrogen (N) cycle and phosphorus (P) cycles—A counterbalance to reduced crop yields leading to dependency of the world's poor on international agrochemical industries (Erenstein and Thorpe 2010).
5. Global freshwater use—It is estimated that by 2030 more than half of the world's population will face water shortages (Ridoutt and Pfister 2010).
6. Land system change, land-use and land-cover change (LULCC)—The current rate, extent, and intensity of LULCC is far greater than at any time in recorded history, driving unprecedented changes in ecosystems and environmental processes on local, regional, and global scales (McAlpine et al. 2009).
7. Loss of biological diversity—The negative effects on natural ecosystem processes and services that benefit and stabilize human society (Xi 2011).
8. Chemical pollution—Human health is directly related to increased levels of atmospheric, terrestrial, and water chemical pollution (Kampa and Castanas 2008).
9. Atmospheric aerosol loading—Increased atmospheric loading and deposition of mineral dust aerosols have important human health implications (Goudie 2009).

Economic Vulnerability

When we look at economic vulnerability, we examine our risk to production, distribution, and consumption of goods and services not only from the private commercial sector but also from the nonprofit and public sectors. The health and vitality of a community's economy is interdependent with the region, nation, and globally. The identification of local, regional, national, and international forces that influence local wages, production, export volume, unemployment, the number and types of jobs are impacted by many external forces. There are many linkages in our economies that shape the robustness of our local, regional, and state economic base. Suggesting that we can predict accurately how to establish a highly productive economy is very different from the examination of a set of economic indicators that will suggest that a local community could withstand or recovery from a natural disaster. Our task then is to identify and examine indicators that will suggest how robust our economy is for a given community and its capacity to contribute in a positive manner to a recovery from a disaster. Economic vulnerability also includes factors that could harm a labor force such as human disease or epidemics. United Nations World Vulnerability Report (Pelling et al. 2004) documents indicators for indexing and monitoring the potential for disasters.

When we assess the economic vulnerability, we evaluate not only jobs and the nature of the local economy but also the status of existing roads, bridges, airports, rail lines, hospitals, prisons, manufacturing plants, shopping areas, utilities, and communication systems to withstand extreme event impacts. It is the potential impact to employee wages, employment, and infrastructure such as electrical, natural gas, communication sectors that impact our community's capacity to recover from a disaster. As Comfort et al. (1999) point out, our vast set of services to our rural and urban communities offer a vital backbone to our commerce and standard of living, the scale of these systems also create dependence and losses, which have vast consequences on our economic stability.

Critical Thinking: Economic indicators that reveal the vulnerability of the economic capital of a community include the diversity of the economic base and the capacity of the critical infrastructure to withstand a disaster. A local community that is heavily dependent on a single type of industry is more vulnerable than another community that enjoys great diversity in local and regional businesses. Communities that enjoy strong transportation, communications, electrical networks are in a position to withstand a disaster than communities that do not have a robust infrastructure.

The infrastructural and economic vulnerabilities are in fact tightly connected but can be clearly separated if we consider two aspects: a physical aspect and a non-physical aspect. Although the built environment and its physical resilience against extreme events may be impacted by the physical forces of a hazard, the economic resilience would deal with pressures and impacts of the global economy. In today's global economy, financial, trading, and policy decisions in other parts of the world may have a significant impact on a local economy.

International agencies judge the size and structure of an economy, exposure to international trade shocks, and extreme natural events to justify loan or aid programs. The U.S. Agency for International Development examines economic vulnerability by determining the frequency and intensity of hazards and conditions such as energy dependency, export characteristics and destinations, and reliance on financing externally (Crowards 1999). Munich Re Group (2002) views disasters from an economic perspective including annual per capita income as a reflection of purchasing power. In the agricultural sector, our economy can measure the production of various goods. But production is highly impacted by external forces such as soil moisture or meteorological forces or geological variables reflecting the hazard itself.

Environmental Vulnerability

Ecological dimensions of vulnerability refer to the capacity of our natural systems to bounce back from disaster or their fragility to harm. It is the inability of these natural systems to deal with stress that may evolve over time and space (Williams and Kaputska 2000). Saltwater intrusion into freshwater marshes can cause the impairment and even the loss of breading grounds for fish and other water creatures, birds, and other coastal animals. Long-term intrusion of saltwater into marshy areas can also impact community surface water systems. Hazardous material contamination that results from flooding, wind, or storm surge can cause immediate and long-term decay of delicate coastal environments. The Gulf of Mexico Oil Spill of 2010 illustrates that the scale of a disaster can be very large, diverse, and has a great impact on families, communities, and a region. Barnett et al. (2008) have concluded that indexes of vulnerability to environmental change cannot hope to be meaningful when applied to large-scale systems and so should focus on smaller scales of analysis. They argue for a context-specific rather than a generic condition.

The development of indicators for environmental sustainability and vulnerability evolved since the late 1990s (Birkmann 2006, 2007; Esty et al. 2005; Polsky et al. 2007; Kaly et al. 1999); complex interactions and the difficulty in developing broad-based indicators pose major obstacles for assessing environmental conditions.

Environmental systems are also significant to the quality of life for a community and its productivity as well as sustainability. Critical views of the rate of deforestation, annual water use as a percentage of total water resources, population density, annual use of water by a household, volume of recycled materials per household, and the relation of coastline to land area. There could be a relationship between threatened species to land area and the ratio of total number of natural disaster to land area (1970–96) (Atkins et al. 2000). The Yale Center for Environmental Law and Policy identified five components for environmental sustainability including:

1. The health of environmental systems
2. Environmental stresses and risks
3. Human vulnerability to environmental impacts
4. Social and institutional capacity
5. Global stewardship (World Economic Forum 2000)

Long-term efforts to monitor environmental change are under way and are intended to broaden our view of hazards. These assessment of long-term climate change anticipate that changes in atmospheric conditions will have significant impacts on energy use and production, public health outcomes, transportation, agriculture, water resources, and ecosystem services (Karl et al. 2009).

Lovins et al. (1999) contend that businesses restore, sustain, and expand our ecosystem so that it can produce vital services and biological resources abundantly. This view suggests that our natural environment as natural capital is to be used but in a conscious manner so as to reduce waste and expand the productivity of our natural resources. They suggest systems thinking so as to reduce energy costs and waste products. Energy savings can be productivity enhancing. This approach suggests that per capita energy and water consumption is a valid indicator of efficient natural resource allocation and consumption. Further, waste minimization also fits within this model and thus per capita waste generation and recycling are good indicators of natural systems sustainability. Waste minimization and pollution prevention are also Environmental Protection Agency's (EPA) recommended risk management strategies. Finally, they recommend that we view our natural environment as natural capital and one where we make an investment that will lead to positive return on our investments.

Hawkins et al. (1999) suggests that a healthy environment provides us with clean air and water, rainfall, productive oceans and water features, fertile soil, and sustainable watersheds. Social, economic, and environmental sustainability are interdependent and you cannot have one without the other.

Hossain (2001) noted the efforts of the Commonwealth Secretariat in using indicators to understand environmental sustainability include annual rate of deforestation, population density, and annual water use as a total water resources (Pantin 1997). The World Bank (1999) approach to environmental analysis is based on climate, water, forest, and pollution. Environmental degradation can result in health and economic losses, poverty, loss of intellectual property rights, loss of natural heritage, and conflict exposure to extreme events. It also might be related to the root causes of a hazard outcome such as disease. As an example, water supply, air pollution (indoor), and sanitation are all related to the highest level of risk from disease. This would suggest that indicators are thus related to specific hazards and may be a strong association to some threats while not to others.

Measuring Vulnerability

An indicator is a quantifiable measurable reflection of a phenomenon. We use indicators to understand a community's capacity to suffer from, cope with, and recover from a disaster. They are measurements that help us to understand key assets in our community. By looking at these indicators over time, they tell us if our community is improving, declining, or just remaining the same.

No set of measures tell us everything that we want to know. The Dow Jones Industrial Average does not include all stocks nor does the Consumer Price Index examine all goods. The key is that we use indicators to give us a barometer of how well something is doing. Quality indicators reflect existing and objective data from well-known sources. The indicators measure something that reflect local conditions or assets that are valued by the community.

Damage measures tangible things that are usually based on built measures such as bridges, homes, commercial or industrial buildings, cars and trucks, or communication towers. Human and cultural measures, such as coping indicators, demonstrate a community's preparedness for a disaster such as persons evacuated or sheltered. The number of people evacuated or shelters could reflect effective warning systems or procedures to help people get out of harm's way. These indicators may reflect a community's preparedness, the effectiveness of hazard mitigation or response strategies, and could explain why some communities recovery more quickly than other suggesting that they are more resilient.

No set of indicators can be all-inclusive as shown by the Dow Jones Industrial Average, which is a widely respected indicator of stock market performance. The Dow Jones does not include every stock traded on the New York Stock Exchange. Nor does the Consumer Price Index measure the prices of all consumer goods. Both indices, like the *Sierra Nevada Wealth Index*, are based on developing and monitoring a sample of indicators that, viewed together, provide a barometer of overall performance at the community level. The 60 indicators included in this web-based version of the *Index* were selected because:

- They are measurable and can be updated with existing and objective data sources.
- They measure the condition of assets of material importance to the Sierra Nevada's wealth.
- They measure the condition of assets, where active public interest exists.

Assessing which indicators are best suited for a hazards analysis is a critical step. Dwyer et al. (2004) provide suggestions on the criteria that help us to select appropriate measures of vulnerability.

- *Support concept:* Any effort to examine the vulnerability of a community is intended to serve an overall purpose. These indicators are tools to help address the overall goals or questions to be addressed.
- *Validity:* Indicators need to reflect accurate and timely data from the community and may be verifiable.
- *Data quality:* The data must be available from a reliable source and should be credible and reproducible.
- *Sensitivity:* A good indicator should measure change in a system or process. Indicators thus should be time sensitive so that they capture change. Data, thus, could be measured on a daily, monthly, quarterly, or annual basis.
- *Simplicity:* Indicators should clearly reflect a social, economic, or environmental condition.
- *Quantitative and Objective:* Indicators must be measurable and clearly reflect the economic, social, or natural condition.

Indicators of Social Conditions

Many studies have identified socioeconomic population characteristics that indicate higher vulnerability (Cutter et al. 2003; Peacock et al. 2000). High-risk groups, such as those with lower incomes, the very young and elderly, the disabled, women living alone, female-headed households, families with low ratios of adults to dependents, ethnic minorities, renters, recent residents, tourists, and the homeless are good social vulnerability indicators. These indicators have been used to demonstrate risk and that "social and economic costs of disasters fall unevenly on [these] different classes of victims and stakeholders" (Bolin and Stanford 1991; Cutter et al. 2000; Heinz III Center for Science, Economics and the Environment 2000; Morrow 1999; Wisner et al. 2004). Economic income limitations impact many people and thus they are not prepared to deal or recover from disasters (Mileti 1999). Age (less than 18 and over 65), gender (females), race, and income (mean household value) are viewed as primary social vulnerability indicators (Cutter et al. 2000).

Other indicators of vulnerability include nonwhite population, household incomes less than $25,000, households who rent, number of individuals over 65, disabled individuals (not including employment disabilities), individuals over age 25 without a high-school diploma, households without a vehicle, renters, single-parent households with children under the age of 18, and households without a phone. These indicators have been used by household or on a per capita basis. The following are examples of indicators that have been used in vulnerability assessments. A brief explanation is provided to suggest why these indicators were selected.

Household earnings less than $25,000: Percentage of household earnings less than $25,000 represents the number of households with earnings less than $25,000 divided by the total number of occupied housing units in the block group. This indicator was chosen to reflect an income threshold instead of households living at or below poverty. This income value reflects the minimum required to qualify for a home mortgage or a home.

No vehicle: Percentage of households by block group without a vehicle (car, truck, or van) was selected as one of our social vulnerability indicators. The Census Bureau defines housing unit as a house, apartment, mobile home, group of rooms, or a single room that is occupied as separate living quarters.

Nonwhite race: Much of the hazards vulnerability research has suggested that a nonwhite race is usually located in the highest hazard areas (Bolin 1993; Peacock et al. 1997; Pulido 2000). The percentage of nonwhite population by block group in New Orleans was selected as a social vulnerability indicator and consists of African-Americans as well as Asian and Latino neighborhoods. Howell (2005) noted that African-Americans were less likely to have evacuated or retreat to a safer place for Hurricane Georges and thus could be at a greater risk for Hurricane Katrina.

Federal Emergency Management Agency (FEMA) prioritizes vulnerability in the order of (1) income distribution, (2) elderly populations, (3) disabled populations, (4) children, (5) minority neighborhoods, and (6) language and cultural barriers. FEMA suggests what populations are vulnerable to specific hazards, and it is likely that vulnerable populations can be assumed as having similar characteristics.

Elderly: The percentage of persons within a block group over the age of 65 was selected as a social vulnerability indicator. The hazards literature sites numerous studies that suggest that the elderly are a particularly vulnerable population to hazards. These studies note that physical, mental, and sensory skills become weaker with age. Age is recognized as an indicator of social vulnerability due to mobility limitations, major dependence on relatives, frequency of respiratory distress, and a lower resilience after the disaster (Cutter et al. 2000; Hewitt 1997; Mileti 1999; O'Brien and Mileti 1992). Howell (2005) found that individuals over the age of 65 were less likely to evacuate or have a plan for evacuation for Hurricane Georges, which threatened New Orleans in the Summer of 1998.

Disability: The percentage of noninstitutionalized individuals with a disability includes persons who have sensory, physical, mental, or other self-care limitation that limits their activities outside the home. This category does not include those with employment disabilities. The percentage is based on the number of disabled individuals in the block group, unlike the other indicators, which use households as the measurement.

Education: This indicator includes the percentage of individuals over the age of 25 with no high-school diploma. Lower education has been suggested as a constraint in understanding hazard warnings (Heinz III Center for Science, Economics and the Environment 2000). This indicator reflects the number of people over the age of 25 in each block group rather than the number of households.

Use of rental housing: The percentage of rented housing units represents the number of occupied households renting divided by the total number of occupied households in a block group. Several studies suggest that renters are vulnerable because of their lack of finances and/or limitations in transportation (Heinz III Center for Science, Economics and the Environment 2000; Morrow 1999).

No phone: The percentage of housing units without a phone represents all households without a phone divided by the total occupied housing units. Telephones are an important means of communication to notify an immediate evacuation. Calls can be made from emergency managers to inform of evacuation and locate those who are willing to answer the phone and respond with key punches that they do not have means to evacuate. Although phone calling as a means of communication has not yet been documented as a warning tool, access to a phone is an important vulnerability criterion. Automated systems for early hazard warning use local phone capabilities (Burby 1998).

Single-parent households: The percentage of single parents who are the head of a household was selected as a social vulnerability indicator. This indicator reflects single parents who are head of household with children under the age of 18.

Indicators of Economic Conditions

Traditional indicators of economic conditions center on employment, housing, business sales, business taxes, and construction. Unemployment rate has been a widely used economic indicator for it reflects each of the sectors of the economy as well as the public and private employment. The level and extent of construction has been used to judge the vitality of a local economy separating permits for industrial or commercial operations from either rental or home construction. Examining the number of housing starts over a long-term period is a good way to determine if the present condition is performing at a higher rate. The level of business taxes by type of business reflects the level of local economic health.

In selecting economic indicators, the measures should be a reflection of the broad basis of the economy and not heavily focused on a single sector. The number of jobs in each of the major economic category as viewed over a long period provides this type of broad measure. Unemployment rates also provide a broad view of the economy. These indicators included the number of employers and employees (by sector industry), overall unemployment rate (especially small business that has one to five employees), the percentage of business failures, household population, and the number of students in schools by type of school or college. Employment estimates provide a good economic indicator for the local economy as reflected in the employment loss in Figure 6.2 for Orleans Parish following the impacts of Hurricane Katrina and Rita in 2005.

Zandi et al. (2006) also provide a breakdown of volume of production as a key indicator of recovery from a disaster including fishing, chemical, retail sales, home and rental prices (average home sale price and the fair market rent for two-bedroom apartment). Comparisons of production volume over multiple years are an excellent

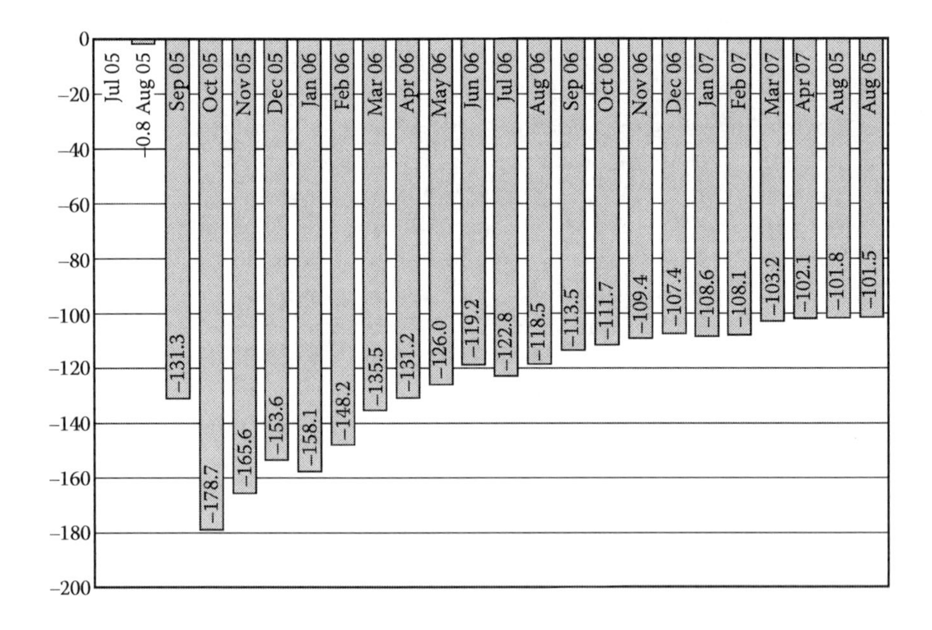

Figure 6.2 Employment loss in metro New Orleans area in thousands.

indicator of the vitality of a local economy. Other economic indicators used in vulnerability assessments include the following:

- Number of residential units destroyed (compare renters and homeowners).
- Number of weeks to restore residential units for use; percent of electrical system shut down.
- Duration of recovery period to restore utilities; percent of businesses closed because of the disaster; the percent of businesses opened for 1 to 6 months after a disaster (recovery period); production level of agricultural commodities or units processed by month; average weekly wage prior to an event and afterwards (by month); percent of homes built prior to 1992; percent of residents in a flood zone with National Flood Insurance Program flood insurance.
- The percent of homes, rental units, or businesses with flood insurance.

Indicators of Environmental Conditions

Environmental capital is a key contributor to the sustainability of communities and human well-being. Figure 6.3 provides an example of a natural forest along the Gulf of Mexico. Forests, wetlands, rivers, and open spaces make significant contributions to recreation commercial enterprises such as fishing, forestry, and agriculture. In a study of the effects of urban design on aquatic ecosystems in the United States, Beach (2004) examined the relationship between land use and the effects of sprawl on both air and water quality. The study shows that as the percent of impervious surfaces increased in a coastal community, the nature of the water runoff into water features changed causing increased levels of nitrogen and phosphorus; organic carbon; trace metals such as copper, zinc, and lead; and pesticides (Schueler and Holland 2000). This dynamic is considered coastal sprawl, which is the expansion of low-density residential and commercial development scattered across large coastal land areas. The study notes that changes in urban growth patterns affected habitat quality, water temperature, pollutants, and aquatic life. In addition, the study noted that as coastal communities expanded using traditional development patterns of sprawl, drivers were forced into longer trips for work, recreation, or just normal shopping. The study used community indicators to show the relationship of development patterns and air quality. Beach used indicators to demonstrate their use in understanding the impact of development practices on ecosystem services, such as air or water quality.

Beach's work centers on a national and even worldwide problem of development patterns in coastal areas. Coastal counties make up 17% of the land area in the United States but just 13% of the nation's acreage. Unfortunately, this coastal zone is home to more than half of the U.S. population. The issue unfortunately is not in the present condition but that this coastal region is where the United States is experiencing population increases. We are continuing to put more people in a small area. Pollution and habitat degradation is the end result of this pattern.

Figure 6.3 Healthy forest, clean water, and soils support abundant fish and wildlife. (Photo courtesy of J. Pine.)

The population density of a community would be part of an effort to determine the burden that people have on the environment. This indicator alone does not reflect the magnitude of human impacts on environmental health. We need further indicators.

Beach cited studies that have demonstrated that when impervious surfaces cover more than 10% of a watershed, water features and estuaries become biologically degraded. A key indicator for a community is the percent of the watershed that is composed of impervious surfaces. When it exceeds 10%, there will be problems according to Beach (2004). The fact is that the ecosystem health including streams, marshes, and rivers impacted by development results in less diverse, less stable, and less productive watersheds. Increases in impervious surfaces lead to higher levels of sediments containing higher concentrations of nitrogen and phosphorus; organic carbon; or metals such as copper, zinc, or lead as well as petroleum hydrocarbons; and pesticides (Schueler and Holland 2000). The study found not only differences in channel erosion but also the health of estuaries. Increased levels of nitrogen lead to algal blooms and fish kills. Increases in fertilizer use in watersheds also reduces water clarity allowing less light to penetrate below the water's surface impacting the health of biologic habitats and aquatic habitats.

The issues related to urban development patterns are limited not only to the productivity of watersheds but also to the changes in the volume of water in streams. The fact is that there is an increased threat of flooding as development increases in a watershed and as impervious surfaces increase (Booth 1991; Booth and Reinelt 1993). Changes in water feature discharge reflect development patterns and provide a clear indicator for potential flooding problems in a community.

Changes in impervious surfaces also result in the rise of water temperature in the water. As the percent of impervious surface area increases in a watershed, the water temperature also increases (Galli 1991). The result may be decreases in oxygen levels resulting in changes in the marine life and environment.

Other studies (Jacob and Blake 2010) monitored variables of environmental conditions that impact ecosystem services. They used measures of mean temperature, days with temperature above a specific value, total precipitation, days with rainfall, and periods of drought in relation to local agricultural production or revenue from local tourism. They also recommend monitoring the impacts of hazards such as frequency and extent of electrical outages, emergency service calls, water quality, flight delays, and various service interruptions. Measuring the impacts of hazardous conditions suggests a tangible way of determining the impact of environmental conditions that are relevant to a community's economy rather than just the condition itself. Gober and Kirkwood (2010) provide an example of the association between climate-induced water shortage and local community economic measures.

Understanding the association between development patterns and environmental conditions is noted in research dealing with hypersprawl. Measures reflecting the expansion of residential development with housing densities of one unit on three acres or less could be associated with local levels of water use or commute time. Measures of housing unit density could show how land-use patterns could impact water use, water quality, or air quality.

Vehicle miles driven by residents could also impact air quality conditions. As the vehicle miles driven per household increases, air quality is impacted. Our urban growth patterns could require that people drive further; the average commuter trip could increase as a result of these development patterns.

Miles of bike trails as noted in Figure 6.4 allow for alternative modes of transportation also reflect community action to enhance environmental conditions and a contribution to human well-being.

Figure 6.4 Healthy landscapes support recreation. (Photo courtesy of J. Pine.)

EPA has adopted measures to determine when water features might be impacted negatively from agricultural or related practices (use of fertilizers on commercial or residential properties). The measure of total maximum daily loads is used by EPA and state environmental agencies to determine how much pollution a body of water can accept without becoming impaired.

The use of these environmental indicators, thus, provides a means of assessing our environment to determine if there is change occurring at the community level.

Methodological Issues

Scale of Analysis

When we assess local conditions as a part of a hazards analysis, we need to view the indicators from the same scale. That is, we would measure each set of indicators from a region, county, city, or neighborhood scale. To ensure that any comparisons that are made are valid, we should make sure that our data reflect the same scale. As an example, if we are assessing the capacity of a community to deal with a disaster, we might obtain data on a county-wide basis and look at Census population data, crime rates, educational attainment data, and public health information for the county level. We would then compare the result with other surrounding counties.

If we have data at one scale and then attempt to compare this with data at another scale, questions can be raised concerning the validity of our analysis. The key is to ensure that the data that we collect is at the same level (county, Census tract, or ZIP code level).

Weighting, Data Availability, and Accuracy

When multiple indicators are part of the analysis, we would need to consider if each indicator is equal and can be used along with others. Are the indicators of equal weight or do some indicators a better reflection of the community's capacity to cope or withstand a disaster event. Most hazards analysis studies use an equal-weighting process, where all the indicators are treated the same. Some, however, assign specific weights to various variables (SOPAC 2000).

Throughout this text, we have stressed the importance of ensuring that data are current, accurate, and available for use in a hazards analysis. As we examine indicators of community vulnerability, the same data quality issues are present and should influence how we judge the capacity of our community to cope with disaster.

Type and Scope of Measures

Twigg and Bhatt (1998) contend that vulnerability is complex and too complicated to be represented by models or frameworks. They see that economic, social, demographic, political, and psychological elements influence how communities are impacted by disasters. Vulnerability is too dynamic and complex to account for the many factors that shape our societal adjustments.

The application of quantitative indicators for community resiliency provides a basis for identifying hazards risk management strategies and hazard mitigation alternatives to reduce vulnerability. In addition, qualitative data can be obtained that reflects the views and impressions of community members as to the causes of vulnerability and potential solutions. Surveys of community members can reveal why some citizens refuse to adopt protective actions where notice of a disaster is present. They can clarify what citizens thought would be the result of a specific hazard event and why they refused to evacuate, take shelter, purchase flood insurance, elevate their home, or take other protective measures. Statistical measures of a community can give us clues as to the causes of vulnerability but only people can explain why they adopted a specific coping strategy and rejected others.

Few studies have assessed the impacts of disasters on communities and examined the relationship between social and economic vulnerability and risk. Hurricane Katrina and the devastating destruction of levees, businesses, and residential property provides an opportunity to examine the relationship between social and economic vulnerability and risk. Did the extent of flooding associated with Hurricane Katrina treat households equally in the City of New Orleans? Did the storm have a far greater physical impact on households that are considered to have a higher social vulnerability?

Interdependence of Social, Economic, and Ecological Capital

Social, economic, and natural systems do not exist independently, but in concert with one another. Some of the indicators from one system might fall into another system such as unemployment has both human and economic dimensions. Educational attainment impacts both socially and economically. Water quality is an indication of environmental capital but may be used for water sports and recreation. Water thus may be used as a critical economic measure, which is associated with tourism. The interrelatedness of these systems suggests that we be more sensitive of what the indicator measures.

If we represent the three types of capital by a circle, then the size of the capital would represent a stronger robust capacity for dealing with disaster. It might also reflect that one type of capital is stronger than another, such as a vital petrochemical industry might have high wages but have a low indicator reflecting the natural environment and higher levels of pollution.

A second view of the relationship between social, economic, and natural environments is their interdependence. A community that relies on the attraction of the natural environment would be drastically impacted by a forest fire, hurricane, flood, earthquake, or other disaster. Since each community is unique in its natural assets, culture, and economic assets, the overlap between them would vary and influence the community's resilience and sustainability. As leaders in the community examine their social, economic, and natural systems, they should not only determine their robustness but also their interrelatedness. Every community is unique and the interrelatedness of their social, cultural, and natural assets determine the robustness of their assets to deal with disaster.

Discussion Questions

Should the vulnerability indicators be hazard specific or appropriate to understand the impacts from any hazard? Should the indicators be selected for a specific geographic area or community?

What criteria may be used to determine if the indicators are a good measure of vulnerability?

What should be the scale of analysis for a hazards assessment? Can I include data that are collected at various times and spatial scales?

Applications

Using one of the sources of capital for a community (sociocultural, economic, or environmental), select six indicators that you believe are a broad reflection of the resilience of a community. For the six indicators, determine an appropriate scale as well as a time period for analysis. The scale could be by county or city, and the basis of analysis could be monthly, quarterly or annually. In addition, determine if the indicators should be expressed in terms of a percentage of households, population, or persons in the workforce. Examine these indicators for your community and determine if you believe they provide a valid basis for assessing the sustainability of your community.

Websites

American Planning Association represents practicing planners, elected and appointed officials, and citizens involved in planning issues. http://www.planning.org

NOAA Coastal Services Center: http://www.csc.noaa.gov

Search their site for Smart Growth

The Smart Growth Network is a coalition of organizations working together to promote smart growth. The website features a calendar of events, legislative news, reports, links to partner organizations, and more. NOAA is a partner of the Smart Growth Network. http://www.smartgrowth.org/

Sierra Nevada Business Council developed the Wealth Index Third Edition (2006). The Sierra Nevada Wealth Index is intended to help business leaders and policy makers understand the assets that sustain our region. The index describes the social, natural, and financial capital, which are the foundation of the Sierra Nevada's economy and thereby provides an integrated understanding of our region's wealth. http://www.sbcouncil.org. Do a search on this site for the Sierra Nevada Wealth Index.

Sustainable Development Indicators: http://www.hq.nasa.gov/iwgsdi/Welcome.html

Sustainable Development Indicators (SDIs) are various statistical values that collectively measure the capacity to meet present and future needs. SDIs will provide information crucial to decisions of national policy and to the general public.

Environmental Indicators: http://www.epa.gov/reg3esd1/data/indicators.htm

Excellent discussion of environmental indicators and resources. See: http://www.epa.gov/nceawww1/roe/pdfs/Region3_Indicators.pdf

EPA's Report on the Environment 2008, Indicators Presenting Data for EPA Region 3 (PDF) (111 pp, 15 MB): This report lists the most reliable indicators with data specific for EPA Region 3, those that EPA believes are of critical importance to its mission to protect human health and the environment.

Key National Indicators Initiative (U.S. Government Accountability Office). http://www.gao.gov/npi/

The Key National Indicators Initiative (KNII), launched in 2003 and currently housed at The National Academies in Washington, DC, has begun work on a comprehensive indicator set that will allow individual citizens, organizations, and elected officials to answer the question, "How are we doing as a nation?". State of the USA™ will gather in one place credible, up-to-date information about the nation's environmental, economic, and social performance. It is currently in development with an anticipated public release at the end of 2008. See: http://www.iisd.org/measure/compendium/DisplayInitiative.aspx?id = 1887.

The State of the Nation's Ecosystems. http://www.heinzctr.org/sites/default/files/pdf/The%20State%20of%20the%20Nation%27s%20Ecosystems%202008.pdf

Example of a local set of "Economic, Social, and Environmental Indicators" 2012.http://www.ciras.iastate.edu/publications/CerroGordoCounty_EconomicSustainabilityIndicators.pdf

National Center for Environmental Economics (NCEE). http://yosemite.epa.gov/ee/epa/eed.nsf/pages/homepage

The U.S. EPA's NCEE offers a centralized source of technical expertise to the agency, as well as to other federal agencies, Congress, universities, and organizations. NCEE's staff specializes in analyzing the economic and health impacts of environmental regulations and policies, and assists EPA by informing important policy decisions with sound economics and other sciences. NCEE also contributes to and manages EPA's research on environmental economics to improve the methods and data available for policy analysis.

United Nations Department of Economic and Social Affairs, Division for Sustainable Development: http://sustainabledevelopment.un.org/index.html

World Bank (2000). *World Development Indicators 2000*. Washington, DC: The World Bank. http://data.worldbank.org/data-catalog/world-development-indicators.

The primary World Bank collection of development indicators compiled from officially recognized international sources. It presents the most current and accurate global development data available and includes national, regional and global estimates.

Environmental Risks to Human Health: New indicators .... The indicators rank countries according to potential environmental threats to human health. http://www.wri.org/publication/content/8349

World Health Organization (2000). *World Health Report 2013*. Geneva: WHO. http://www.who.int/whr/en/

Report Highlights Importance of Chesapeake Bay Health to Human Health. Johns Hopkins Bloomberg School of Public Health. http://www.jhsph.edu/public healthnews/press_releases/PR_2004/Burke_Bay_indicators.html

References

Adger, W. N. (2006). Vulnerability. *Global Environmental Change*, *16*(3), 268–281.

Adato, M., Carter, M. R., and May, J. (2006). Exploring poverty traps and social exclusion in South Africa using qualitative and quantitative data. *The Journal of Development Studies*, *42*(2), 226–247.

Atkins, J. P., Mazzi, S., and Easter, C. D. (2000). *A Commonwealth Vulnerability Index for Developing Countries: The Position of Small States*. Commonwealth Secretariat, Economic Paper 40.

Barnett, J., Lambert, S., and Fry, I. (2008). The hazards of indicators: insights from the environmental vulnerability index. *Annals of the Association of American Geographers*, *98*(1), 102–119.

Beach, D. (2003). Coastal sprawl: The effects of urban design on aquatic ecosystems. *In of the United States, Pew Oceans Commission 2002*.

Berkes, F., and Folke, C. (1998). Linking social and ecological systems for resilience and sustainability. *Linking social and ecological systems: management practices and social mechanisms for building resilience*, *1*, 13–20.

Birkmann, J. (2006) *Measuring Vulnerability to Natural Hazards: Towards Disaster Resilient Societies*. New York: United Nations Publications.

Birkmann, J. (2007). Risk and vulnerability indicators at different scales: applicability, usefulness and policy implications. *Environmental Hazards*, *7*(1), 20–31.

Bolin, R. and Stanford, L. (1991). Shelter, housing and recovery: a comparison of US Disasters. *Disasters*, *15*, 24–34.

Booth, D. B. (1991). Urbanization and the natural drainage system—impacts, solutions, and prognoses. *Northwest Environmental Journal*, *7*(1), 93–118.

Booth, D. B. and Reinelt, L. E. (1993). *Consequences of Urbanization on Aquatic Systems: Measured Effects, Degradation Thresholds, and Corrective Strategies*. Proceedings of Watershed '93, A National Conference on Watershed Management, Alexandria, VA.

Boyd, K. A. (2005, March). Assessing storm vulnerability using the Coastal Risk Atlas. In *Geological Society of America Abstracts with Programs*, (Vol. 37, No. 2, p. 50).

Bruneau, M., Chang, S. E., Eguchi, R. T., Lee, G. C., O'Rourke, T. D., Reinhorn, A. M., Shinozuka, M., Tierney, K., Wallace, W. A., and von Winterfeldt, D. (2003). A framework to quantitatively assess and enhance the seismic resilience of communities. *Earthquake Spectra*, *19*(4), 733–752.

Burby, R. J. (1998). *Cooperating With Nature*. Washington, DC: Joseph Henry Press.

Cannon, T., Twigg, J., and Rowell, J. (2003). Social vulnerability, sustainable livelihoods and disasters. Report to DFID Conflict and Humanitarian Assistance Department (CHAD) and Sustainable Livelihoods Support Office.

Comfort, L., Wisner, B., Cutter, S., Pulwarty, R., Hewitt, K., Oliver-Smith, A., Wiener, J., Fordham, M., Peacock, W., and Krimgold, F. (1999). Reframing disaster policy: the global evolution of vulnerable communities. *Environmental Hazards, 1*(1), 39–44.

Crowards, T. (1999). *An Economic Vulnerability Index for Developing Countries, with Special Reference to the Caribbean.* Barbados: Caribbean Development Bank.

Cutter, S. L. (1996). Vulnerability to environmental hazards. *Progress in Human Geography, 20*(4), 529–539.

Cutter, S. L., Mitchell, J. T., and Scott, M. S. (1997). *Handbook for Conducting a GIS-Based Hazards Assessment at the County Level.* Columbia, SC: University of South Carolina.

Cutter, S. L., Barnes, L., Berry, M., Burton, C., Evans, E., Tate, E., and Webb, J. (2008). A place-based model for understanding community resilience to natural disasters. *Global Environmental Change, 18*(4), 598–606.

Cutter, S. L., Boruff, B. J., and Shirley W. L. (2003). Social vulnerability to environmental hazards. *Social Science Quarterly, 84*(2), 242–261.

Cutter, S. L., Burton, C. G., and Emrich, C. T. (2010). Disaster resilience indicators for benchmarking baseline conditions. *Journal of Homeland Security and Emergency Management, 7*(1).

Cutter, S. L., Emrich, C. T., Webb, J. J., and Morath, D. (2009). *Social Vulnerability to Climate Variability Hazards: A Review of the Literature.* Final Report to Oxfam America, pp. 1–44.

Cutter, S. L., Mitchell, J. T., and Scott, M. S. (2000). Revealing the vulnerability of people and places: a case study of Georgetown County, South Carolina. *Annals of the Association of American Geographers, 90*(4), 713–737.

Dow, K. (1992). Exploring differences in our common future(s): the meaning of vulnerability to global environmental change. *Geoforum 23*, 417–436.

Dwyer, A., Zoppou, C., Nielsen, O., Day, S., and Roberts, S. (2004). *Quantifying Social Vulnerability: A Methodology for Identifying Those at Risk to Natural Hazards.* Canberra, Australia: Geoscience Australia.

Erenstein, O. and Thorpe, W. (2010). Livelihoods and agro-ecological gradients: a meso-level analysis in the Indo-Gangetic Plains, India. *Agricultural Systems, 104*(1): 42–53.

Esty, D. C., Levy, M., Srebotnjak, T., and deSherbinin, A. (2005). *2005 Environmental Sustainability Index: Benchmarking National Environmental Stewardship.* New Haven, CT: Yale Center for Environmental Law and Policy.

Füssel, H. M. (2007). Vulnerability: a generally applicable conceptual framework for climate change research. *Global Environmental Change, 17*(2), 155–167.

Galli, J. (1991). *Thermal Impacts Associated with Urbanization and Storm Water Management Best Management Practices.* Washington, DC: Metropolitan Washington Council of Governments, Maryland Department of Environment.

Gober, P. and Kirkwood, C. W. (2010). Vulnerability assessment of climate-induced water shortage in Phoenix. *Proceedings of the National Academy of Sciences, 107*(50), 21295–21299.

Goudie, A. S. (2009). Dust storms: recent developments. *Journal of Environmental Management, 90*, 89–94.

Heinz III Center for Science, Economics and the Environment (2000). *The Hidden Costs of Coastal Hazards: Implications for Risk Assessment and Mitigation*. Washington, DC: Islands Press.

Hewitt, K. (1997). Regions of Risk: A Geographical Introduction to Hazards. Harlow, Essex: Addison Wesley, Longman.

Hoffman, S. M. (2003). *The Hidden Victims of Disaster*, Hazard Workshop Presentation, Boulder, CO.

Holand, I. S., Lujala, P., and Rød, J. K. (2011). Social vulnerability assessment for Norway: a quantitative approach. *Norsk Geografisk Tidsskrift-Norwegian Journal of Geography, 65*(1), 1–17.

Hossain, S. M. N. (2001). *Assessing human vulnerability due to environmental change: concepts and assessment methodologies*. Master of Science Degree Thesis, Royal Institute of Technology, KTH, Stockholm, Sweden.

Howell, S. E. (2005). *Citizen Hurricane Evacuation Behavior in Southeastern Louisiana: A Twelve Parish Survey*. New Orleans, LA: Survey Research Center, University of New Orleans.

Hufschmidt, G. (2011). A comparative analysis of several vulnerability concepts. *Natural Hazards, 58*(2), 621–643.

IPCC (2007). Climate change 2007: the physical science basis. In S. Solomon, M. Qin, M. Manning, Z. Chen, K. B. Marquis, K. B. Averyt, M. Tignor, and H. L. Miller. (eds.), *Contribution of Working Group 1 to the Fourth Assessment Report of the Intergovernmental Panel on Climate Change*. Cambridge: Cambridge University Press.

Jacob, K. and Blake, R. (2010). Chapter 7: indicators and monitoring. *Annals of the New York Academy of Sciences, 1196*, 127–142.

Rosenzweig, C., Solecki, W. D., Blake, R., Bowman, M., Faris, C., Gornitz, V., Jacob, K., and Zimmerman, R. (2011). Developing coastal adaptation to climate change in the New York City infrastructure-shed: process, approach, tools, and strategies. *Climatic Change, 106*(1), 93–127.

Kaly, U., Briguglio, L., McLeod, H., Schmall, S., Pratt, C., and Pal, R. (1999). *Environmental Vulnerability Index (EVI) to Summarize National Environmental Vulnerability Profiles*. SOPAC Technical Report 275. Suva. Fifi. South Pacific Applied Geoscience Commission.

Kampa, M. and Castanas, E. (2008). Human health effects of air pollution. *Environmental Pollution, 151*(2), 362–367.

Karl, T. R., Melillo, J. M., and Peterson, T. C. (eds.) (2009). *Global Climate Change Impacts in the United States*. Cambridge: Cambridge University Press.

Kasperson, R. E., Renn, O., Slovic, P., Brown, H. S., Emel, J., Goble, R., Kasperson, J. X., and Ratick, S. (1988). The social amplification of risk: a conceptual framework. *Risk Analysis, 8*, 177–187.

Kasperson, J. X., Kasperson, R. E., and Turner, B. L. (1995). Regions at Risk. Tokyo: United Nations University Press.

Khazai, B., Merz, M., Schulz, C., and Borst, D. (2013). An integrated indicator framework for spatial assessment of industrial and social vulnerability to indirect disaster losses. *Natural Hazards, 67*, 1–23.

Kumpulainen, S. (2006). Vulnerability concepts in hazard and risk assessment. In: Philipp Schmidt-Thorne (ed.), *Natural and Technological Hazards and Risks Affecting the Spatial Development of European Regions*. Geological Survey of Finland, Special Paper 42, pp. 65–74.

McAlpine, C. A., Etter, A., Fearnside, P. M., Seabrook, L., and Laurance, W. F. (2009). Increasing world consumption of beef as a driver of regional and global change: a call for policy action based on evidence from Queensland (Australia), Colombia and Brazil. *Global Environmental Change, 19*(1), 21–33.

Morrow, B. H. (1999). Identifying and mapping community vulnerability. *Disasters, 23*(1), 1–18.

Munich Re Group (2002). *Topics 2002.* Report of the Geo-science Research Group, Munich Reinsurance Company, Munich.

Nakagawa, Y. and Shaw, R. (2004). Social capital: a missing link to disaster recovery. *International Journal of Mass Emergencies and Disasters, 22*(1), 5–34.

National Research Council (2000). *Ecological Indicators for the Nation (2000), Commission on Geosciences, Environment and Resources.* Washington, DC: National Academic Press.

Ni-Bin, C., Rui, F., Zhiqiang, G., and Wei, G. (2010). Skin cancer incidence is highly associated with ultraviolet-B radiation history. *International Journal of Hygiene and Environmental Health, 213*(5), 359–368.

O'Brien, K., Leichenko, R., Kelkar, U., Venema, H., Aandahl, G., Tompkins, H., Javed, A., Bhadwal, S., Barg, S., Nygaard, L., and West, J. (2004). Mapping vulnerability to multiple stressors: climate change and globalization in India. *Global Environmental Change, 14*(4), 303–313.

O'Brien, P. W. and Mileti, D. S. (1992). Citizen participation in emergency response following the Loma Prieta earthquake. *International Journal of Mass Emergencies and Disasters, 10*(1), 71–89.

Pantin, D. (1997). *Alternative Ecological Vulnerability Indicators for Developing Countries with Special Reference to SIDS.* Report prepared for the Expert Group on Vulnerability Index. UN DESA, December 17–19, 1997.

Peacock, W. G., Betty H. M., and Gladwin, H. (eds.) (2000). *Hurricane Andrew and the Reshaping of Miami: Ethnicity, Gender, and the Socio-Political Ecology of Disasters.* Miami, FL: International Hurricane Center, Florida International University.

Peacock, W. G., Morrow, B. H., and Gladwin, H. (eds.) (1997). *Hurricane Andrew: Ethnicity, Gender, and the Sociology of Disasters.* New York: Psychology Press.

Pelling, M., Maskrey, A., Ruiz, P., Hall, L., Peduzzi, P., Dao, Q. H., and Kluser, S. (2004). *Reducing Disaster Risk: A Challenge for Development.* New York: United Nations Development Program, Bureau for Crisis and Recovery.

Polsky, C., Neff, R., and Yarnal, B. (2007). Building comparable global change vulnerability assessments: the vulnerability scoping diagram. *Global Environmental Change, 17*(3–4), 472–485.

Popovski, V. and Mundy, K. G. (2012). Defining climate-change victims. *Sustainability Science, 7*(1), 5–16.

Pulido, L. (2000). Rethinking environmental racism: white privilege and urban development in Southern California. *Annals of the Association of American Geographers, 90*, 12–40.

Richmond, B. M., Fletcher, C. H., Grossman, E. E., and Gibbs, A. E. (2001). Islands at risk: coastal hazard assessment and mapping in the Hawaiian Islands. *Environmental Geosciences, 8*(1), 21–37.

Ridoutt, B. G. and Pfister, S. (2010). A revised approach to water foot printing to make transparent the impacts of consumption and production on global freshwater scarcity. *Global Environmental Change, 20*(1), 113–120.

Schueler, T. and Holland, H. K. (2000). *The Practice of Watershed Protection*. Ellicott City, MD: Center for Watershed Protection.

SOPAC (2000). *Environmental Vulnerability Index: Development and Provisional Indices and Profiles for Fiji, Samoa, Tuvalu and Vanuatu*, EVI Phase II Report.

Turner II, B. L., Kasperson, R. E., Matson, P. A., McCarthy, J. J., Corell, R. W., Christensen, L., Eckley, N., et al. (2003). A framework for vulnerability analysis in sustainability science. *Proceedings of the National Academy of Sciences of the United States of America, 100*, 8074–8079.

Turvey, R. A. (2000). *Methodology for Vulnerability Assessment of Developing Countries with Relevance to Small Islands and Least Developed Countries*. Tokyo: United Nations University Institute of Advanced Studies.

Twigg, J. and Bhatt, M. (1998). *Understanding Vulnerability: South Asian Perspectives*. London: Intermediate Technology Publications on behalf of Duryog Nivaran.

United Nations (1997). *Report of the Secretary-General on the Development of a Vulnerability Index for Small Island Developing States*. New York: Commission for Sustainable Development.

UNCTAD (1997). *The Vulnerability of Small Island Developing States in the Context of Globalisation: Common Issues and Remedies*. Report prepared for the Expert Group on Vulnerability Index, New York.

Veron J. E., Hoegh-Guldberg, O., Lenton, T. M., Lough, J. M., Obura, D. O., Pearce-Kelly, P., Sheppard, C. R., Spalding, M., Stafford-Smith, M. G., and Roger, A. D. (2009). The coral reef crisis: the critical importance of < 350 ppm CO_2. *Marine Pollution Bulletin, 58*(10), 1428–1436.

Villa, F. and McLeod, H. (2002). Environmental vulnerability indicators for environmental planning and decision-making: guidelines and applications. *Environmental Management, 29*(3), 335–348.

Wallis, A. D., Aguelles, E., Lampe, D., and Meehan, M. (2001). *Imaging the Region: South Florida via Indicators and Public Opinions*. Miami, Florida: Florida Atlantic University/Florida International University Joint Center for Urban and Environmental Problems.

Wells, J. (1997). *Composite Vulnerability Index: A Revised Report*. London: Commonwealth Secretariat.

Williams, L. and Kaputska, L. (2000). Ecosystem vulnerability: a complex interface with technical components. *Environmental Toxicology and Chemistry, 19*(4), 1055–1058.

Wisner, B., Blaikie, P., Cannon, T., and Davis, I. (2004). *At Risk: Natural Hazards. People's Vulnerability and Disasters*. London: Routledge.

World Bank (1999). *Environment Matters, Annual Review*. Washington, DC: The World Bank.

World Economic Forum (2000). *Pilot Environmental Sustainability Index: An Initiative of the Global Leaders Tomorrow Environment Task Force*, Annual Meeting 2000, Davos, Switzerland, in collaboration with Yale University and Columbia University.

Xi, J. (2011). Ecological accounting and evaluation of urban economy: taking Beijing city as the case. *Communications in Nonlinear Science and Numerical Simulation, 16*(3), 1650–1669.

Yohe, G. and Tol, R. S. (2001). Indicators for social and economic coping capacity—moving toward a working definition of adaptive capacity. *Global Environmental Change, 12*(1), 25–40.

Zandi, M. S., Cochrane, R., Ksiazkiewicz, R., and Sweet, R. (2006). Restarting the economy. In E. Birch and S. Wachter (eds.), *Rebuilding Urban Places after Disaster: Lessons from Hurricane Katrina*. Philadelphia, PA: University of Pennsylvania Press.

Chapter 7

Risk Communication

John C. Pine and Stephen L. Guillot, Jr.

Objectives

The study of this chapter will enable you to:

1. Define risk communication and the communication process.
2. Examine communication barriers in discussing risk with the public or other community stakeholders.
3. Examine the target audience in the risk communication process.
4. Discuss tools for risk communication including maps, figures, and community engagement.
5. Explore strategies for managing risk communication including community engagement, ethics, decision making, and legal issues.
6. Explore how organizations learn through risk communication.

Key Terms

Adaptive behavior
Credibility
Dialogue
Precautionary actions
Risk
Risk communication
Social amplification
Social media
Trust

Issue

Doubt, skepticism, and uncertainty have become part of individual views of our capacity to deal with disasters. The response to Hurricane Katrina in 2005 raised many questions about our risks associated with natural hazards. People have doubts about government policies and processes, as well as the priorities of both businesses and community-oriented agencies. Many may question agency representatives especially when someone states that they know what is best. We must be sensitive to how we communicate about risk and how discussions of risks associated with hazards may be received by others. Understanding how individuals perceive risk to hazards and how the communication of risk impacts individual and organization actions is essential in reducing vulnerability to hazards. People no longer see that hazards and our vulnerability are associated with chance and factors outside human control. People appreciate that our systems are not perfect, have limitations, and may be vulnerable to human and organizational failures. Our goal is to enhance the public's understanding of the risks associated with hazards and improve organizational adaptive measures by increasing our communication of risks (Senate Committee on Homeland Security and Governmental Affairs 2006). What can be done to strengthen our individual, community, and organizational resilience?

Introduction

Hazards identification and risk analysis provide a basis for profiling hazards that might impact a public jurisdiction, business, or agency. These processes clarify when and where a disaster might occur and the impacts that it could result. Information from the assessment process can be used in many ways to help us adapt to our risks including short-term hazard warnings or in the development of long-term mitigation strategies to reduce adverse consequences from disasters. A jurisdiction might initiate communication strategies to help the community know how vulnerability might be reduced through the use of sheltering or evacuation protective actions. Many decisions associated with hazards are made on an individual, family, community, regional, and national basis. Hazards analysis does not conclude with the risk analysis but is a tool to reduce vulnerability and strengthen individual, organizational, and community resilience to hazards. Understanding the role of risk communication in individual and organizational decision making is critical in establishing and sustaining resilient communities. Our goal is to enhance our decision-making capacity through conscious communication strategies at all levels.

Risk Communication

Risk communication involves the process of sharing information about hazards, risks, vulnerability, assets, and adaptive mechanism within organizations or with the public. The process is intentional and goal directed including sharing information about a hazard or to identify appropriate strategies to reduce vulnerability to a specific hazard. We see that risk communication involves more than just talking about the hazards, but a process that provides a framework for enhancing our capacity to understand hazards and foster constructive adaptive strategies at the individual, community, or organizational levels to foster sustainability and resilience.

Individuals are concerned about their own safety and security and have the capacity to protect their welfare; public, private, and nonprofit agencies have the capacity to build a culture of trust and credibility to ensure that their expertise is used to support sound decision making. But failures occur at all levels and we acknowledge that our organizations are not perfect. The key is to realize that sound decision making is an intentional action by individuals and organizations. The risk communication process has a critical role in supporting sound decision making and the adoption of strategies to cope and deal with hazards.

Hundreds of miles of levees were constructed to defend metropolitan New Orleans against storm events. These levees were not designed to protect New Orleans from a Category 4 or 5 monster hurricane. The original specifications of the levees offered protection that was limited to withstanding the forces of a moderate hurricane. Once constructed, the levees were turned over to local control, leaving the United States Army Corps of Engineers to make detailed plans to drain New Orleans should it be flooded (U.S. House of Representatives 2006).

> The Local sponsors - a patchwork quilt of levee and water and sewer boards - were responsible only for their own piece of levee. It seems no federal, state, or local entity watched over the integrity of the whole system, which might have mitigated to some degree the effects of the hurricane. When Hurricane Katrina came, some of the levees breached-as many had predicted they would - and most of New Orleans flooded to create untold misery. (A Failure of Initiative, Final Report of the Select Bipartisan Committee, U.S. House of Representatives, 109th Congress [2006])

Risk communication is more than just talking with people; it is an intentional process to gather information to further explain the nature and extent of hazards and disasters as well as to provide input into the decision-making process. Risk communication can thus be viewed as a tool in the hazards risk management process. This tool is helpful to organizations and communities to deal with risks and reduce vulnerability to hazards and disasters.

Risk Communication Process

McGuire (1969) provides us with a lasting approach to understand the communication processes that are persuasive in nature and built on who says what, by a medium, to whom, and with what desired intensions. Figure 7.1 provides

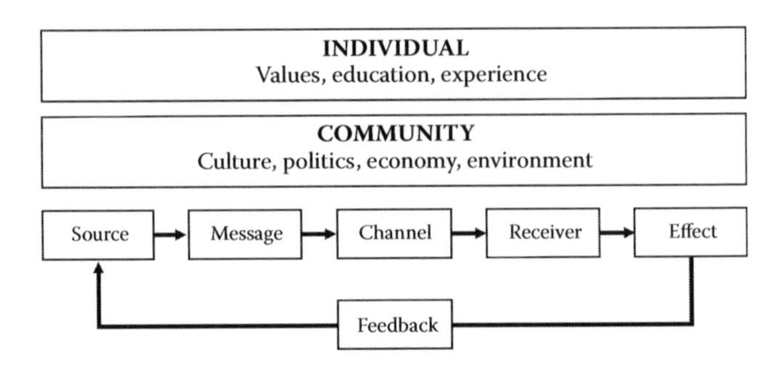

Figure 7.1 The classical persuasion model.

a diagram of this communication process. The key is to appreciate the source of the communication including its credibility, trust, and authority. The nature of the message itself involves both the hard and soft characteristics including the style, words, pace, complexity, as well as the scientific or technical nature of the content. The medium includes how we send the message, which might involve written or oral communication through the Internet, radio or television, video-conference, phone, or in person. For the person whom we are communicating with, the receiver may be old or young, educated, of a different culture or ethnic background, speak a different language, and may have an interest in the subject of our dialogue. Finally, our intent in the communication process may simply be to just inform, obtain compliance with some official order, reach agreement with some future action, raise a question for discussion and exploration, or simply just form the basis for an ongoing dialogue.

The context of our risk communication may involve diseases; natural hazards such as floods, earthquakes, fires, or drought; and target employees, citizens, legislators, or business representatives. Our communication message may involve short-term warnings of hazards or long-term awareness initiatives or efforts to raise support for changes in codes or hazard mitigation programs.

Critical Thinking: Risk communication is person centered in a social, cultural, and environmental context. We must acknowledge that when we are talking about hazards, disasters, and risk, we are dealing with complex issues that affect our way of life and our community. Risk communication may naturally involve conflicts between parties and we must acknowledge how we intend on dealing with differences in our communities and within our own organizations.

As one views risk communication within a risk management context, we could characterize this communication process as persuasive in nature since it is intentional to bring about some desired action. However, this view of risk communication places great emphasis on one-way communication that results in a planned outcome. Risk communication in a risk management context changes the desired outcomes from just one-way communication to an open exchange of information and mutual understanding of complex issues. The process, in this way, becomes more of a two-way exchange of information that can lead to further clarification of issues, identification of possible alternatives to reducing the impacts of a hazard or strategies that individuals, families, organizations, or communities could take to enhance resiliency.

Blaikie et al. (1994) note that disasters are more than just a natural event and are the product of social, political, and economic factors. Hurricanes that strike coastal areas cause extensive destruction because of development practices and the desire to build, live, and vacation in coastal areas. Recent hurricanes demonstrate that floods and wind damage are more than just a physical event; it was a political, economic, social, and environmental crisis that is human influenced. Therefore, risk is more than an objective phenomena but one that includes social and emotional reactions of people. Individual perception of risk is thus a critical part of the risk communication process; we should be sensitive to individual perceptions if we want to communicate with others about disasters and our social, economic, and environmental exposures. The human aspects of risk, and how people interpret information concerning hazards and disasters, must be recognized as a critical part of risk communication. We must examine the social aspects of risk and ensure that it is included in the risk communication process.

Barriers in Risk Communication

How one views a hazard is influenced by one's own values and dynamics of power, conflict, and trust in organizations. Risk is, thus, highly subjective and can be perceived very differently by citizens, public officials, businesses, and agency personnel. Risk is a concept that is impacted by how we understand and cope with the dangers and uncertainties of life. A scientist may view risk in light of model outputs, data limitations, and assumptions. Nonscientists have their own decision rules that may be highly intuitive and subjective (Kraus et al. 1992; Morgan et al. 2002).

Our perception of risk is influenced by what we believe are the immediate direct impacts and their longer-term indirect impacts. One could face small-scale immediate damage or longer-term financial repercussions including lawsuits or unexpected recovery costs. The ripple effect of a disaster event could thus be long term and far-reaching. Our perspective and how we perceive these possible effects will impact how we regard the adverse impacts of a disaster. The concept of social amplification of risk is demonstrated in limited direct impacts of events that trigger major indirect impacts (Slovic and Weber 2002). The concept of social amplification suggests that the communication process may result in amplifying or attenuating information associated with disaster risk. It can also generate ripple effects from the social or economic impacts of disasters to include ongoing mental stress, impacts on business sales or property values, and conflict in the community. Button (2013) provides an example of our selective perception by showing that those who are concerned about climate change are much more likely to pay attention to messages that are related to their preferences.

Terpstra (2011) found that emotions, trust in public agencies, and perceived risk influence flood preparedness behavior. The results showed that individual emotions, trust, and perceived risk affect behavioral intentions. Specifically, negative emotions associated with a past disaster event play a role in behavioral intentions. Near misses and milder threats also contribute to one's affective responses to risk. Further, if people were aware of environmental issues and feel threatened, they were more likely to pay attention to climate change issues. Property owners who consider themselves at risk are more willing to adapt but may not know how to adapt (Button 2013).

Lindell and Hwang (2008) also saw the contribution of personal experience on perceived risk and response actions. Their study confirmed the importance of hazard experience, gender, and income in affecting perceived personal risk. They stressed that hazard managers need to carefully identify their target audiences so the right messages can be communicated through the right channels so as to increase household adoption of prevention actions.

Grothmann and Reusswig (2006) suggest that risk communication concerning exposures to a potential hazard can result in either an individual precautionary action or adaptive behaviors after a disaster event. How one perceives information associated with a specific risk may result in adaptive behavior that significantly reduces vulnerability. People who live in high-hazard zones often fail to act to reduce their vulnerability (Peek and Mileti 2002), but some households do take action to avoid damage and reduce their vulnerability (Rogers and Prentice-Dun 1997). The difference may be that a minimum level of threat is perceived before a preventive action is taken. The key is that communities have an opportunity through risk communication initiatives to influence individual and household behavior and the value of adaptive behavior in light of local risks.

Critical Thinking: So, what do we believe and whom do we trust? How do we decide whom to believe? Slovic and Weber (2002) notes that people respond to hazards that they perceive are real. If these perceptions are faulty, then their actions could be ineffective or misdirected. If we use complex scientific concepts or statistics to make our case for protective action, and people do not understand their meaning, then the likely outcome will be distrust, conflict, and ineffective actions.

We have learned from past disasters that we should define risk within an intergovernmental framework with a focus on protecting citizens. Further, we should engage citizens in this communication process in an examination risk. "Authorities for catastrophic risk management should ensure that those vulnerable have sufficient and timely information regarding their condition and a reciprocal ability to respond to requests for their informed consent especially regarding tradeoffs of safety for cost. The public needs to be encouraged to actively and intelligently interact with its development of local plans" (Moteff 2005).

Kirkwood (1994) contends that there is often a difference between the object evaluation of risk and the public perceptions of the risk. This gap is explained by "experts" who suggest that the public just cannot understand complex scientific knowledge. Unfortunately, scientists may believe that their examination of risk is rational, objective, and nonjudgmental and that risk must be explained based on technical grounds. To do otherwise would lead to gross oversimplification of risk. Kirkwood notes that unfortunately this view of expert opinion does not fit with the reality that two different experts who examine the same problem may conclude differing estimations of risk. In addition, he clarifies that the expert and the public look at risk very differently; the expert examines risk based on a rational-documented process and the public by looking at potential injury, death, or loss. The public is making decision based on their rules of thumb and subjective judgments for avoiding danger rather than a complex examination of data.

Fischhoff et al. (1982) observe that many experts believe that people are so poorly informed that they require institutions to defend them. Furthermore, these uninformed citizens might be better off surrendering some political rights to technical experts. He further explains that some experts justify their unwillingness to discuss complex risks with the layman because they believe that information would make people anxious and that they could not use the information wisely if provided. Fischhoff attempts to explain this difference by noting that people are very different and that we should avoid generalizations. Some are risk takers and others avoiders. Some are cautious where others rash; it is just part of an individual's personality. He stresses that people's perceptions about risk may sometimes be erroneous but they are seldom stupid or irrational; an individual citizen may have a different way of processing risk and the possibility of harm.

Critical Thinking: Tierney et al. (2006) suggest that the mass media plays a significant role in promulgating erroneous beliefs about disaster behavior such as looting and lawlessness as part of a disaster. They note that the media made unconfirmed reports of widespread civil unrest and "urban warfare" following the flooding of New Orleans from Hurricane Katrina. What do you see are the adverse consequences of publishing unconfirmed reports of lawlessness in a disaster on actions that individual citizens might take or that public officials might take?

The expert and the layperson are different from one another in education level and the level of knowledge at their disposal but not in the way they think. One should ensure that the communication between the expert and the public be respectful and balanced. Effective hazard risk communication and management requires cooperation of many laypeople and results in an informed citizenry. We benefit from carefully examined judgments including quantitative and qualitative data. We must recognize our own cognitive limitations and temper our assessments of risk with a respectful eye to the public and openness to other views.

The best way of getting a good assessment of risk is from diverse and independent views. When decisions are made from limited perspectives, the results often reveal many unexpected outcomes that were not considered. We need to be prepared for a wider discussion and address other points of view so as to a common mistrust of public institutions.

Cook et al. (2004) examined the discourse between experts and the public and came to the conclusion that scientists fall into three groups including knowledgeable experts (scientist themselves), the public, and opponents (including the press). The public under this framework is categorized as uninformed (ignorant) and emotional with no understanding of risk. This view of the public allows the scientist to be free of having to engage with the public in dialogue that would be pointless since the uninformed have nothing to contribute to a decision-making process. Opponents as a group have something at stake, are unconcerned with truth, and have nothing to gain by a dialogue.

Vasterman (2005) suggests that the media can have an influence on how the public perceives risk associated with hazards. He refers to the concept of "media-hype" as a form of self-inflating media coverage in which various forms of public media use exaggeration or distortion during news coverage. He sees that this form of coverage of disasters and the threats posed by hazards provides a process for framing issues and amplifies elements of the coverage provided. He examines this form of self-inflating coverage from an international perspective and the role that media plays in society.

In contrast to the analysis of the media provided by Vasterman (2005), Nelson et al. (2009) examined the role of the media in the aftermath of the Minneapolis bridge collapse. They note that mediated learning is important during a crisis in reducing the negative emotional consequences of crisis events as well as to prompt the learning of information that can be importing in future emergencies. The study documents the Interstate 35 collapse of a bridge over the Mississippi River and the capacity of individuals to learn from the media during a crisis. A crisis from their perspective provides opportunities for learning and the acquisition of knowledge that is helpful in the determination of appropriate future actions in a crisis. Nelson suggests that crisis events provide the opportunity to highlight critical lessons for society stressing the value of preparation, overreliance on technology, or just lack of foresight by individuals to a potential risk. Individuals may not have a chance to prepare for all life-threatening events, but these events may provide a unique learning opportunity. Nelson's study of how radio, broadcast television, and cable as well as Internet news services can contribute to learning from a crisis event and reducing the psychological consequences of mass media coverage.

Critical Thinking: How one frames a position is critical in any dialogue. Scientists may view a situation from empirical objectivity and consider this as the only legitimate perspective. On the other hand, many issues may be framed in other ways such as morally, economically, socially, politically, aesthetically, or even scientifically. Is it justifiable? What does it cost? Who benefits? Who controls it? Does it make things more pleasing? Is it safe? Many nonscientists see many of these perspectives as very legitimate ones.

Risk Communication Tools

Risk communication associated with natural- and human-caused hazards is a challenging process when one attempts to explain the complex scientific elements of hazards and disasters. Communication concerning climate change and its links to potential extreme events illustrates the difficulties in understanding possible adverse events as well as those that have a low probability. We often look for tools and aid to help explain complex phenomena such as a map or chart depicting risk.

Communicating Risks with Maps

Hazard maps are one of our best tools to help communicate the nature and extent of risks associated with natural- and human-caused disasters. The National Flood Insurance Program (NFIP) has used flood insurance rate maps as a means of communicating risks associated with flooding hazards. These maps provide a visual image of risks and identify areas that are vulnerable to 100-year or longer flooding events. The base construction elevations for new construction or changes in existing structures in floodplain areas may be noted on a map.

Hazard maps come in all shapes and sizes. We use these maps to help us describe the nature and characteristics of a specific hazard (wind speed, size of storm, intensity, and related hazards) by a specific location. We use maps to identify the location and vulnerability of the local population.

A map can be an excellent tool to support our communication of risks and should include a title, a mapped area, a legend, and any credits and should provide a perspective on direction, symbols, and a scale. The map title should be short and concise. It should precisely say what is displayed in the map. The map title is usually placed above the mapped area. It is better to use a main and a subtitle instead of one long main title. The map title should have the largest type size of any text on the map. It can be all in upper case or in upper and lower case letters.

The mapped area should show a graphic representation of the cultural and physical environment and contain graphic information about a hazard or our vulnerabilities. What we represent in the map provides the content for communicating what we want someone to understand.

The map should include sources for information, the map producer, publishing date, data collection methods, information about the map projection, and other explanatory notes. This information is also referred to as metadata. The legend explains all graphic representations from the mapped area. Symbols in the legend should look exactly as they appear in the mapped area (same size, color, etc.). We also include a symbol that provides direction; maps are usually oriented with north being up. As part of the content of the map, we also use symbols to represent:

1. *Point features:* Vulnerable population sites, high-risk facilities, such as a nuclear power plant, damaged areas, and so on.
2. *Linear features:* Highways, bridges, canal, hurricane track, and so on.
3. *Areal features:* Wildfire-damaged neighborhoods, flooded areas, and landslides.

Finally, we include a distance scale on the map to show the relationship between distance on the map and the ground. This relationship is usually expressed in the form of a ratio relating one unit on the map (numerator) to many units on the ground (denominator). The smaller the denominator, the larger the map scale. A larger scale map covers a smaller area, which is shown with more detail. In addition, a larger scale map shows features from the physical and cultural environment that are less generalized. Smaller map scales allow one to display a larger area on the map. For maps that are provided in larger scale, more features may be shown but the physical and cultural environment are generalized.

Figure 7.2 provides an example of a map that was prepared to communicate risk to local residents, business owners, and local officials in coastal Louisiana following Hurricane Rita in 2005. The map was displayed in a local library, and meetings were held with small groups so as to facilitate communication about risks in their community. Planning for the event and what would be displayed in the map was done with local emergency management officials and representatives of Louisiana State University hazards research laboratory. The map provided an exceptional visualization of the coastal environment including landmarks, land elevation, political boundaries, and risk zones. Viewers of the map could easily find property of interest, geographic features that might influence their level of risk, and hurricane surge zones.

It is also an illustration of a type of map generally referred to as a thematic map, which consists of a geographic base map and various thematic overlays. This type of map is ideal for communicating relationships between hazards and known features such as roads, public buildings or parks, or water features. The

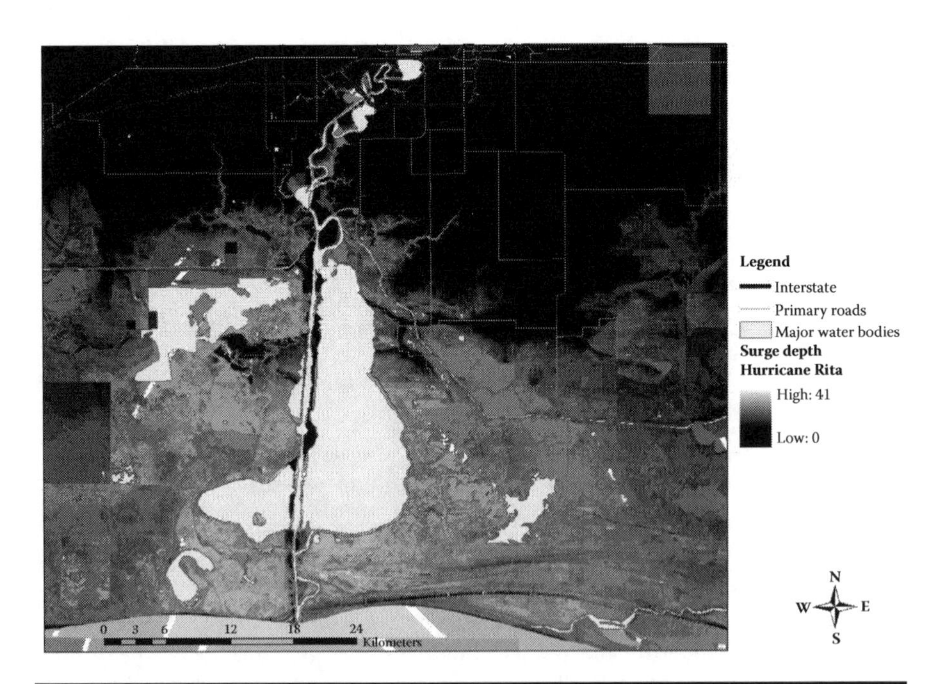

Figure 7.2 Base elevation map of St. Mary Parish, LA.

Map of St. Mary Parish is also a qualitative thematic map since it shows where something is located.

Reference maps as the digital flood map shown in Figure 7.3 customarily display both natural and man-made objects from the geographical environment. The emphasis is on location, and the purpose is to show a variety of features such as coastal elevations in this map (Robinson and Petchenik 1976).

The flood insurance rate map shown in Figure 7.3 is also a thematic map. Its purpose is to provide information on flood zones in an area and shows base

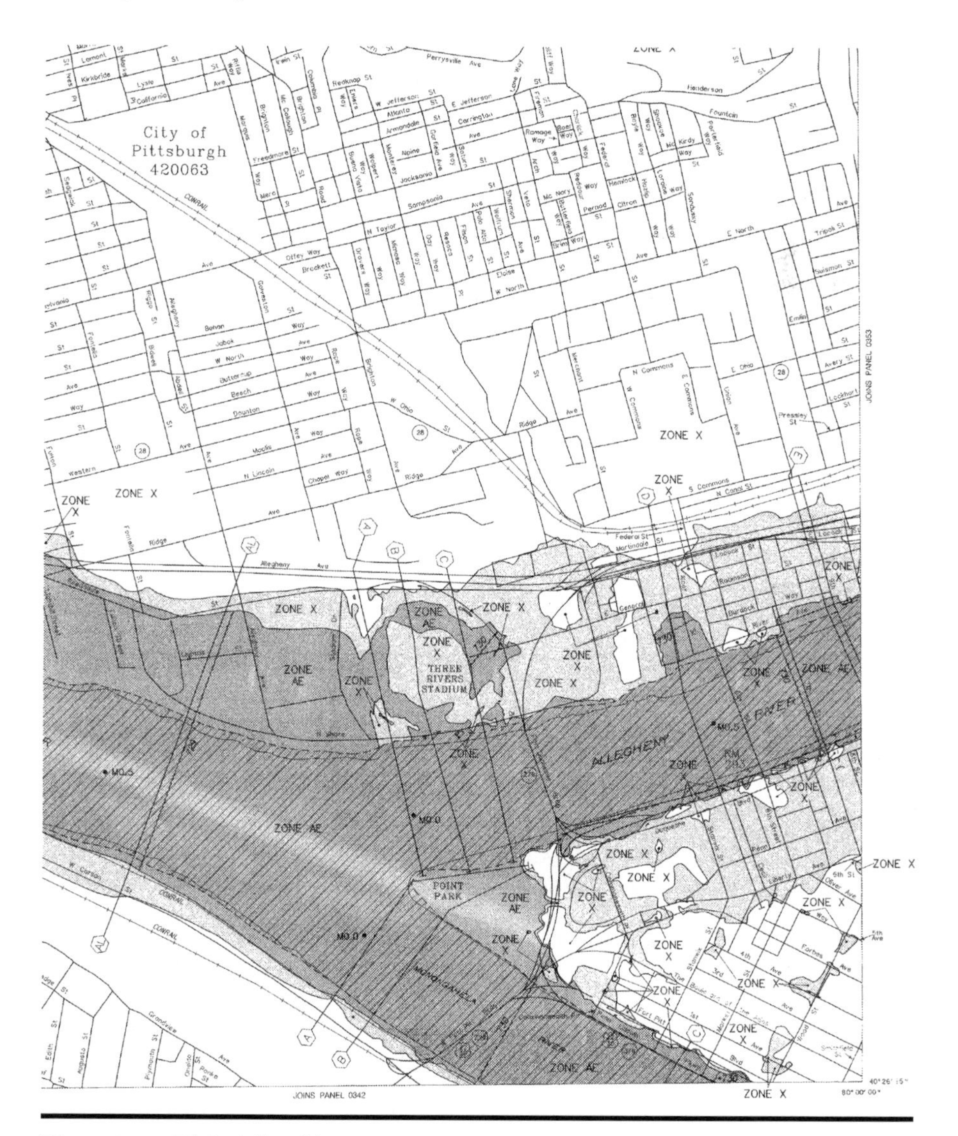

Figure 7.3 Digital flood insurance rate map.

elevations throughout his risk areas. Transportation features are shown as an aid in orienting one to the area covered by the map. The key to all thematic maps is that they illustrate "structural characteristics of some particular geographical distribution" (Robinson 1975).

We use quantitative thematic maps to show how much of something is found at a location. Census Bureau data are often shown in quantitative thematic maps for a community showing population for a Census Block, Block Group, Track or County boundary. A bivariable quantitative disaster map could display a dispersion plume model from a chemical incident and nighttime population density for the same area.

Maps are able to convey information concerning hazards that may not be communicated through words. Figure 7.4 shows the location of 100-year flood zones along with graphics that reflect nighttime population counts. This is a useful graphic that conveys potential population vulnerability. One is able to see community vulnerability from inland riverine flooding, hurricane storm surge, or heavy sustained winds. A map along with added discussion of the nature of potential natural hazards brings the nature and scope of a specific risk to the viewer.

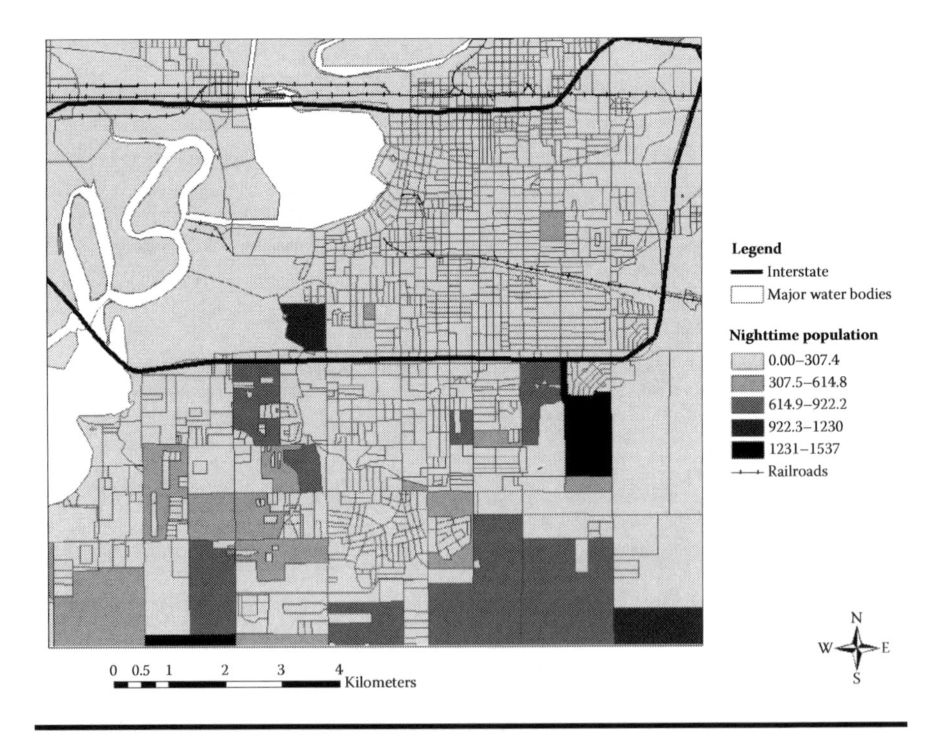

Figure 7.4 Nighttime population with flood zone.

Use of Figures

Figures may be used to convey information and provide a basis for explaining hazard exposures. Federal Emergency Management Agency (FEMA) helps citizens to appreciate risks by clarifying many different situations when examining the probability and consequences that hazards present. Figure 7.5 is used by FEMA to acknowledge the many situations that exist when we examine the advantages in the purchase of flood insurance under the NFIP (National Flood Insurance Program). If we owned property in a low-lying area near a water feature that floods often but we have no structures on this property, then we have high probability (10) but low consequence (1). There would be no need for insurance. However, if we have a structure on the property and as the consequence of flooding increases, then flood insurance would be beneficial.

The figure helps us to judge risk by examining when the risk value is 12 for someone with a higher chance of flooding (high probability but a low consequence), has the same risk as someone with lower probability but higher consequences. We can see that if we purchase flood insurance where the risk value is 40, then through consequence management we reduce the adverse impact from flooding. The final example from Figure 7.5 shows that when work on a levee makes it higher but

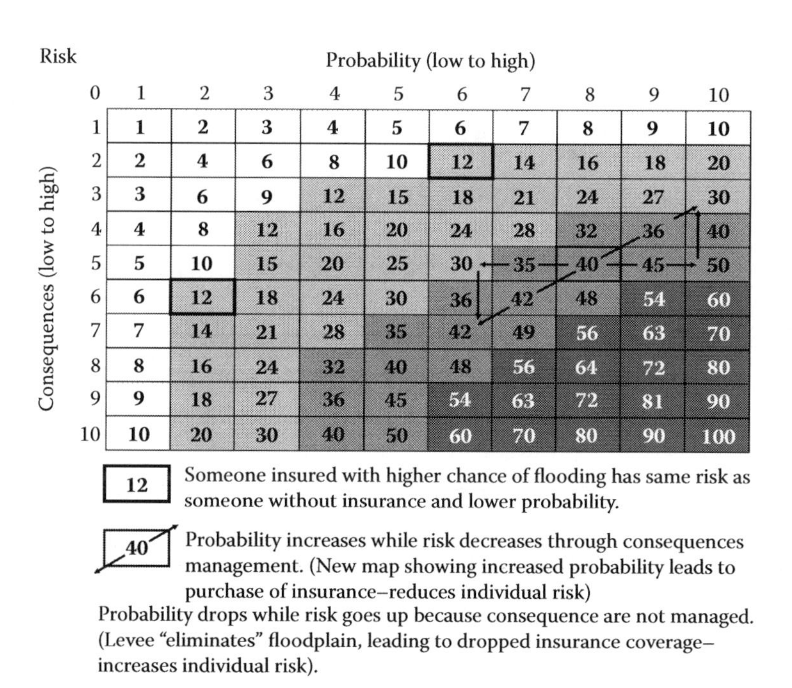

0	1	2	3	4	5	6	7	8	9	10
1	1	2	3	4	5	6	7	8	9	10
2	2	4	6	8	10	12	14	16	18	20
3	3	6	9	12	15	18	21	24	27	30
4	4	8	12	16	20	24	28	32	36	40
5	5	10	15	20	25	30	35	40	45	50
6	6	12	18	24	30	36	42	48	54	60
7	7	14	21	28	35	42	49	56	63	70
8	8	16	24	32	40	48	56	64	72	80
9	9	18	27	36	45	54	63	72	81	90
10	10	20	30	40	50	60	70	80	90	100

Figure 7.5 Risk probability consequence chart.

a homeowner drops their flood insurance, then probability is lowered but consequence is increased by dropping the flood insurance.

Figures may facilitate communication by providing a basis for exploring different situations that a business owner, homeowner, or a renter may have about their situation. The chart thus becomes a problem-solving tool that enhances understanding of complex information.

Social Media

The expansion of cell phones and microcomputers such as tablets have provided opportunities to communicate prior to and during disaster events. Liu et al. (2008) examined the role of online photo sharing in times of disaster. The expanded use of photo technology and file sharing has increased the use of eyewitness photography in disaster response and recovery efforts (Kaigo 2012). They note that photo-sharing websites such as Flickr provide a forum for disaster-related grassroots activity.

Several studies have examined the use of social media in disasters (Abbasi et al. 2012; Yates and Paquette 2011) and found that blogs, discussion forums, chat rooms, wikis, YouTube Channels, LinkedIn, Facebook, and Twitter all have been used in disasters. These tools have been used by individuals and communities to warn others of unsafe areas or situation and as a fundraising tool for disaster relief. Yates and Paquette (2011) determined that there were two categories of social media for emergencies and disasters. They first used a passive approach to disseminate information and get feedback through messages, wall posts, or polls. The second approach uses a systematic strategy to issue warnings, use social media to obtain requests for assistance, and use postings to increase situational awareness or obtain images of an area affected by an event. The study notes the adoption by the Federal Communications Commission and FEMA of the Personal Localized Alerting Network in 2011 of a mobile device application of the commercial Mobile Alert System. The study concludes that social media has great potential to enhance communications between citizens, first responders, volunteers, the private sector, and government agencies at all levels.

Merchant et al. (2011) note that social media is changing the way people communicate in their day-to-day lives and how they communicate during and after a disaster. Specifically, the use of social media enhances a person's situational awareness. Individuals can use global positioning system capabilities of mobile phones to find relief sites or the location of family members. Cottle (2011) noted that social media was used as a communication tool in the Arab uprisings of 2011 to suggest that media systems and communication networks facilitated events and communicated them around the world.

Targeting Specific Audiences

There are many groups who may be engaged in the risk communication process prior to a disaster as well as victims following a disaster. Other groups could include different public agencies, nonprofit groups, business, first responders, or volunteers.

A strategy in communicating with different groups may change given that our goals might shift depending on who is targeted and if the communication is prior to or following a disaster. Clarifying the audience who we will engage is critical in determining the content and the process of communication. Some to whom we communicate may be involved in ongoing emergency preparedness, while others in emergency response. The role of the audience in the emergency management process will impact our strategy in engaging them in an ongoing process to understand and deal with risks.

General public: The largest audience of which there are many subgroups such as the elderly, the disabled, minority, low income, youth, or members of their families.
Disaster victims: Those individuals impacted by a specific disaster event.
Business community: A key ally in disaster recovery, preparedness, and mitigation activities.
Media: An audience but also a critical partner effectively communicating with the public.
Elected officials: Governors, Mayors, County Executives, State Legislators, and Members of Congress.
Community officials: City/County Managers, public works, or department heads.
First responders: Police, fire, and emergency medical personnel.
Volunteer groups: American Red Cross, Salvation Army, the religious organizations who are critical to first response to an event.

Critical Thinking: Identify groups in your community who might want to discuss hazards and risks. Determine how they might be engaged in the hazard risk communication process. How do their roles differ? How might we engage and communicate with community groups and help them to understand risks and the advantages in adopting appropriate action strategies to minimize or avoid the adverse impacts of a disaster?

Risk Communication Myths

A myth is something that is widely believed but in truth is not supported by facts. Unfortunately, widely believed risk communication myths interfere with the way we deal with others (Covello 2002) and our attempts to help communicate information about hazards and risks.

1. Communicating openly and directly with people is more likely to alarm than calm people.

   Truth: The fact is that this is not the case if information about hazards and risks associated with disasters is done properly. We need to educate and inform people and not simply alert them. Do not just show an image of a flood zone, but give people the chance to express their concerns, ask questions, and receive accurate answers.

2. Many issues associated with hazards and risks are too technical and too difficult for the public to understand.

   Truth: The fact is that a technical explanation of a hazard can confuse people. Our job is to communicate the issues no matter how complex they may be but be conscious of how we convey information and encourage feedback and welcome questions. The public may not make technical decisions, but their opinions deserve consideration by those who are making those decisions.

3. Risk communication is not my job.

   Truth: We can find many opportunities to communicate information about risks to others, and it is our public duty to look for ways to help others understand risk.

4. If we listen to the public, we may divert limited resources to concerns that are not important.

   Truth: Listening to and communicating with the public does allow for opportunities for people to have input into organizational policy decisions. But we do not set organizational agendas and priorities based solely on prevailing public concerns. Our job is to help organizations manage issues and expectations. The public's concerns cannot be ignored, but neither can they necessarily dictate policy. The better-informed people are, the more likely it will be that the public's and your opinions on priorities are aligned. Providing for opportunities for public input into policies may be a positive factor in decision making rather than a waste of time as is suggested by this myth.

Slovic and Weber (2002) believe that most individuals overestimate their capacity to deal with disaster and or any emergency. This is illustrated by the widely held belief by many that, "they are better than average drivers, more likely than average to live past 80, less likely than average to be injured by tools they operate and so on (Svenson 1979)." Although such beliefs are obviously unrealistic, people still base their perceptions of risk on their own experience and that risks look very small. People still speed, run red lights, or tailgate without mishaps. Our personal experience teaches us that we are safe despite our unsafe driving practices. Our experience is shaped by daily news shows leading us to believe that when accidents happen, they happen to others. Unfortunately, we see that people do not see beyond their own perceptions, information, and experience.

Another common myth held by some scientists is that the public is ignorant of science and that they have no understanding of risk. It is perceived by many scientists that the public demands that they be exposed to "no risk." This is an assertion that is not supported by the evidence. Studies show that most people understand very well that nothing is risk free and are able to "live with" uncertainty and the lack of control that it entails (Wynne 2002).

Cook et al. (2004) suggests that despite these perceptions by many scientists, the public understands that life is not risk free and that risk is present in our daily lives. As individuals representing agencies, we use risk communication to help the public to see that something is not risk free or safe but that we are reasonably safe. A few mistakes occur and accidents do happen despite our best efforts to prevent them.

We can state that our hazard models show that there is a very low possibility of a disaster but cannot say that we are without risk. We can only express that our models have not shown anything that we need to be worried about. Cook et al. (2004) see that the public are concerned about these kinds of statements. Unfortunately, many scientists believe that the public cannot understand risk and that if a disaster is to happen, it will not impact them.

Covello (2002) stresses that we must earn trust and build credibility with the public. We need to establish constructive communication that is based on our audiences perceiving us as trustworthy and believable. This is built on how they perceive us including caring, antagonistic, or competent. If we are seen as honest and open, and our perceived commitment to common goals, Covello suggests that we:

- Look at others as partners and work with them to inform, dispel misinformation, and, to every degree possible, allay fears and concerns.
- Examine their concerns. Statistics and probabilities do not necessarily answer all questions. Be sensitive to their fears and worries on a human level.
- Be honest and open. Once lost, trust and credibility are very difficult to regain.
- Work with other credible sources, build alliances, and establish mutual interests.
- Conflicts and disagreements among organizations create confusion and breed distrust by the public.
- Address the needs of the media. The media's role is to inform the public, which will be done with or without your assistance.

Critical Thinking: Conflict is a normal part of dealing with hazards and risks. We should expect it and understand how to cope with it in a positive manner. Conflict should not be viewed in a negative manner but an opportunity to reveal the complex nature of the issues associated with disasters and our communities. Why is conflict such an ongoing part of our lives?

Covello (2002) provides some insights into overcoming conflict and how to deal with potential distractions using positive approaches to coping with hazards. He identifies common problems and then provides us with a suggestion for dealing with conflicts (Figure 7.6).

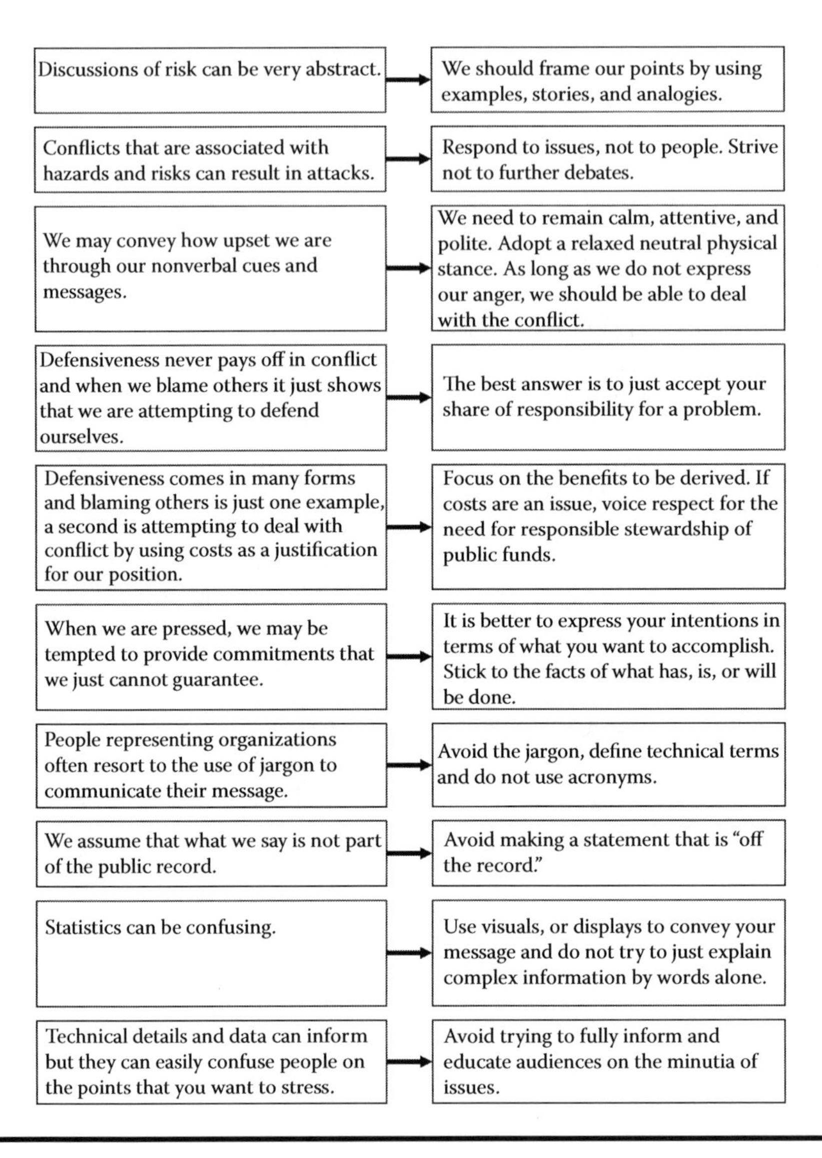

Figure 7.6 Conflict situations and coping strategies.

Managing Risks

We have stressed that risks may be viewed from two perspectives including the likelihood of something happening and the consequences if it did happen (Beer and Ziolkowski 1995). We have explained that the concept of hazards evolved from the notion of games of chance. This view stresses that risks are a phenomena outside individual control and are associated with involuntariness and unanticipated consequences. He goes further to stress that risk is also associated with voluntariness through individual adventurousness or individual decisions to take a chance. Risk may be managed only if we appreciate the nature of hazards and how we might be impacted economically, politically, socially, and environmentally (Young 1998). A key element of anyone to cope with a disaster is the information that is provided by public, private, and nonprofit organizations through risk communication and education. What we do with our understanding of hazards and risks impacts our capacity to deal with disasters on an individual, organizational, or community level. We, therefore, cannot manage risks that we do not understand.

An appreciation of our social, economic, and environmental vulnerability is critical in selecting and implementing adaptive strategies to reduce the adverse impacts of disasters. Vulnerability involves the capacity to withstand a hazard and for individuals, families, and communities the capacity to cope with hazards. Hazards risk management involves strategies to strengthen community building codes and their enforcement, adopt planning guidelines, purchase a flood insurance policy, invest in elevating an existing or new structure above anticipated flood levels, relocating critical storage areas, or developing operational plans if transportation routes are closed. Through risk management, we see that we can adopt and implement specific actions that can reduce vulnerability and enhance our resilience to disasters. When disaster strikes, we are caught up in reacting to the damage rather than looking for ways to mitigate or prevent the damage from occurring.

Decision Making

We must acknowledge that there are no guarantees that our decisions and actions will reduce all damages. Unfortunately, there is always a possibility that failure can occur and that risk cannot be completely eliminated (Barnes 2002). The key is that we must build a process for enhancing our understanding of risk and that information about hazards and their impacts can support short-term and long-term decision making that acknowledges our vulnerabilities and risks to hazards.

Organizations must adopt a position of risk management as a strategy to understand the nature of hazards. Risk communication must be part of the hazards identification process where we share our community assets, resources, and the nature of hazards that could impact us. It is also part of risk analysis where we explore our vulnerabilities through spatial analysis tools and techniques. Risk analysis thus becomes the basis for us to adopt sound decisions and systems.

Risk communication within our organizations and with the public is a critical part of our attempts to manage hazards and to deal with risk. The intent is ensure that everyone understands hazards and what can be done to reduce our vulnerability on an individual, community, region, or national level from warning systems to forecasting or crisis decision-making processes (Plate 2002). We thus need to combine the best science and decision support systems with an ongoing dialogue with our community.

Community Engagement

We have stressed that risk is associated with our actions or inactions, from natural events, and from our use of technology. Although each disaster event is unique, a disaster involves human decisions that may increase or decrease our vulnerability. Risk communication throughout the hazards analysis process facilitates our capacity to identify and implement sound strategies to manage risks.

One of the key questions is who should be involved in the hazards analysis process and in making decisions that result from this process. This question is best resolved by examining the goals of the hazards analysis and who is needed to provide you with a quality effort. To attain the goals, you may need to involve a variety of people from throughout the organization and even outside.

Another look at the hazards analysis process in Figure 7.7 shows that it involves risk communication and community engagement throughout the process from hazard identification, risk analysis, and the identification and adoption of coping mechanisms. If stakeholders are to be involved, it needs to begin from the start and have them help explain the unique human and cultural capital of the community, economic and critical infrastructure, and especially what elements of the ecosystem capital make the community unique.

A key element of public risk communication is community engagement. Too often we see public initiatives that attempt to "sell" or "persuade" the public on a policy or action such as the purchase of flood insurance, adopting a family evacuation plan, securing home water heaters from earthquake hazards, or just stocking up on essentials such as water and nonperishable foods. The efforts might include the production of informative brochures, coloring books for children, incentives from utility companies to secure home gas water heaters, or newspaper flyers advocating a specific emergency preparedness suggestion or hazard mitigation initiative. In most cases, these programs to sell the public have marginal effects and we wonder why we cannot get people to do what is smart. An alternative approach is to engage the community at various levels in a dialogue to help inform, sell, educate, or persuade families to adopt hazard mitigation efforts. We have learned in many community initiatives that involvement and engagement is the key and that including the public in the process can not only inform citizens but may also shape the form of emergency preparedness, hazard mitigation, and emergency response activities.

In a study of the European Union's Flood Risk Management Directive (2007), Fleischhauer et al. (2012) examined two questions, including: (1) How can stakeholder and public involvement be improved in risk communication? and

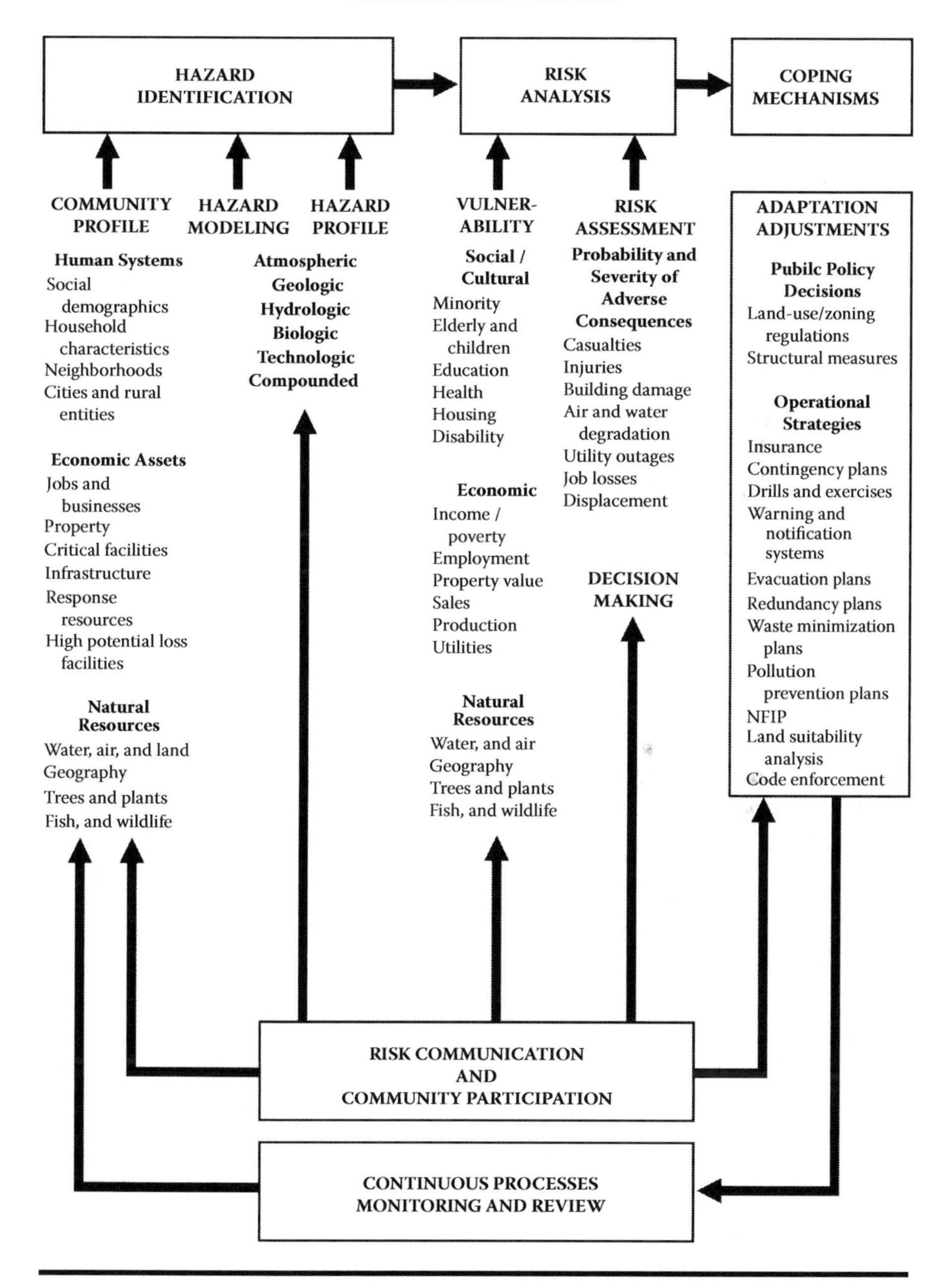

Figure 7.7 Community involvement in the hazards analysis process.

(2) How can the quality and fairness of the flood risk management processes be guaranteed? The study stressed that not only scientific knowledge but also local knowledge as well as stakeholder and public risk perceptions could be used to form the decisions. It also provided new tools for communication, participation, and optimization of a flood risk governance process at the local level.

Recent disasters have provided examples of how social media can also be used by agencies attempting to send messages to the public. The approach to communication is a powerful tool for engaging a distributed self-organizing population and can be used in disaster response and recovery (Palen et al. 2010).

Community planners have known that citizens who have a stake in the neighborhood will work side by side with public employees and officials to solve problems and help identify solutions to complex problems in the given opportunity. This engagement, however, involves a two-way dialogue, where issues are identified and potential solutions are examined. The result is often surprising for a wide group of community stakeholders surface that may not have been included in the past and may be essential to implementation of a community-wide plan.

Critical Thinking: Public agencies appreciate that citizen support is critical in any implementation of a public policy. Why do we spend so much time and effort on the development of the policy and program but little time in engaging the community in a dialogue on the policy or program? Give an example of a community that has successfully engaged stakeholders in a dialogue and explain why the effort was so successful. How could a public agency use the Internet and social media to solicit input into community planning and recovery initiatives?

Community stakeholders may include small business owners, financial institutions, the school district, healthcare providers, local community agencies and religious organizations, manufacturing, homeowners and renters alike, and a broad range of service industry representatives. Each stakeholder has an interest and a unique perspective to view risks. Identifying these stakeholders and encouraging their participation can be the key in obtaining broad-based community engagement. Then the role of the public sector is to facilitate and enable the stakeholders to become and sustain their involvement.

Critical Thinking: As hazard profiles are prepared for a community, would engagement strategies be the same as with the identification of community assets? Chances are that the identification and description of community hazards needs to involve skilled community planners, geographic information systems specialists, modelers from the local university, or extension outreach specialists as well as emergency management staff at the local or state level. Are there other stakeholders who have specialized knowledge of either geologic, hydrologic, climate, or hazardous materials hazards and can be called in to help. Even the local school system might be engaged to help explain the nature and extent of a local hazard as part of a science project. This engagement can enlighten the school district on hazards that had not been fully appreciated and showcased in the local media.

One of the key elements of decision making is determining whom if anyone is involved in the process. Given that we live in an open society and that there are many interested parties in the results of a hazards analysis, involving stakeholders in the process is not uncommon. The question then centers on who is involved and the process that best fits with their role in dealing with organizational or community risk from hazards. To what extent and how should residents, representatives from business and nonprofit organizations be involved in analyzing risks, defining the problem, and decision making?

There are many special interests associated with any examination of risks and hazards, and how should these be considered? To what extent should they be discussed in an open arena? How to include stakeholders in the hazards analysis process in the problem-solving process is essential if acceptance and understanding of the output of the analysis is a priority. Risks associated with hazards are multidimensional and stakeholders from numerous public, private, and nonprofit agencies. The process of decision making is thus as important as the final results of the analysis.

Unfortunately, we do not anticipate the degree of conflict that may arise in attempting to find ways of reducing the adverse impacts of disasters. It is, therefore, crucial that we manage the conflicting forces that are stirred up by the alternatives that we are considering. There are many reasons that we encounter conflict in decision making; one involves failing to reach agreement on our goals, priorities, and essential elements of a solution. It is crucial that we take our time in examining alternatives and adequately account for the unknown nature of many disaster impacts. Many outcomes are possible.

Petak (1985) notes that conflicts in determining acceptable mitigation or risk management solutions center on the fact that hazards always involve risks and to resolve disputes, one must develop and integrate as much information as possible about existing conditions and potential future outcomes. He observes that current approaches to problem solving place a great deal of power in the hands of the technical expert and professional administrator. He suggests that elected officials must be included and assume their responsibility for protecting public interests. He stresses that it is not possible to be risk free for there are no zero risks to natural hazards. What makes this process difficult is that there are two sides of dealing with risks, one that is empirically verifiable and associated with facts and the other is based on individual judgment. Acknowledging these two perspectives contributes to a more balanced discussion of risks and reaching acceptable solutions to problems presented by disasters.

Stakeholders Involvement

One of the key issues in the hazard analysis process is who is involved in the problem-solving process. Technical versus democratic approaches to analysis provide two very different ways of determining the level of risk that an organization or community face. One is based on experts and scientific fact where the other includes a more democratic and participatory model involving the public along with the experts. If our goal is to have analysis that is acted on by the public, then the latter approach is one that should be considered. Including stakeholders from the organization or the community provides benefits associated with different perspectives examining complex problem. An alternative may be an adversarial model that provides opportunities for the public to react, respond, or give input into the discussion.

Because the process of determining risk acceptability (including mitigation spending and regulatory practices) is one that is influenced by many different interests, it is possible to include a broad range of groups to participate in defining what is acceptable risk for a local community or organization. Increased public participation in the process can broaden the discussion as to what constitutes acceptable risk. Unfortunately for organizations and communities, it is becoming more difficult today to deal with hard decisions relating to hazards and our vulnerability. We must ensure that our mitigation policies and priorities reflect a clear understanding of potential risk. Determining what is acceptable risk is difficult to resolve because each interest group in a community or organization has his or her own priorities, biases, social values, and resources. More importantly, a conflict over what is acceptable is often the result of relevant facts that allow us to view complex problems. The problem is further complicated because even experts often disagree on the degree of risk that we face from a hazard. Techniques to overcome conflicts associated with acceptable risk may result in common approaches to risk management. The key is to have a process in which to review the nature of risks associated with natural hazards.

Guillot (2013) observed in a study of emergency alert and warning systems that messages must be tailored to accommodate all facets of the diverse population within communities. He found that it was critical to construct and convey clear, concise messages that identified and encourages adequate appropriate protective actions to be taken. He stressed that effective risk communication could be accomplished through concerted efforts by communities to enhance preparedness through public outreach and education programs. Collaboration among the broadcast media, broadcast meteorologists, and emergency management was essential. His research confirmed the need for broad community outreach and education relating to severe weather.

Fischhoff (1984) suggests that we consider the following when attempting to deal with acceptable risk.

- Recognize the complexities of the nature of acceptable risk.
- Acknowledge many approaches and perspectives in dealing with risk.
- Use more than one approach in examining risks and what is acceptable.
- Determine a basis for quantifying risk.
- Develop a decision-making process that matches the expectations of those involved.
- Identify the role of government in risk.

One of the fundamental barriers in dealing with acceptable risk is that one or more of our alternatives to dealing with the problem may threaten the life, health, welfare, or property of others.

There is a real benefit in examining our thinking, for it drives us to look deeper at our options in addressing a problem. Developing a structure for comparing alternatives is seen in a cost–benefit analysis of our choices. This analysis may need to provide for future values and for estimating indirect costs of a solution. The process provides a constructive quantitative means in comparing one approach to another. But more will likely need to be considered such as what does our selected approach represent in terms of our values and priorities, and does this solution provide a precedent that is unwelcomed?

Engaging community stakeholders in the risk analysis process is also essential for they have their own views of the probability of a hazard event and the significance of the consequences of a disaster. Soliciting stakeholders in the risk analysis process not only improves the quality of the analysis but it also leads to a more thorough understanding of risk and potential strategies that might address them.

A key means of engaging stakeholders in the hazards analysis process is present in every community as a result of the passage of the Community Planning and Right-to-Know legislation in 1986. This statute evolved from national efforts to enhance planning for human-caused, technological hazards associated with the threat presented by hazardous chemicals. State emergency preparedness committees were created by this national legislation and local community planning committees (LEPCs) were formed. Membership on local LEPCs included representation of stakeholders from public, private, and nonprofit sectors. Each person on the LEPC has something to contribute and can be a link to the broader community. The membership of the LEPC usually includes emergency responders representing police, fire and emergency medical, health care, media, chemical processing businesses, and education. The LEPC provides a natural means of engaging stakeholders in the hazards analysis process for not only human-caused disasters but also for natural hazards.

Barnes states that many communities experience a common conflict over known or suspected links between chemical hazards and perceived health impacts. He sees that there is a public disbelief when public officials or business representatives state that there is minimal risk from hazardous substances.

Barnes notes that the public may have fears of large-scale industrial disasters or long-term chronic exposures to hazardous substances in the community. He suggests that many citizens perceive that there is inequity in exposure to harm. Trust and credibility are the core issues impacting the perception of risk in the community, and how strategies and policies to address risks from hazards should be constructed. He sees that there may be a technical rationale concerning a hazard along with a cultural perception of risk in the community and unless addressed could cause extensive conflict.

The potential for conflict is observed by Krimsky and Plough (1998) as a difference in technical and cultural views of risk. For example, a technical perspective views risk through scientific methods, evidence, and explanations, whereas a cultural rationale views risk from a political one. Technical perspectives use organizational authority and professional expertise, whereas cultural perspectives appeal to one's wisdom or cultural traditions. Technical understanding evolves from quantitative methods, whereas cultural ones evolve from personal interviews or experience. One view suggests that it is the role and responsibility of the expert and regulatory agencies for decisions about, and regulation of risks associated with health hazards. This is a view that the professionalized bureaucracy and the scientific community know best. This perspective may be in contrast with a citizen's view of risk and may be shaped by cultural influences, personal knowledge, or unarticulated views of hazards.

Unless these differences in how one perceives risk are surfaced and articulated in an open setting and one that promotes communication and trust, conflict or polarization can develop in the community. The issues presented by Krimsky and Plough concerning health risks from chemical hazards could easily evolve from risks from flood, wildfire, earthquake, or other technical complex hazards. Without broad-based community engagement in an open dialogue concerning risk, it is easy to see that conflict will evolve between the technical elite and the community.

Our attempts to manage risks may evolve into measures to contain it as well as to open communication about risks. We find that the past can illuminate failures, their causes, and their control as lessons for engaging new issues and threats. The future commands the exercise of foresight, an imaginative process involving possible scenarios stirred by such questions of what it, or would could happen if, or might this occur under these circumstances. These questions help us to see options and the possibilities and allow us to identify alternatives to dealing with risks. Risk communication thus facilitates opening our perspectives and horizons to how we might manage and deal with the future.

Ethics and Decision Making

Ethics raises issues concerning how we are to act and under what standards we should base our decisions. In making decisions, we reflect on what we believe is right or wrong and the standards that these judgments are based. Our decisions and the manner in which they are made reflect what we believe is important and what groups, issues, or facts that we ignore. Ethics is then the study of our standards that we adopt and serve as a basis for what we consider right, suitable, and appropriate (Velasquez et al. 1987).

When we are considering ethics in making decisions, we take a look at how others may be impacted or how the environment (including property, animals, the air, or water) might be impacted. These concerns may be viewed at a personal, institutional, or societal level.

Ethical considerations are not limited to legal concerns of what is right or wrong for we look at the rights of citizens, employees, investors, and our institutions. The key question in decision making and ethics is what standards of right and wrong are we including in making judgments. These standards may relate to obligations and benefits that result or the degree of fairness to parties associated with the decision.

We make ethical judgments throughout the hazards analysis process from determining what we consider when we characterize the local community and what is at stake when a hazard could be present. We judge what is suitable and appropriate when we select one methodology over another when our approach could result in some form of harm. Our standards of right and wrong may also influence the degree that we consider potential hazard impacts on various environments, social groups, or who wins or benefits from our analysis in the commercial sector. Central to ethical decision making is to determine when our standards of right or wrong influence our judgments and have impacts on people, enterprises, society, and the environment.

Critical Thinking: Take a look at the potential impacts of our decisions in the hazards analysis process and determine who benefits or who might suffer some loss. Are all stakeholders treated equally or are some benefiting from your assessment of hazards and their associated risks? Reflect on the outcomes of your hazards analysis and determine if all interests have had the opportunity for input equally. Do your recommendations and decisions reflect a common good? Given that many stakeholders in the hazards analysis process have very different positions and interests, do your recommended risk management strategies or hazard mitigation initiatives suggest that some parties will benefit at the price of others? How could your conclusions, recommendations, and decisions support the common good and be seen as impartial?

Legal Issues in Decision Making

Emergency management does not exist in a vacuum but part of a broader context of federal and state statutes, court decisions, administrative rules, and constitutional provisions. In addition, local jurisdictions adopt ordinances and administrative procedures. Agency statutes or regulations may require that a comprehensive hazards analysis be conducted as a part of emergency planning or hazard mitigation efforts. A comprehensive hazards analysis for a community or organization may support zoning revisions, changes in building codes, or other public policies.

Depending on state law, public entities may have enjoyed immunity from civil claims for financial compensation because of the negligence of governmental agencies, officials, employees, or volunteers. This protection known as sovereign immunity was based on a belief that neither the King nor his representatives could do any wrong. Under this doctrine, individuals or businesses that were harmed by the actions of public entities are prohibited from recovering damages from the government and were thus immune from suit (Reynolds 1982). Public agencies covered by sovereign immunity may pay claims, but they are not required to do so.

State laws have changed over the past 40 years making public entities at the state and local governments more accountable for their actions (Pine 2005). Most states adopted tort claims acts that provide some immunity for agencies and employees but no longer provide total protection for state and local public officials, employees, or volunteers. In the case of state or local officials participating in a hazards analysis, state law governs their potential liability and almost all states provide statutory immunity for being involved in emergency management activities. It should be noted that this statutory immunity also covers volunteers who assist in emergency management activities such as a hazards analysis.

A citizen who believes that they have been harmed, may file a claim against a public agency or employee (including volunteers), files a tort claim against the jurisdiction sponsoring the hazards analysis. The claim would state that they suffered some harm because of the negligence of public officials or employees who participated in the hazards analysis or who made decisions based on the analysis. The claim of negligence would state that the public official or employees did not exercise an appropriate degree of care, skill, and diligence, which the reasonable or prudent person would exercise under similar circumstances. The injured party would state that they suffered property damage or personal injury as a result of decisions that were based on the hazards analysis and that those doing the hazards analysis were negligent.

This rule, as applied to governmental entities and those who act on their behalf (volunteers), must be understood in terms of the essential elements of negligence. All claims of negligence include the following:

1. *Duty:* Failing to conform to a defined standard of care owed to others and established by statute, common law (based on judicial decisions), or a policy by the governmental entity.
2. *Breach:* Failing to conform to that standard of care or failure to carry out the required duty.
3. *Damage:* Actual loss or damage to the injured party.
4. *Causation:* There must be a connection between the act of the governmental employee, official, or agency body and injury to a third party.

All negligence cases have these elements in common and absence of proof of any one element will defeat a finding of liability (American Law Institute 1965).

Many state emergency management acts require local jurisdictions to prepare a comprehensive emergency response plan. Given that state law imposes a duty to plan on local entities, failing to prepare a plan could leave the community open to a claim of negligence. Emergency plans that are out of date have not been based on realistic assumptions, inflate capabilities, do not involve all operations areas and functions of governmental entities, or are not fully developed could be the basis for a claim. The court would have to determine if the local jurisdiction met their statutory duty to plan.

Critical Thinking: A local public entity that does hazards analysis that forms the basis of the local emergency plan is taking a positive step in satisfying the duty to plan. The court would examine the work of the public entity and judge if their actions were reasonable and prudent under the circumstances. Doing a comprehensive and complete hazards analysis is a positive action that would demonstrate an intent to meet the duty imposed by the emergency management statute. Involving trained people in the hazards analysis process, using accepted hazard models and risk analysis techniques, having capable people review the results of the hazards analysis, and testing the assumptions of the analysis would all be symbols of responsible actions. Having a realistic, effective, and current hazards analysis and plan is one of the best ways for a community to reduce its liability exposure.

Note that the intent of the statutory immunity provision in state laws is intended to encourage local officials to do their jobs. Legislatures recognize that emergency management is complex and that it involves discretionary judgments. Immunity is intended to encourage local officials to do their best and not be distracted by the treat of lawsuits. Further, public officials are asked to use their judgment and execute their jobs in a professional manner.

State law provides extensive protection to public officials and employees; however, there are limitations to the protections. The first exception is for willful misconduct, which includes actions that are intentional or completely unreasonable under the circumstances or actions that area consider gross negligence. Gross negligence is not mere thoughtlessness, inattention, or a mere mistake resulting from inexperience or confusion but actions that are completely unreasonable under the circumstances.

The best approach in dealing with potential liability is to include trained staff in conducting the hazards analysis, get internal and external feedback on the results of the analysis, involve stakeholders, and share the results with the public. Document how you conduct your hazards analysis and note why you did and what you did throughout the process.

Indemnification

Public officials and representatives of governmental agencies involved in a hazards analysis are generally entitled to protection against personal financial loss or indemnification if they are subject to a claim of negligence while working on the community hazards analysis. Almost all states recognize that the governmental unit is responsible or liable for the negligent acts or omissions of its agents or employees who are acting within the scope of their duties as public employees. Employees in this context include not only paid staff but also volunteers who participate in the hazards analysis process. The public entity is thus vicariously liable for the actions of government officials and employees.

Critical Thinking: One final note on liability and claims of negligence should be examined. Local governments retain the services of legal counsel who has a clear understanding of the law of torts under state law. The General Counsel is a key person to know for they know state laws, which apply to local government and its employees. Public employees and volunteers involved in a community hazards analysis should communicate with legal counsel and explain what they do and how they operate. Legal counsel can clarify state law and if needed additional steps in the hazards analysis might be taken.

Acknowledging Risk as a Part of Risk Communication

Examining the past and benchmarking is one way to understand and evaluate our hazard policies and practices. Our past policies and practices have influenced how communities have been impacted by disasters. Our past beliefs, practices, policies, and how our organizational systems interfaced all form the basic elements of community functioning. The fact is that many of our losses from disasters are imbedded root causes that are part of our public policies, business practices, and social values in communities. Katrina revealed more fundamental problems in New Orleans that were part of the community educational, medical and mental health, emergency management, public safety, and public housing systems. Reversing root causes that are part of very complex systems is not a simple fix such as adapting a building, bridge, or water feature. Patterns of behavior, practice, and policy that help to create vulnerability are often pervasive throughout the society and invisible to our normal view of a community. The root causes of vulnerability can be reversed by recognizing problems in our systems and approaching them in an ongoing systematic way.

Our work needs to begin with a judicious and honest assessment of threats, followed by investments in prevention and mitigation and by construction of response systems that will be equal to a larger of class of disturbances than we have previously allowed ourselves to contemplate (Leonard and Howitt 2006).

We know that technological failures virtually always occur within the context of management failures, and there is a growing body of literature that describes management implementations designed to reduce large-scale failure (e.g., Dekker 2002; Roberts and Bea 2001; Weick and Sutcliffe 2001).

Learning as a Part of Risk Communication

Learning from documented failures is a powerful method for reducing risks of repeated losses. Another method is to learn from close shaves. Many dangerous events fortunately culminate in only an incident rather than an accident, but the repetition of similar incidents can serve as early warnings of danger. Indeed, the logging and analysis of such events on the nation's airways partially accounts for commercial aviation's impressive safety record. A system for reporting close encounters of aircraft was installed decades ago. Anticipating the possibility that perpetrators of high-risk events might be reluctant to blow the whistle on themselves, many years ago, the Federal Aviation Administration arranged for National Aeronautics and Space Administration (NASA) to collect incident data and to sanitize it to protect the privacy of the incident reporter. NASA also screened reports to identify patterns as early warning of dangerous conditions. Similar systems are in place for reporting nuclear power plant incidents.

With the growing recognition of human factors in accidents or in failures to limit damage, a class of situations entailing uncommonly high risks but conspicuously good safety records was examined. In the Navy, for example, high risks are a part of daily operations of submarines and aircraft carriers. Yet accident rates are paradoxically low. Careful analysis of these situations showed that certain qualities of leadership and organizational culture foster integrity, a sense of responsibility among all participants, a tolerance by authority figures for dissent, and consensus on common goals of safe performance. High safety performance is associated with an institutional culture that is bred from the top of the management pyramid. The most critical element of that culture is mutual trust among all parties (e.g., Roberts 1990).

Discussion Questions

What is the link between hazards analysis and risk communication?

Why do experts and the public have such a difficult time in communicating?

Why is conflict such an ongoing element of the risk communication process?

Risk may be highly subjective and perceived very differently by citizens, public officials and officers, the press, and others. How might each view risks associated with natural hazards?

Why is it that some people take precautionary actions when faced with hazards and others do nothing?

Why are hazard maps such a useful tool in helping to communicate the nature and extent of risks associated with natural- and human-caused disasters?

How do you get the public engaged in a discussion of risk? How should the community be involved in the risk communication process?

What barriers inhibit the communication of information about hazards?

Most states provide immunity to state and local emergency management agencies and officials for carrying out emergency management functions. Do you believe that this is appropriate? Why were laws established to provide immunity?

What might be an example of "willful misconduct" by a local official or employee involved in a hazards analysis?

Applications

Search for maps with different scales on the Internet and find examples of cultural and physical features that are represented with different categories of symbols (point-like, linear, or real) at different scales.

What are some examples of qualitative and quantitative thematic (disaster) maps. For example, a qualitative thematic (disaster) map could show the different types of disaster (flood, forest fire, hurricane, etc.) for an entire country or a continent. A quantitative thematic (disaster) map could visualize the amount of damage one or more disasters caused in each state of the United States.

There are numerous examples of the use of social media in disasters such as the coverage by Fresh Air host Terry Gross.

http://www.wnyc.org/articles/wnyc-news/2013/apr/23/blog-social-media-and-disasters/

or the Public Broadcasting Service program News Hour

http://www.pbs.org/newshour/bb/media/jan-june13/dd_04-16.html

Find your own example of an assessment of the use of social media in disasters.

Websites

Risk Communication: Public Health Communication during a pandemic flu. http://www.pandemicflu.gov/news/rcommunication.html

ATSDR—Health Risk Communication Primer: http://www.atsdr.cdc.gov/risk/riskprimer/index.html

Excellent Reference for Risk Communication websites (From Risk World): http://riskworld.com/websites/webfiles/ws5aa014.htm

CDC Emergency and Risk Communication: http://www.bt.cdc.gov/erc/

University of Washington, Institute for Risk Analysis and Risk Communication: http://depts.washington.edu/irarc/

University of Maryland, Center for Health and Risk Communication http://www.comm.riskcenter.umd.edu/

Michigan State University. Health & Risk Communication Center, College of Communication Arts and Sciences: http://hrcc.cas.msu.edu/

World Health Organization Risk Communication: http://www.who.int/foodsafety/micro/riskcommunication/en/

References

Abbasi, M. A., Kumar, S., Andrade Filho, J. A., and Liu, H. (2012). Lessons learned in using social media for disaster relief—ASU crisis response game. In S. J. Yang, A. M. Greenberg, and M. Endsley (eds.), *Social Computing, Behavioral-Cultural Modeling and Prediction*, pp. 282–289. Springer-Verlag Berlin Heidelberg.

American Law Institute (1965). *Restatement of the Law: Torts*, Second Edition. St. Paul, MN: American Law Institute.

Barnes, P. (2002). Approaches to community safety: risk perception and social meaning. *Australian Journal of Emergency Management*, *1*(1), 15–23.

Beer, T. and Ziolkowski, F. (1995). *Environmental Risk Assessment: An Australian Perspective.* Report to Environment Protection Agency, AGPS, Canberra.

Blaikie, P., Cannon, T., Davis, I., and Wisner, B. (1994). *At Risk: Natural Hazards, People's Vulnerability and Disasters.* London: Routledge.

Button, C. D. (2013). *Coastal vulnerability and climate change in Australia: Public risk perceptions and adaptation to climate change in non-metropolitan coastal communities.* Ph.D. Dissertation, The University of Adelaide, South Australia.

Cook, G., Pieri, E., and Robbins, P. T. (2004). 'The scientists think and the public feels:' expert perceptions of the discourse of GM food. *Discourse & Society*, *15*(4), 433–449.

Cottle, S. (2011). Media and the Arab uprisings of 2011: research notes. *Journalism*, *12*(5), 647–659.

Covello, V. (2002). Myths, principles, and pitfalls. In *Communicating in a Crisis: Risk Communication Guidelines for Public Officials.* Rockville, MD: U. S. Department of Health and Human Services Substance Abuse and Mental Health Services Administration.

Dekker, S. (2002). *The Field Guide to Human Error Investigations.* Aldershot, England: Ashgate.

Fischhoff, B. (1984). Setting standards: A systematic approach to managing public health and safety risks. *Management Science*, 30, 823–843.

Fischhoff, B., Slovic, P., and Lichtenstein, S. (1982). Lay foibles and expert fables in judgments about risk. *The American Statistician*, *36*(3), 240–255.

Fleischhauer, M., Greiving, S., Flex, F., Scheibel, M., Stickler, T., Sereinig, N., Koboltschnig, G., Malvati, P., Vitale, V., Grifoni, P., and Firus, K. (2012). Improving the active involvement of stakeholders and the public in flood risk management: tools of an involvement strategy and case study results from Australia Germany and Italy. *Natural Hazards and Earth System Sciences*, *12*, 2785–2798.

Grothmann, T. and Reusswig, F. (2006). People at risk of flooding: why some residents take precautionary action while others do not. *Natural Hazards*, *38*, 101–120.

Guillot, S. L. Jr. (2013). *Emergency warning system: Factors influencing citizen decision-making.* Ph.D. Dissertation, University of South Wales, Pontypridd, Wales, UK.

Kaigo, M. (2012). Social media usage during disasters and social capital: Twitter and the Great East Japan earthquake. *Keio Communication Review*, *34*, 19–35.

Kirkwood, A. S. (1994). Why do we worry when scientists say there is no risk? *Disaster Prevention and Management*, *3*(2), 15–22.

Kraus, N., Malmfors, T., and Slovic, P. (1992). Intuitive toxicology: expert and lay judgments of chemical risks. *Risk Analysis*, *12*, 215–232.

Krimsky, S. and Plough, A. (1988). *Environmental Hazards: Communicating Risks as a Social Process.* Auburn House, MA: Praeger.

Leonard, H. B. and Howitt, A. M. (2006). Katrina as prelude: preparing for and responding to future Katrina-class disturbances in the United States—Testimony to U. S. Senate Committee, March 8, 2006. *Journal of Homeland Security and Emergency Management*, *3*(2).

Lindell, M. K. and Hwang, S. N. (2008). Households' perceived personal risk and responses in a multihazard environment. *Risk Analysis, 28*(2), 539–556.

Liu, S. B., Palen, L., Sutton, J., Hughes, A., and Vieweg, S. (2008). *In Search of the Bigger Picture: The Emergent Role of On-line Photo Sharing in Times of Disaster.* Proceedings of the Fifth International ISCRAM Conference. Washington, DC. F. Fiedrich and B. Van de Walle (eds.).

McGuire, W. J. (1969). The nature of attitudes and attitude change. *The handbook of social psychology, 3*(2), 136–314.

Merchant, R. M., Elmer, S., and Lurie, N. (2011). Integrating social media into emergency-preparedness efforts. *New England Journal of Medicine, 365*(4), 289–291.

Morgan, M. G., Fischhoff, B., Bostrom, A., and Atman, C. J. (2002). *Risk Communication: A Mental Models Approach.* New York: Cambridge University Press.

Motef, J. (2005). *Risk Management and Critical Infrastructure Protection: Assessing, Integrating, and Managing Threats, Vulnerabilities and Consequences.* Washington, DC: Congressional Research Service Report for Congress, , Updated on February 4, 2005.

Nelson, L. D., Spence, P. R., and Lachlan, K. A. (2009). Learning from the media in the aftermath of a crisis: findings from the Minneapolis bridge collapse. *Electronic News, 3*(4), 176–192.

Palen, L., Anderson, K. M., Mark, G., Martin, J., Sicker, D., Palmer, M., and Grunwald, D. (2010). *A Vision for Technology-mediated Support for Public Participation & Assistance in Mass Emergencies & Disasters.* Proceedings of the ACM-BCS Visions of Computer Science 2010 International Academic Research Conference, April 14–16, 2010, p. 8, The University of Edinburgh.

Peek, L. A. and Miletik, D. S. (2002). The history and future of disaster research. In R. B. Bechtel and A. Churchman (eds.), *Handbook of Environmental Psychology.* New York: John Wiley & Sons.

Petak, W. J. (1985). Emergency management: a challenge for public administration. *Public Administration Review, 45*, 3–7.

Pine, J. C. (2005). *Tort Liability Today: A Guide for State and Local Governments,* 10th Edition. Washington, DC: Public Risk Management Association.

Plate, E. J. (2002). Flood risk and flood management. *Journal of Hydrology, 267*, 2–11.

Reynolds, O. M. (1982). *Handbook on Local Government Law.* Second Edition. St. Paul, MN: West Publishing.

Roberts, K. H. (1990). Some characteristics of one type of high reliability organization. *Organization Science, 1*, 160–176.

Roberts, K. H. and Bea, R. G. (2001). Must accidents happen: lessons from high reliability organizations. *Academy of Management Executive, 15*, 70–79.

Robinson, A. H. (1975). *Mapmaking and Mapprinting: The Evolution of a Working Relationship.* Chicago, IL: University of Chicago Press.

Robinson, A. H. and Petchenik, B. B. (1976). *Nature of Maps.* Chicago, IL: University of Chicago Press.

Rogers, R. W. and Prentice-Dunn, S. (1997). *Protection Motivation Theory.* New York: Plenum Press.

Senate Committee on Homeland Security and Governmental Affairs (2006). *Hurricane Katrina: A Nation Still Unprepared.* Washington, DC: Senate Committee on Homeland Security and Governmental Affairs.

Slovic, P. and Weber, E. U. (2002). *Perception of Risk Posed by Extreme Events.* This paper was prepared for discussion at the conference "Risk Management Strategies in Uncertain World," Palisades, New York, 12–13 April, 2002.

Svenson, O. (1979). Process descriptions of decision making. *Organizational behavior and human performance, 23*(1), 86–112.

Terpstra, T. (2011). Emotions, trust and perceived risk: affective and cognitive routes to flood preparedness behavior. *Risk Analysis, 31*(10), 1658–1675.

Tierney, K., Bevc, C., and Kuligowski, E. (2006). Metaphors matter: disaster myths, media frames, and their consequences in hurricane Katrina. *The Annals of the American Academy of Political and Social Science, 604*, 57–81.

U. S. House of Representatives (2006). *A Failure of Initiative.* Final Report of the Select Bipartisan Committee to Investigate the Preparation for and Response to Hurricane Katrina. 109th Congress, Second Session. U.S. Government Printing Office, Washington, DC.

Vasterman, P. L. M. (2005). Media-Hype: self-reinforcing news waves, journalistic standards and the construction of social problems. *European Journal of Communication, 20*(4), 508–530.

Velasquez, M., Andre, C., Shanks, T. S. J., and Meyer, M. J. (Fall 1987). What is Ethics? Issues in Ethics, *1*(1).

Weick, K. E. and Sutcliffe, K. (2001). *Managing the Unexpected.* San Francisco, CA: Jossey Bass.

Wynne, B. (2002). Risk and environment as legitimatory discourses of technology: reflexivity inside out? *Current sociology, 50*(3), 459–477.

Yates, D. and Paquette, S. (2011). Emergency knowledge management and social media technologies: A case study of the 2010 Haitian earthquake. *International Journal of Information Management, 31*(1), 6–13.

Young, E. (1998). Dealing with hazards and disasters: risk perception and community participation in management. *Australian Journal of Emergency Management, 13*(2), 14–16.

Chapter 8

Hazards Risk Management Process

Greg Shaw

Objectives

The study of this chapter will enable you to:

1. Establish the role/value of a Hazards Risk Management process.
2. Define key terms associated with Hazards Risk Management.
3. State the essential questions of Hazards Risk Management.
4. Describe the Government Accountability Office framework for risk management and its inherent limitations.
5. Describe an overall framework for accomplishing the comprehensive Hazards Risk Management process.
6. Explain the content and importance of each component within the Hazards Risk Management framework.
7. Explain how the Hazards Risk Management process supports Comprehensive Emergency Management.

Key Terms

Comprehensive Emergency Management
Hazards
Hazards Risk Management
Management
Prevention

Risk
Risk analysis
Risk assessment
Risk communication
Stakeholders

Issue

What process will best inform decision makers in their efforts to balance safety and security expenditures with the myriad challenges, requirements, and opportunities facing all organizations and communities?

Critical Thinking: Billions of dollars are spent in organizations from all sectors (private, public, and nonprofit) and all levels of community from individuals and their families to the federal government on measures to manage risk from natural, technological, and intentional hazards. Perfect hazards risk management is unobtainable, and decisions must be made to consider and formulate hazards risk management interventions in the context of overall organizational/community priorities. As presented and explained in this chapter, can the Hazards Risk Management process inform decision makers in establishing priorities, which balance competing needs while devoting limited resources to the most effective and efficient risk management interventions?

Introduction

Chapter 1 describes the nature, purpose, and application of hazards analysis as a process and a tool that supports the phases of Comprehensive Emergency Management (CEM) (Preparedness, Mitigation, Response, and Recovery). This chapter takes a step back from hazards analysis as an activity undertaken to understand hazards and the risks they pose. It focuses on the larger Hazards Risk Management (HRM) philosophy and framework as an iterative and ongoing process that is intended to inform decisions dealing with safety, security measures, and sustainability at all levels of organizations and communities. The structure of the HRM framework described in this chapter is adapted from the Emergency Management Australia Emergency Risk Management process set forth in the *Emergency Risk Management Applications Guide.* A much more detailed description and discussion of HRM can be found in 1000 plus pages of the FEMA EMI Emergency Management Higher Education Program Hazards Risk Management course available on the Higher Education Program website (Federal Emergency Management Agency [FEMA] 2004).

Terminology

As discussed in Chapter 1, there are multiple and often conflicting definitions of terms associated with hazards and HRM. These definitions may change over time to reflect certain areas of emphasis and are not necessarily consistent, even within a particular discipline or organization.

For example, The National Fire Protection Association issued the 2004 document, Standard on Disaster/Emergency Management and Business Continuity Programs, which defines mitigation as "Activities taken to eliminate or reduce the probability of the event, or reduce its severity or consequences, either prior to or following a disaster/emergency." (National Fire Protection Association 1600 2004, p. 4). The 2007 edition of this document redefines mitigation as "Activities taken to reduce the severity or consequences of an emergency," (National Fire Protection Association 1600 2007, p. 4) and introduces the new term, prevention, which is defined as "Activities to avoid an incident or to stop an emergency from occurring." (National Fire Protection Association 1600 2007, p. 5). Most recently, the 2010 and 2013 editions have changed the definition to "Activities taken to reduce the impact from hazards." (National Fire Protection Association 1600 2010, p. 6; National Fire Protection Association 1600 2013, p. 6). Following from these definitions, mitigation, as a widely understood and accepted phase of the long-established framework of CEM, has been bifurcated into the two phases of prevention and the newly defined meaning of the term mitigation, which focuses on consequence management.

To complicate matters further, mitigation was defined in the Department of Homeland Security (DHS) issued National Response Plan of December 2004 in the more traditional manner as "Activities designed to reduce or eliminate risks to persons or property or to lessen the actual or potential effects or consequences of an incident." (United States Department of Homeland Security 2004, p. 68). The successor National Response Framework documents of 2008 and 2013 do not include a definition of mitigation, while the 2010 DHS Risk Lexicon does provide a similar definition consistent with the traditional scope of mitigation: "Ongoing and sustained action to reduce the probability of, or lessen the impact of, an adverse incident." (United States Department of Homeland Security 2010, p. 21).

Considering this example, differing definitions for the HRM process and terms contained within the process are to be expected and accepted. To avoid confusion, these terms should be defined and used consistently. Accordingly, the following terms related to the HRM process are presented and defined along with the rationale for selecting the chosen definition for use in this chapter.

Lacking a widely accepted definition for the term HRM, the term is defined based on its three component words: hazard(s), risk, and management. Consistent with a definition of hazard included in Chapter 1, the definition from the 1997 FEMA publication *Multi Hazard Identification and Assessment* is selected for developing a definition of HRM: "Events or physical conditions that have the potential to cause fatalities, injuries, property damage, infrastructure damage, agricultural loss, damage to the environment, interruption of business, or other types of harm or loss." (p. xxv). Defining hazards in this manner is purposeful since it is inclusive of all sources of hazards and does not necessarily emphasize any one category of natural, technological, or human-induced (intentional/terrorist) events.

Risk and the more expansive concept of risk management are also subject to multiple definitions and are often misunderstood or confused with other terms such as risk identification, risk assessment, risk analysis, and risk communication. As discussed later in this chapter, risk management is a function composed of several subfunctions that work together for the purpose of informing decision making at all levels of organizations and communities. Risk, as the foundational term for risk management, has differing meanings in different disciplines, such as medicine, finance, safety, and security. The selected definition for risk derived from Ansell and Wharton (1992) is general in nature and applies across these disciplines: Risk is the product of probability (likelihood) and consequences of an event. Defining risk in this manner implies that risk can be managed by influencing either or both the probability (through mitigation and preparedness actions) and consequences (through mitigation, preparedness, response, and recovery actions).

The chosen definition for the term "manage" comes from the Merriam Webster Dictionary: "To work upon or try to alter for a purpose." Other definitions of manage include words like direct, govern, and succeed, which imply achieving control. Although a manager of risk strives to achieve control over risks, this is generally not totally achievable due to uncertainties, unknowns, and other intervening concerns. As stated by Borge, "Risk management is not, and will never be, a magic formula that will always give you the right answer. It is a way of thinking that will give you better answers to better questions and by doing so helps you shift the odds in your favor" (2001, p. 4). In dealing with risk, one is seldom or ever in complete control and the best one can do is work to influence future events in a manner that is perceived as favorable.

Therefore, combining these three definitions with the author's personal bias, HRM is defined as: A process that provides a general philosophy and a defined and iterative series of component parts that can be used to establish goals and objectives and inform decisions (strategic and tactical) concerning the risks associated with all hazards facing an organization and/or community. This definition of HRM is intended to emphasize each of the three component terms and the application of the process to all hazards and all phases of CEM. HRM, as an iterative process, is thus intended to provide an understanding of hazards and risks and a rational, inclusive, and transparent process for identifying, assessing, and analyzing hazard risks across all sectors and at all levels of community to inform decision makers as they allocate limited resources to the myriad and often competing priorities of their organization/community.

As discussed in the following section, Risk Management (a more commonly used term that can be used synonymously with HRM) has gained prominence in the post-9/11 environment, particularly as a tool for dealing with human-induced (intentional/terrorist) hazards. This predominantly terrorism-focused application of Risk Management has evolved to a more HRM all-hazards focus, particularly with the fallout from Hurricane Katrina and the perceived failures of all levels of government to adequately mitigate against, prepare for, respond to, and recover from the catastrophic events resulting from natural and technological hazards.

Risk Management

In the post-9/11 environment, the term risk management has gained prominence, particularly in the vernacular and practice of Homeland Security. The Homeland Security Act of 2002 requires DHS to conduct comprehensive assessments of vulnerability (a component of risk) to the United State's critical infrastructure and key resources (The White House 2002). Homeland Security Presidential Directives 7: Critical Infrastructure Identification, Prioritization, and Protection, and 8: National Preparedness, both issued in December 2003, endorse risk management as a way of allocating resources (The White House 2003a; The White House 2003b). The National Infrastructure Protection Plan (NIPP) issued in July 2006 is based on three foundational blocks including a "Risk management framework establishing processes for combining consequence, vulnerability, and threat information to produce a comprehensive, systematic, and rational assessment of national or sector risk." (United States Department of Homeland Security 2006, p. 35). Within the 2006 NIPP, Chapter 3 is titled *The Protection Program Strategy: Managing Risk* and Chapter 7, titled *Providing Resources for the CI/KR Protection Program*, includes a section titled "The Risk-Based Resource Allocation Process." Within the 2009 version of NIPP, the content remains largely unchanged with Chapter 3 now titled *The Strategy: Managing Risk* and Section 7.1 now titled "The Risk-Informed Resource Allocation Process" (United States Department of Homeland Security 2009). This seemingly minor change from risk based to risk informed reflects the increased understanding that risk information is just one of myriad considerations and inputs to the decision-making process.

The commitment to a risk management-based approach within DHS was further demonstrated by the newly appointed Secretary Michael Chertoff in the months following his confirmation. In his April 26, 2005 address to government and business leaders at New York University, Secretary Chertoff stated "Risk management is fundamental to managing the threat, while retaining our quality of life and living in freedom. Risk management can guide our decision-making as we examine how we can best organize to prevent, protect against, respond and recover from an attack ....

For that reason, the Department of Homeland Security is working with state, local and private sector partners on a National Preparedness Plan to target resources where the risk is greatest." Although terrorism-focused, Secretary Chertoff's remarks can and should be extended to all hazards and clearly emphasize the importance of risk management in "guiding" decision making supporting CEM.

The experiences observed in the next year and a half and the lessons learned during the 2005 hurricane season only strengthened Secretary Chertoff's commitment to risk management as a foundation of Homeland Security. In his December 14, 2006 address at The George Washington University, Washington, DC, Secretary Chertoff stated "Probably the most important thing a Cabinet Secretary in a department like this can do as an individual is to clearly articulate a philosophy for leadership of the department that is intelligible and sensible, not only to the members of the department itself, but to the American public. And that means talking about things like risk management, which means not a guarantee against all risk, but an intelligent assessment and management of risk; talking about the need to make a cost benefit analysis in what we do, recognizing that lurching from either extreme forms of protection to total complacency, that's not an appropriate way to build a strategy; and finally, a clear articulation of the choices that we face as a people, and the consequence of those choices."

Taken together, Secretary Chertoff's remarks, though separated by time and events by over 18 months, emphasize several very important points concerning the purpose and application of risk management:

1. Risk management can "guide" (inform) decision making across the phases of CEM.
2. Risk management is applicable to and across all levels of government (local, state, federal), all sectors (public, private, and nonprofit), and to the American public.
3. Decisions based on risk management should include a cost–benefit analysis (not just monetary costs and benefits but all costs and benefits, such as social, political, and public relations).
4. Communication (clear articulation) is a necessary component of risk management.
5. Risk management should support strategic planning and management.

Critical Thinking: Stephen Flynn in his 2004 book *America the Vulnerable* makes the very profound statement concerning understanding and dealing with risk in the post-9/11 environment: "What is required is that everyday citizens develop both the maturity and the willingness to invest in reasonable measures to mitigate that risk." (p. 64). How do we, as everyday citizens, gain this maturity and willingness to understand the risks facing our organizations/communities and to accept as reasonable the measures taken to mitigate that risk? What are our roles and responsibilities as members of our organizations and communities to engage in a process for managing risk and what are the roles and responsibilities of our organizational and community leaders to include us in that process?

To address these key points, a widely distributed, understood, and accepted framework for risk management is needed. Recognizing this need, the Government Accountability Office developed and distributed a Risk Management Framework displayed in Figure 8.1 (Government Accountability Office, 2007, p. 9).

The Government Accountability Office report from which this framework was extracted makes the point that "Risk management, a strategy for helping policymakers make decisions about assessing risks, allocating resources, and taking actions under conditions of uncertainty, has been endorsed by Congress and the President as a way to strengthen the nation against possible terrorist attacks." (2005, p. 5). The report goes on to state that the "GAO developed a framework for risk management based on industry best practices and other criteria" (2005, p. 6). This framework, shown in Figure 8.1, divides risk management into five major phases: (1) setting strategic goals and objectives, and determining constraints; (2) assessing the risks; (3) evaluating alternatives for addressing these risks; (4) selecting the appropriate alternatives; and (5) implementing the alternatives and monitoring the progress made and results achieved.

Given that the GAO has provided an authoritative and relatively widely-accepted framework and approach to risk management, why is an alternative HRM framework and process required? The GAO framework as presented is in

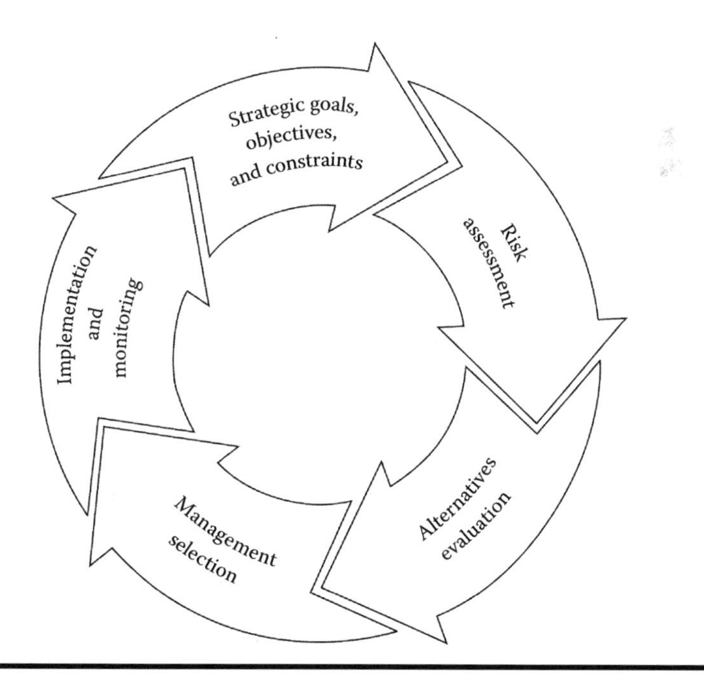

Figure 8.1 Risk management framework.

fact inclusive of certain components of the HRM process but goes beyond the intent of HRM to include risk-informed decision making and the implementation and monitoring of these risk management decisions. The HRM process, as described in the following sections, provides a context for risk-informed decision making and the identification, assessment, analysis, and presentation of hazard risk data and information. HRM is intended to support CEM as one input to informed decision making that attempts to balance safety and security expenditures with the myriad challenges, requirements, and opportunities facing all organizations and communities. The GAO framework also implies that the component steps are sequential, which they are not. The steps influence each other throughout the process and later steps may necessitate the revisiting of earlier steps and revisions of the results of each step.

A major shortfall of the GAO framework is that it largely ignores the necessity of continuous risk communication and monitoring and review throughout the overall process that can doom the overall process to failure. The point of emphasis here is that HRM is an ongoing process that continually examines the impact of organizational activities to ensure that risks are identified, considered, and understood to support decisions impacting our vulnerability to those risks. To maximize effectiveness, any risk management process must continuously communicate strategies and tactics to manage the adverse impacts of risks throughout the impacted organization/community.

To improve the risk management process, a set of framing questions and a framework for HRM are presented and described as a recommended philosophy and approach to informing safety and security decision making in any sector and at all levels of organizations and communities.

Hazards Risk Management Framing Questions

Before embarking on the HRM process, and particularly before starting any risk assessment, the following questions should be asked and answered in a manner generally understood and acceptable to the audiences impacted by the HRM process results.

What are the organization's/community's strategic goals and objectives and considering those goals and objectives:

- What is the scope of our HRM effort?
- What is an acceptable level of risk?
- Who determines what an acceptable level of risk is?
- Can risk be managed?
- For those risks that can be managed, what are the interventions (controls/countermeasures) available to manage risk?
- What combination of risk management interventions (controls/countermeasures) make sense in terms of nonrisk-specific considerations (economic, environmental, social, political, legal)?

Framework for Hazards Risk Management

Figure 8.2 displays the process for risk management as adapted from the Emergency Risk Management process set forth in the 2004 Emergency Management Australia, *Emergency Risk Management Applications Guide* (Emergency Management Australia 2004). The HRM framework includes the general format of the Emergency Risk Management framework but meets a different purpose as described in this section of the chapter. The HRM framework includes six steps: (1) Establish the context; (2) Identify the hazards; (3) Assess the hazards risk; (4) Sort the hazards by risk magnitude; (5) Analyze the risks from each hazard; and (6) Group and prioritize risks (with two continual components: Communicate and Consult, and Monitor and Review). Roughly categorized, steps 1 and 2 accomplish hazards identification, steps 3 and 4 hazards risk assessment, and steps 5 and 6 hazards risk analysis. Note that chapters in this book examine hazards identification and characterization,

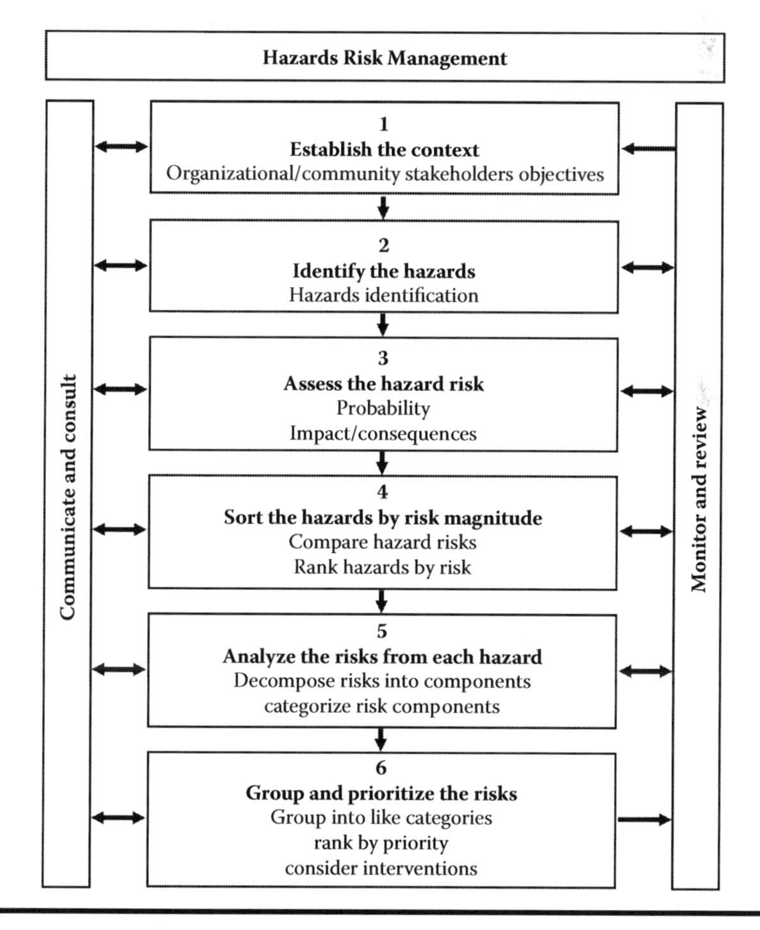

Figure 8.2 Hazards Risk Management process.

modeling, spatial analysis, and risk and vulnerability analysis. We thus view the hazards analysis process in the context of HRM and as a process to generate information for selecting appropriate hazard mitigation strategies.

The HRM framework is constructed to define an inclusive, iterative, and continuous process that addresses the HRM framing questions listed above and provides a foundation for the four phases of CEM: Preparedness, Mitigation, Response, and Recovery. Inherent in each of the phases of CEM is the goal of effectively and efficiently managing the hazards that may adversely impact an organization/community and its ability to achieve its strategic and tactical goals and objectives. Following the HRM process is intended to provide the "needs assessment" for CEM and as such establishes a focus and steering direction. Understanding the HRM process is a key to developing a risk-informed, all-hazards emergency management program.

Each component of HRM process is discussed in this section of the chapter. Much of the content in this discussion is adapted from *Emergency Management Principles and Practices for Healthcare Systems* (Second Edition) published by the Institute for Crisis, Disaster, and Risk Management Research Team at the George Washington University for the Veterans Health Administration/U.S. Department of Veterans Affairs, Washington, DC, June 2010 (Barbera et al. 2010).

Components of the Hazards Risk Management Process

Communicate and Consult

Continual communication and consultation within and without an organization/community provides a means of inclusion and the establishment and management of realistic expectations for the HRM process and its eventual incorporation into the organization's/community's overall Emergency Management program. Step 1 in the HRM process calls for establishing the organizational/community context, involving stakeholders, and setting objectives. Communication and consultation does not stop there. Repeating the statement on risk management of Secretary Chertoff of DHS from earlier in this chapter, he views his responsibilities as a Cabinet Secretary to include the need to "clearly articulate a philosophy for leadership of the department that is intelligible and sensible, not only to the members of the department itself, but to the American public" (Chertoff 2006). This role is shared by all leaders of organizations and communities with risk management responsibilities. To meet this responsibility, the entire HRM process should be open, accessible, and intelligible to the impacted public.

As a matter of guidance and emphasis for continuous communication and consultation throughout the HRM process, the 1989 National Research Council Report, "Improving Risk Communication" provides the following statement that should guide all risk communication. "Risk communication is a process, the success of which is measured by the extent that it, first, improves or increases the base of information that decision makers use, be they government officials, industry managers, or individual citizens, and second, satisfies those involved that they are adequately informed within the limits of available knowledge" (p. 74).

It should also be noted that we stress the role of communication and stakeholder participation in the hazards analysis process in both Chapters 1 and 7 of this book. The public has a critical need to know and understand the nature of risks in the community, and risk communication should be an intentional part of hazard risk management and the hazards analysis process.

Monitor and Review

The HRM process is never actually finished, as it is subject to reanalysis and revision when changes occur in the internal and external environments. Continuous monitoring and review of findings from all steps should be conducted to keep the overall process relevant and on track with the Emergency Management program. Drills, exercises, and actual events will test the Emergency Management program, and both the positive and negative observations related to system vulnerabilities should be noted and analyzed. The HRM process also constitutes a major means of monitoring and reviewing any findings related to reduced as well to newly recognized hazard risks. For example, an exercise could examine whether a new process or procedure has effectively reduced a previously recognized hazard risk. Similarly, an exercise, a threat, or an actual event may prompt the recognition of a previously unidentified hazards and/or hazard risk.

Step 1: Establish the context

Context refers to the external environment in which the organization/community exists and functions, and the internal characteristics of the organization/community itself. Therefore, establishing the context for the HRM process (and, essentially, for an overall Emergency Management program) is the logical starting point for the process. To accomplish this, the organizational/community context, the stakeholders, and the objectives for the HRM process must be defined.

Organizational/community context: The organizational/community context for the HRM process is established based on the organization's/community's responsibilities; the economic, environmental, social, political, and legal realities; and the review and input of stakeholders. This step in the process begins with hazard identification, where an organization or community is characterized from a social, economic, political, geographical, structural, and ecological context. In short, the starting point of HRM is to clearly describe the nature of our organization/community.

The organization's/community's boundaries to include strategic and tactical goals and objectives and legal and moral roles and responsibilities are delineated. Additionally, economic, environmental, social, political, and legal constraints on the organization/community for resource allocation supporting emergency management requirements and initiatives should be identified and recognized. This step helps to answer all framing questions. For example, a community should consider its demographics, economic state, and strategic and tactical goals for growth and development and its roles and responsibilities to its population and to surrounding communities to determine the scope and constraints of its HRM effort.

Stakeholder involvement: Identifying and engaging a stakeholder group is the second critical step in establishing the context, and particularly can assist in identifying the social, economic, environmental, political, and legal realities and constraints that impact the HRM process. The individual/group responsible for the HRM process (HRM Committee) accomplishes this by identifying all appropriate stakeholders, both internal and external to the organization/community, that should be included and considered in all steps of the HRM process. Stakeholders are defined as key people, groups of people, or institutions that may significantly influence the success of the process.

Stakeholder analysis is a technique that is increasingly used in private industry to identify and assess the importance of stakeholders and thereby judge that the stakeholder group is balanced and comprehensive. To ensure that multiple perspectives are adequately considered and represented in the overall HRM process, the following steps help define a successful stakeholder analysis (Management Sciences for Health and United Nations Children Fund website):

- Identify people, groups, and institutions that will influence your HRM process.
- Develop strategies to build the most effective support possible for the process and reduce any obstacles for the successful implementation of an effective emergency management program. For example, simply by inviting outsiders such as representatives from business, public safety, Local Emergency Planning Committee, and impacted community groups into the HRM process may help resolve misconceptions and miscommunications.

Objectives: Establishing the specific objectives for the HRM process follows from defining the organizational/community context and the involvement of the appropriate stakeholders. Realistic and measurable objectives based on observable outcomes for both strategic (long-term) and tactical (short-term) activities are essential for all other steps and components of the HRM process, particularly communication and consultation, and continuously monitoring and reviewing the process and the results.

Step 2: Identify the hazards

This component involves the listing of all possible hazard types that could significantly impact the organization/community.

Comprehensive hazard identification: The full range of hazards must be captured. The list includes hazards that do not directly or physically impact the organization/community, but could generate demand on the organization's/community's resources from other organizations/communities. Hazard identification should also include hazards that, if they occur, could cause major impacts on the goods and/or services provided by the organization/community, such as loss of customers, litigation, liability payments, and poor publicity. A comprehensive examination of the hazard identification process is provided in Chapter 2.

Hazard identification strategy—organization/community resources: Although multiple resources are available to assist in identifying hazards, an essential consideration is coordination with outside organizations/communities, including nearby organizations and municipalities, regional leadership, and state authorities. Hazards that could potentially impact an organization/community are commonly hazards that may impact the larger area, and therefore have already been identified or are being defined by other organizations/communities at the local, regional, and state levels.

Hazard identification strategy—web resources: Other resources are available to assist in hazard identification. These include local, state, and national websites; FEMA and National Oceanic and Atmospheric Administration publications; and the 15 National Preparedness Scenarios established within the 2007 *National Preparedness Guidelines* (United States Department of Homeland Security 2007). Examples include the following: The New York City Office of Emergency Management provides a list and description of the "NYC Hazards" expected to impact the New York City area (New York City Office of Emergency Management website); the State of Virginia Department of Emergency Management provides a list and description of potential hazards (Virginia Emergency Management Agency website); FEMA provides historical data and links to hazard-related websites (FEMA website). Many of these resources provide important historical data on hazards and useful guidance for individuals, organizations, and communities.

Categorizing hazards: As hazards are identified, it useful to group the hazards according to the following categories, where commonalities predominate in both cause and actions necessary to address the hazard risk. These categories are further explored as part of the hazard identification process within a hazard modeling context in Chapter 3.

- *Natural hazards:* Hazards that primarily consist of the forces of nature.
  - For example, hurricane, tornado, storm, flood, high water, wind-driven water, tidal wave, earthquake, drought, lightning-caused wildfire, infectious disease epidemic.

- *Technological hazards:* Hazards that are primarily caused by unintentional malfunction of technology, including human and system actions.
  - For example, industrial, nuclear, or transportation accidents; power and other utility failure; information technology failure; hazardous materials release; and building collapse.
- *Intentional hazards:* Hazards that are caused primarily by deliberate human threat or executed action. These are usually criminal, civil disobedience, or terrorist in nature.
  - For example, civil strife, terrorism, or criminal attacks on the organization/community.

Step 3: Assess the hazard risk

Hazard risk assessment is conducted in the next two HRM steps. Risk, as previously defined in this chapter, is a product of probability and consequences. Each hazard identified by the organization/community should, therefore, be assessed individually according to its probability of occurrence and its impact (consequences) on the organization/community as a means of approximating each hazard's level of risk.

Hazard risk assessment strategy: How the hazard risk assessment is presented and accomplished varies among sources, but all share the common purpose of establishing the relative importance of and between hazard risks. For that reason, most hazard risk methods use a ranking system that assigns a quantitative value to each individual hazard to allow a preliminary method of sorting by numeric value. It must be remembered, however, that the assignment of quantitative values to probability and consequence is often subjective and can be based on information with inherent uncertainties.

For example, the exact probability of any event occurring as a result of a natural, technological, or intentional hazard is not necessarily determined by past occurrences and is subject to continual changes in the organization's/community's internal and external environments. Determining the probability of an intentional hazard (terrorist strike) impacting an organization/community is particularly perplexing due to the general lack of historical data and the fact that an intelligent hazard vector, such as a terrorist, can adjust the location and nature of the attack based on the hazard risk management controls in place.

Similarly, an exact determination of consequence is not possible since it requires an a priori understanding of the hazard event scenario prior to the occurrence and a complete understanding of the organization/community impacted. Complicating the matter further is the necessity to define the categories (people, property, vital services, business loss, etc.) to be included in the measure of consequence and a common metric for combining the categories. Also, looking beyond the immediate consequences of an event, long-term consequences such as environmental damage and financial impacts of a localized event may permanently alter the environment and cascade through the larger economy. These long-term consequences may dwarf the immediate consequences and are not necessarily quantifiable (e.g., the immediate consequences on a maritime container port in terms of physical damage and loss of life due to a hurricane, as compared to the longer-term impact on the environment, local economy, and possibly the overall national economy).

For these reasons, it is essential that the purpose of the hazards risk assessment be emphasized throughout the HRM process—to establish the relative importance of hazard risk between the identified hazards for sorting—and that the stakeholders providing input to the assessment have a common understanding of the rating measures in order for them to be consistent across all inputs. A discussion of the myriad hazards risk assessment methods and templates currently in use go far beyond the scope of this chapter. The key point is that the method and template selected must be useful and appropriate for the scope of the particular HRM effort.

Step 4: Sort the hazards by risk magnitude

This step continues the hazards risk assessment and consists primarily of assigning a relative level of importance to each hazard risk value from step 3, thereby placing each hazard risk in the context of the overall cohort of the identified hazard risks.

Sorting strategies: This sorting of hazard risks entails the comparison of the assigned risk values (established in step 3) associated with each hazard and the designation of each hazard to one of the broad categories (high risk, moderate risk, and low risk) via mathematical or expert judgment methods. Although simple numerical values are commonly used to represent probability and consequence, the comparative value of the selected metrics must be fully understood for this ranking system to have merit. In other words, is a value of 3.92 calculated in the risk assessment essentially equal to a value of 4.15 considering the inherent uncertainties in assigning probability and consequence ratings? If so, separating hazards with these two values into different categories may be misleading and cause judgment errors later in the HRM process. To address this, presenting the score for all identified hazards on a single graph or spreadsheet may allow appropriate grouping of hazards, with less reliance on specific (but relatively uncertain) numerical assignments.

The use of expert judgment: Expert judgment should enter into this hazard sorting and may result in a rearrangement of the results based on specific intelligence related to probability and/or consequence. For example, a terrorist attack using biological agents may have almost unimaginable consequences that totally dwarf the probability considerations, thus elevating such an event to the top of the priority list. In the absence of specific intelligence, however, the rank of this hazard may be moved to below that of an event with a better defined probability, such as a hurricane in a rural coastal community.

Conversely, a lower ranked terrorist hazard may suddenly be elevated to the top of the hazard list based on new threat information, thereby overriding the earlier expert judgment. The dynamic nature of the natural, technological, and intentional hazard environment necessitates an expert level review and judgment beyond mere numerical sorting. A complicating factor in the use of expert judgment is the definition and identification of true "experts." In the post-9/11 environment, there appear to be many "terrorism experts," ready and willing to express their views on the topic. What exactly qualifies them as experts may be subject to debate and honest disagreements.

Step 5: Analyze the risks from each hazard

Hazard risk analysis is accomplished in the next two steps. Hazards are considered individually during the earlier steps of the HRM process, and the risk (probability and consequences) of each hazard is compared only at a very macro level. The final analysis step of the HRM process should allow decision makers to look across all hazards to identify components of hazard risk that are common to multiple hazards. At this point, the components of risk refer to the consequences of an event resulting from the occurrence of a hazard. This approach promotes the identification of options that reduce or eliminate components of risk from multiple hazards through a single intervention (see step 6) and therefore supports the most effective and efficient allocation of resources to HRM.

Hazards risk analysis strategy: To accomplish this, the components (consequences) of risk from each hazard should be "decomposed" into significant elements that can be compared and/or grouped across the range of identified hazards. A relatively common grouping of hazard risk consequences is in the categories of human, property, and operational consequences. The groupings should refer to the processes and resources that are disrupted (i.e., so that they can later be grouped across hazards according to the "all-hazards" processes and resources that are affected). For example, a generic organization experiencing a hurricane would expect to experience consequences in each category with specific consequences that impact the resources and operational processes of the organization. The specific consequences and disruptions could include:

- Human consequences
  - Inability for staff to reach or remain at work: child/elder care responsibilities, transportation disruption, concern about personal property, loss of personal property, and others
  - Injury/death to staff (at work or at home) and customers/clients/guests within the facility due to high winds and debris causing window/door glass failure
- Property consequences
  - Flooding, roof failures, and other water effects
  - Wind and debris damage to buildings, outside equipment, vehicles, and other property on facility premises

  - Storm surge effects if relevant
  - Maintenance problems due to failure of personnel to report for work
- Operational consequences
  - Inability to provide products and/or services
  - Inability to meet contractual agreements
  - Interruption of cash flow
  - Damage to reputation

Categorizing the hazard risk: For each significant hazard, the hazard risk is analyzed, decomposed into elements, and grouped in a format that will allow like elements to be identified across all hazards.

Step 6: Group and prioritize the hazard risks and consider risk management interventions

This step completes the hazards risk analysis by sorting and comparing the hazard risk elements determined in step 5.

Grouping and prioritizing strategy: It is very likely that some identical hazard risk elements (defined in step 5) will be present in a wide range of hazards that cross the natural, technological, and intentional hazard categories. For example, nearby hazardous materials releases with explosive potential (technological or intentional), an approaching tornado (natural), and a realistic truck bomb threat (intentional) all expose an organization's employees, staff, and guests to injury and death due to the physical damage to the building unless prevented/mitigated by structural measures, immediate protective actions, and the availability of an emergency response capability. Guarding against this hazard risk element (injury and death) would be prioritized with other hazard risk elements according to a prioritization scheme which generally places life and safety issues first with property protection and continuity of operations as lower priorities.

Consider hazard risk element interventions: The individual hazard risk elements are further analyzed to develop potential interventions for consideration in the development of formal hazard mitigation and preparedness plans. Following from the above example, some potential interventions could include:

- Mitigation/prevention interventions to reduce the risk of injury or death by reducing the likelihood of physical damage to the building:
  - Structural measures to increase the strength of the building
  - Removing windows
  - Covering windows with protective coatings
  - Relocating personnel work spaces away from outer walls
  - Standoff barriers to keep vehicular traffic away from the building's perimeter
  - Building security measures to check vehicles entering the parking area

- Mitigation/consequence management preparedness interventions to reduce the risk of injury or death by reducing the consequences of physical damage to the building:
  - Emergency action plans covering sheltering in place and evacuation
  - Communication capabilities (e.g., general announcement, alarms, computer alert) that deliver relevant and actionable information to building occupants
  - Communication capabilities that receive and deliver relevant and actionable information from and to outside public safety organizations
  - Awareness, training, and exercises
  - Internal emergency response organization and capability
  - Mutual aid agreements with other organizations/communities
  - Coordination with public safety organizations

If the identified interventions are applicable across multiple hazards, this may prompt further grouping of risk elements. At this step in the process, the risk elements are grouped together. Selection of interventions for implementation, however, is an activity that occurs later, during formal Emergency Management planning. Consideration of potential interventions is used only to prompt the grouping of risk elements in a manner where they may be addressed through economy of scale or in a manner that provides greater benefit than if each element is individually addressed.

Obviously, each potential intervention has resource implications (costs—both tangible and intangible), which must be considered in the context of contribution (benefits) to the HRM goals and objectives. In general, costs, particular monetary and time costs, are easy to quantify. However, intangible costs such as reduced employee morale; decreased accessibility to a building for employees, customers and guests; and disruption of previous policies and procedures may be significant and should not be ignored. Actual benefits derived from risk interventions are much more difficult to identify and quantify. Particularly, in the absence of a hazard event, benefits may be largely invisible to decision makers who must allocate limited resources to multiple organizational/community priorities that may have little or nothing to do with HRM.

Clearly, some of the preparedness interventions such as planning, awareness, training, exercising, coordination, and mutual aid are of relatively low cost, can be implemented easily, and provide some contributions to achieving HRM goals and objectives. Others, such as internal and external communication and the development of internal response organization and capability have higher costs, but are recognized as applicable across all hazards, and as such, may be judged as cost effective.

Application of the Hazards Risk Management Process

Overview

An example of processing an identified hazard through the six-step HRM process may serve to illustrate the value of this approach. The organization/community selected for this example is a fictional private urban university, which is both an organization (a nonprofit business) and a distinct community within the larger urban city community (8,000 resident students, 12,000 nonresident students, 8,000 faculty and staff, and a capital plant valued in excess of $1 billion). The university can be impacted across the spectrum of hazard categories (natural, technological, and intentional) as described in step 2 of the HRM process. One identified hazard, civil strife, is selected for this example. The university is located in proximity to highly visible and controversial organizations/entities that are subject to protest events coinciding with scheduled and widely publicized meetings. Previous protest events have been marked by violence, injuries, property damage, disruption of local transportation, restricted access to areas of the community, and arrests of protesters and bystanders.

Step 1: Establish the context

Organizational/community context: The university's context is described in terms of its stated strategic goals and objectives.

A review of the university's strategic plan provides the overall goal of "Providing a safe and supportive environment for students, staff, and faculty to pursue education, research, personal and professional growth." Supporting this goal are specific objectives:

- Maintaining the safety and well-being of all members of the university's community
- Protecting the physical and intellectual property of the university
- Maintaining continuity of essential university operations following disruption
- Preserving the university's reputation as a high-quality institution of higher learning
- Supporting the surrounding community and the overall urban community as a good neighbor

For any and all hazards, the university is focused on the above listed objectives in the order of priority as listed. The highest priority is to afford protection for all members of the university community. The HRM process should look across all the phases of CEM (Mitigation, Preparedness, Response, and Recovery) to assess and analyze the hazard risks and to identify possible hazard risk management interventions for the university's leadership consideration. The safety and well-being of community members demands a primary focus on mitigation (prevention and consequence management) interventions.

The university's physical location, legal and moral roles, responsibilities, and authority determine the scope and constraints of the HRM effort. For the example hazard, civil strife, reduction of risk to a zero level is not obtainable short of removing the hazard, prohibiting the members of the university community from coming into any contact with the hazard and/or cordoning the university off from the hazard. None of these measures are possible due to the physical setting of the university, the commitment of the university to academic and intellectual freedom, and the rights of the university and surrounding community members. Therefore, there is some level of risk associated with the example hazard, and the university can use the HRM process to establish an acceptable level of risk and identify the potential interventions to be considered for implementation.

Stakeholder involvement: Stakeholders include representatives of the university's student body (all categories of students including resident and nonresident and undergraduate and graduate), faculty, staff, students' parent groups, alumni, police, fire department, urban government authorities, surrounding communities, other area universities, and so on. The level of involvement of each stakeholder group depends on the nature of the identified hazards as the HRM process proceeds. For the example hazard, all these stakeholders have some level of involvement with public safety and urban government authorities as the primary source of relevant information and required coordination.

Objectives: Establishing the specific objectives for the HRM process follows from defining the organizational/community context and the involvement of the appropriate stakeholders.

Step 2: Identify the hazards

This component involves the listing of all possible hazard types that could significantly impact the organization/community. For this example, civil strife is identified as a primary hazard. A comprehensive hazards identification strategy extending beyond this example would include research, expert consultation, and stakeholder input and review to identify all hazards that may impact the university as an organization and as a community.

Step 3: Assess the hazards risk

Identified hazards are considered individually to determine their probability and impact for the purpose of assigning a level of risk.

Hazard risk assessment strategy: For the example hazard, the assigned probability is very high and approaching certainty based on the projected schedule for meetings in the future and the history of demonstrations for past meetings. On an ordinal numerical scale of 1 to 5 of probability ranging from highly unlikely (rating of 1) to almost certain (rating of 5), the example hazard would be assigned a probability rating of 5.

Revisiting the university's strategic objectives identified in step 1, the consequences of civil strife can have significant immediate and long-term consequences impacting each of the objectives. Members of the university community may be injured, university physical property may be damaged, and university operations may be disrupted. Based on these consequences and the university's perceived actions across the phases of CEM, the university's reputation and stature in the larger community may be damaged impacting future enrollments, faculty recruitment and retention, and research funding. For these reasons, the consequences of civil strife are assigned a rating of 5 on an ordinal numerical scale from minimal (rating of 1) to significant (rating of 5).

These ratings are combined by a selected mathematical algorithm (usually multiplying or adding) to establish the relative level of hazard risk between the identified hazards for further sorting. In this example, any method of combining the probability and consequences would lead to a determination that civil strife poses a high level of hazard risk to the university.

Step 4: Sort the hazards by risk magnitude

During this step, a relative level of importance is assigned to each hazard risk value determined in step 3 for the purpose of placing each hazard risk in the context of the overall cohort of the identified hazard risks. The numeric ratings of probability and consequence and their combination are subject to uncertainties and potential rating bias. For that reason, the hazards risks should be sorted into general categories such as low, moderate, and high risk. The civil strife hazard would obviously be placed in the high-risk category, whereas other hazards such as a hurricane might be placed in the moderate-risk category and an accidental spill of a toxic chemical in the low-risk category.

In addition to sorting by mathematical ratings, expert judgment should be applied to the sorting process to account for the dynamic nature of hazard risk and specific intelligence related to the probability and/or consequences of the hazard risk. In this example, the meetings resulting in demonstrations are scheduled for specific dates, and outside of those dates the probability of civil strife declines to an unlikely or highly unlikely level. Additionally, intelligence may indicate changes in the size and motivation of the demonstrations for different meetings, which could change the consequences rating. These considerations can and should rearrange the sorting of hazard risk based on the current situation and the available information.

Step 5: Analyze the risks from each hazard

In this and the following step, each hazard risk is considered in the context of all identified hazard risks to determine commonalities across multiple hazards. For the example of civil strife, the components (consequences) of the hazard risk are decomposed into significant elements that can be compared and/or grouped across the range of identified hazards. Considering human, property, and operational consequences, civil strife could result in the following:

- Human consequences
  - Injury/death to students, faculty, staff, visitors, and transients at the site of the demonstrations and within the confines of the university due to localized or spreading violence
  - Detention/incarceration of students, faculty, and staff in the proximity of demonstrations
  - Inability of students, faculty, and staff to reach or remain at the university due to transportation disruptions or physical security measures
- Property consequences
  - Damage to university buildings from the outside due to vandalism
  - Damage to university buildings from the inside due to access to the buildings and vandalism
  - Damage to other university property such as buses, signage, and outside areas due to vandalism
  - Maintenance problems due to inability of personnel to report for work
- Operational consequences
  - Inability to conduct classes
  - Inability to conduct scheduled events
  - Inability to receive and/or ship supplies
  - Damage to reputation

Step 6: Group and prioritize the hazard risks and consider risk management interventions

For each significant hazard, the decomposed hazard risk elements are grouped in a format that will allow like elements to be identified across all hazards. For example, a hurricane would share many similar hazard risk elements with civil strife such as the potential for injury/death, inability to reach or remain at the university, damage to university buildings and property, inability to conduct classes, conduct scheduled events, and receive and/or ship supplies.

Grouping and prioritizing strategy: The grouped hazard risk elements are then prioritized with life and safety issues first with property protection and continuity of operations as lower priorities.

Consider hazard risk element interventions: The individual hazard risk elements are further analyzed to develop potential interventions for consideration in the development of formal hazard mitigation, preparedness, response, and recovery plans. For the civil strife hazard example, some (for the sake of this example, the list of interventions is not intended to be exhaustive) potential interventions include the following:

- Mitigation
  - Establishing policy that prohibits university community members from entering specified areas where civil strife is expected
  - Cordoning off the university area with police enforcement
  - Limiting access to university buildings (dorms, classroom and office buildings, and administrative buildings) and allowing only university community members with valid identification cards
  - Closing common buildings that are normally open to the public (student union, food courts, and athletic venues)
  - Implementation of instant messaging technologies for connectivity with all university community members
  - Training/education for university community members on the nature of the hazard
  - Information and updates passed to university community members and stakeholder groups (particularly, the parents of resident students) encouraging hazard avoidance and what to do if involved in the consequences of the hazard
  - Shutting down university operations during the expected period of the civil strife
- Preparedness
  - Review of mitigation, preparedness, response, and recovery plans.
  - Training for personnel with mitigation, preparedness, response, and recovery responsibilities.
  - Liaison with public safety agencies and community emergency management agencies.
  - Information updates to all university community members and stakeholders as appropriate.
  - Increased medical response capabilities on campus.
  - Increased police patrols on campus.
  - Increase levels of essential supplies (food, water, medical).
- Response
  - Activate the university emergency response organization and plans.
  - Continue information updates to all university community members and stakeholders as appropriate.
  - Shut down university operations consistent with the situation.
- Recovery
  - Resume operations as the situation permits.
  - Assess and deal with the consequences of the event.

- Provide counseling services for university community members as needed.
- Communicate with all university community members and stakeholders as appropriate.
- Conduct an after-event review to capture lessons learned and to identify corrective actions.
- Review and revise all emergency management-related plans, policies, and procedures based on the after-event review.

Obviously, many of the potential interventions such as planning, community-wide training and education, passing information, instant messaging technologies, and liaison with public safety and emergency management agencies are applicable across all categories of hazards and should be considered in the context of this widespread utility. Other interventions such as shutting down operations, increased police presence, increased medical response capabilities, and increased university building security may be particular to the specific hazard or a limited group of hazards. Some interventions such as prohibiting university community members from being in the area of the civil strife and cordoning off the university may be highly effective for hazard risk deduction processes, but are not feasible within the scope and constraints of the HRM effort as described in step 1.

This concludes the specific steps of the HRM process. The outputs of the process including the hazard risk groupings and priorities and potential interventions are communicated to the university's decision makers for their consideration in determining what can and will be done to manage hazard risks. The HRM outputs are an essential input to informed decision making that attempts to balance safety and security expenditures with the myriad challenges, requirements, and opportunities facing the university.

Hazards Risk Management and Comprehensive Emergency Management

The true measure of utility and success of the HRM process is its application to informing decisions within the organization's/community's overall Emergency Management program. The HRM process serves to identify hazards risk management interventions across the phases of CEM: mitigation, preparedness, response, and recovery. From that point on, it is up to the decision makers to answer the last of the HRM framing questions to decide what combination of risk management interventions (controls/countermeasures) make sense (economic, environmental, social, political, and legal) for their organization/community. As stressed throughout this chapter, HRM as a process and its component steps are not the sole input to answering this question. They are just one input to informed decision making that attempts to balance safety and security expenditures with the myriad challenges, requirements, and opportunities facing all organizations and communities.

Discussion Questions

What is the real value of the HRM process to you as an individual and as a member of an organization and/or community?

What are the measures of success of the HRM process?

As quoted from the 1989 National Research Council Report "Improving Risk Communication," "Risk communication is a process, the success of which is measured by the extent that it, first, improves or increases the base of information that decision makers use, be they government officials, industry managers, or individual citizens, and second, satisfies those involved that they are adequately informed within the limits of available knowledge." Do you agree with this statement, particularly as it applies to risk communication in the post-9/11 environment? Do you feel that the risk information available to you in the context of your personal- and professional-related responsibilities is adequate?

How could you increase your involvement in the HRM process for your organization/community?

Applications

Consider an organization you are affiliated with or your community (local, state, or federal) and apply the HRM process to identify hazard risk management interventions that should be considered. To assist in following the HRM process, answer the following questions.

1. Are there nonrisk-specific considerations (economic, environmental, social, political, and legal) that should be considered when setting goals and objectives for HRM?
2. What are the goals and objectives for the HRM process?
3. Who are the stakeholders that should be included in the HRM process? Are there varying levels of inclusion for different stakeholders?
4. Who determines what is an acceptable level of risk?
5. What is an acceptable level of risk?
6. What sources of information would you consult to identify the hazards that may directly or indirectly impact your organization/community?
7. How would you assess (quantify) the risk of the hazards identified? How would you sort the hazards by the risk magnitude? Based on your personal knowledge of your organization/community and available information (e.g., information from your local/state Emergency Management Agency websites), assess representative hazards by their probability and consequences.
8. How would you analyze the hazard risk from the identified, assessed, and sorted hazards?

9. How would you group and prioritize the hazard risk from the identified, assessed, sorted, and analyzed hazards?
10. How would you present (communicate) the results of the HRM process to stakeholders and decision makers?

Websites

Department of Homeland Security: http://www.dhs.gov/index.shtm
Department of Veterans Affairs: *Emergency Management (EM) Principles and Practices for Healthcare Systems* (Second Edition), 2010: http://www.gwu.edu/˜icdrm/projects/VHA2/index.htm#download
Emergency Management Australia: http://www.ema.gov.au/
Federal Emergency Management Agency: http://www.fema.gov
FEMA Higher Education Program: http://training.fema.gov/EMIWeb/edu/
FEMA State Offices and Agencies of Emergency Management: http://www.fema.gov/about/contact/statedr.shtm
Government Accountability Office: http://www.gao.gov
Ready America: http://www.ready.gov/america/index.html
The National Fire Protection Association: http://www.nfpa.org

References

Ansell, J. and Wharton, F. (1992). *Risk: Analysis, Assessment, and Management.* Chichester, New York: John Wiley & Sons.

Barbera, J., Macintyre, A., Shaw, G., Westerman, L., Seefried, V., and de Cosmo, S. (2010). *Emergency Management (EM) Principles and Practices for Healthcare Systems* (Second Edition). Washington, DC: The Institute for Crisis, Disaster, and Risk Management (ICDRM), George Washington University (GWU). http://www.gwu.edu/˜icdrm/projects/VHA2/index.htm#download. Accessed August 18, 2013.

Borge, D. (2001). *The Book of Risk.* New York: John Wiley & Sons.

Chertoff, M. (2005). *Remarks by Secretary of Homeland Security Michael Chertoff at the Center for Catastrophic Preparedness and Response and the International Center for Enterprise Preparedness.* Released by New York University. http://www.nyu.edu/intercep/events/Chertoff%20remarks%204.05.htm. Accessed July 23, 2013.

Chertoff, M. (2006). *Remarks by Homeland Security Secretary Michael Chertoff on Protecting the Homeland: Meeting Challenges and Looking Forward.* Released on December 14, 2006, George Washington University, Washington, DC. http://www.tsa.gov/press/releases/2005/12/20/remarks-homeland-security-secretary-michael-chertoff-dhs-accomplishments. Accessed July 23, 2013.

Emergency Management Australia (2004). *Emergency Risk Management: Applications Guide.* Manual 5. http://www.em.gov.au/Documents/Manual%2005-ApplicationsGuide.pdf. Accessed July 23, 2013.

Federal Emergency Management Agency, *Types of Disasters*. http://www.fema.gov. Accessed July 23, 2013.

Federal Emergency Management Agency (1997). *Multi-Hazard Identification and Risk Assessment: The Cornerstone of the National Mitigation Strategy*. http://www.fema.gov/library/viewRecord.do?id=2214. Accessed July 23, 2013.

Federal Emergency Management Agency (2004). *Hazards Risk Management course, EMI Emergency Management Higher Education Project*. http://training.fema.gov/EMIWeb/edu/. Accessed July 23, 2013.

Flynn, S. (2004). *America the Vulnerable*. New York: Harper Collins.

Government Accountability Office (2005). *Risk Management: Further Refinements Needed to Assess Risks and Prioritize Protective Measures at Ports and Other Critical Infrastructure* (Report No. GAO-06-91). http://www.gao.gov/. Accessed July 23, 2013.

Government Accountability Office (2007). *Applying Risk Management Principles to Guide Federal Investments* (Report No. GAO-07-386T). http://www.gao.gov/. Accessed July 23, 2013.

Management Sciences for Health and United Nations Children Fund, *The Guide to Managing for Quality*. http://erc.msh.org/quality/ittools/itstkan.cfm. Accessed July 23, 2013.

National Fire Protection Association (2004). *Standard on Disaster/Emergency Management and Business Continuity Programs*. Quincy, MA: NFPA.

National Fire Protection Association (2007). *Standard on Disaster/Emergency Management and Business Continuity Programs*. Quincy, MA: NFPA.

National Fire Protection Association (2010). *Standard on Disaster/Emergency Management and Business Continuity Programs*. Quincy, MA: NFPA.

National Fire Protection Association (2013). *Standard on Disaster/Emergency Management and Business Continuity Programs*. Quincy, MA: NFPA.

National Research Council Report (1989). *Improving Risk Communication*. Washington, DC: National Academy Press.

New York City Office of Emergency Management, *NYC Hazards*. http://www.nyc.gov/html/oem/html/hazards/hazards.shtml. Accessed August 18, 2013.

United States Department of Homeland Security (2004). *National Response Plan*. http://purl.access.gpo.gov/GPO/LPS56895. Accessed August 18, 2013.

United States Department of Homeland Security (2006). *National Infrastructure Protection Plan*. http://purl.access.gpo.gov/GPO/LPS71533. Accessed August 18, 2013.

United States Department of Homeland Security (2007). *National Preparedness Guidelines*. http://www.dhs.gov/xprepresp/publications/gc_1189788256647.shtm. Accessed December 18, 2007.

United States Department of Homeland Security (2009) *National Infrastructure Protection Plan*. https://www.dhs.gov/national-infrastructure-protection-plan. Accessed August 18, 2013.

United States Department of Homeland Security (2010). *DHS Risk Lexicon*. http://www.dhs.gov/xlibrary/assets/dhs-risk-lexicon-2010.pdf. Accessed August 18, 2013.

Virginia Department of Emergency Management Agency, *Hazards and Treats*. http://www.vaemergency.gov/readyvirginia/busines-toolkit/hazards. Accessed April 29, 2014.

The White House (2002). *The Homeland Security Act of 2002*. https://www.dhs.gov/homeland-security-act-2002. Accessed July 23, 2013.

The White House (2003a). *Homeland Security Presidential Directive 7: Subject: Critical Infrastructure Identification, Prioritization, and Protection.* http://www.dhs.gov/homeland-security-presidential-directive-7. Accessed July 23, 2013.

The White House (2003b). *Homeland Security Presidential Directive 8: Subject: National Preparedness.* http://www.dhs.gov/presidential-policy-directive-8-national-preparedness. Accessed July 23, 2013.

Chapter 9

Planning for Sustainable and Disaster-Resilient Communities

Gavin Smith

Objectives

1. Understand the nexus between the topics discussed in Chapters 1 through 9 and their application to hazard mitigation planning.
2. Understand the connection between the following concepts: hazards management, sustainability, disaster resilience, planning, and climate change adaptation.
3. Understand the hazard mitigation planning process, including key components.
4. Understand the role of planners, the plan-making tools and participatory processes they use, and their potential to create disaster-resilient communities.
5. Understand existing hazard mitigation planning policies and programs, including their connectivity to climate change adaptation.

Key Terms

Adaptive capacity
Administrative capability
Advocacy planning
Background studies

Capability assessment
Climate change adaptation
Collaborative planning
Comprehensive plan
Disaster Mitigation Act of 2000
Disaster recovery
Disaster recovery planning
Disaster resilience
Dispute resolution techniques
Emergent groups
Facilitation
Fiscal capability
Focusing events
Hazard analysis
Hazard mitigation
Hazard mitigation
Hazard mitigation committee
Hazard Mitigation Framework
Hazard Mitigation Grant Program
Hazard mitigation plan
Hazard mitigation planning
Hazard mitigation policies
Hazard mitigation projects
Hazard mitigation strategy
Hazards base maps
Hazards management
Land suitability analysis
Land-use planning
Land-use planning tools
Legal capability
Local government paradox
Mediation
Multiobjective planning
Negotiation
Physical planning
Plan adoption and implementation
Plan monitoring, evaluation, and modification
Planning process
Police power
Policy dialogue
Policy planning
Political capability
Post-Katrina Emergency Management Reform Act

Pre-Disaster Mitigation
Public involvement
Robert T. Stafford Disaster Relief and Emergency Assistance Act
Safe development paradox
Sustainable development
Technical capability
Vulnerability assessment

Introduction

This chapter will discuss the nexus between hazards analysis and planning, emphasizing disaster resilience and how it fits within the concepts of hazards risk management, sustainable development, hazard mitigation, and climate change adaptation. The hazard mitigation plan provides a tool to link the concepts discussed throughout the text, such as the identification and analysis of hazards; the use of techniques to assess social, economic, and environmental vulnerability (i.e., spatial analysis, modeling, and economic loss estimation); and the development of risk management or hazard mitigation strategies and their application to at-risk individuals, groups, and institutions. This chapter will conclude with a discussion of land-use planning tools and processes and their potential to achieve disaster-resilient communities.

Sustainability, Disaster Resilience Climate Change Adaptation, and Hazard Mitigation Planning

The concept of sustainability has gained widespread recognition among scholars and practitioners as a sound principle to guide development practices. Sustainable development emphasizes attempts to live in harmony with the natural environment in a manner that provides improved social, environmental, and economic conditions for current and future generations (World Commission on Environment and Development 1987). The complimentary aims of hazard mitigation (Beatley 1998; Berke and Beatley 1992; Godschalk et al. 1999; Schwab et al. 1998; Smith and Wenger 2006) and disaster resilience (Beatley 1995; Berke 1995; Burby 2001; Schwab, et al. 1998) have been added to this conceptual framework (see Figure 9.1).*

A number of international and national commissions and boards have led efforts to link natural hazards mitigation and sustainability, including more recently, the addition of climate change adaptation. The United Nations hosted the Rio Summit, which produced one of the first definitions of sustainability that included hazard

* For a summary of hazard scholarship addressing sustainable development themes, see Smith and Wenger (2006, p. 236).

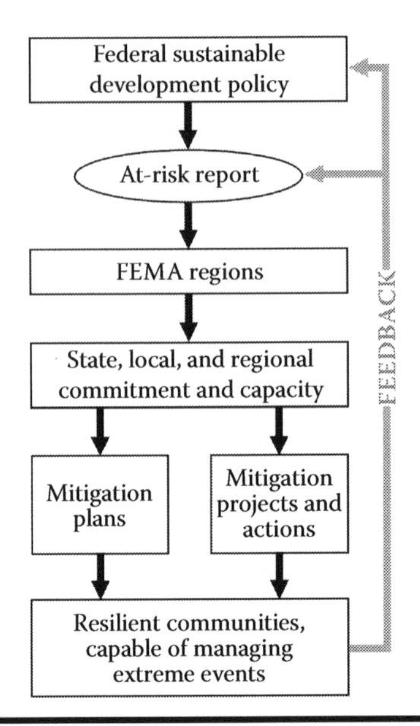

Figure 9.1 A sustainable mitigation policy system. (From Godschalk, D., et al. *Natural Hazard Mitigation: Recasting Disaster Policy and Planning,* Island Press, Washington, DC, 1999, p. 531. With permission.)

mitigation, and declared an International Decade of Natural Hazard Reduction, beginning in 1990. *The President's Council on Sustainable Development* describes specific action items targeting a reduction in governmental subsidies that encourage unsustainable development in known hazard areas (Beatley 1998, pp. 237–238). More recently, the International Panel on Climate Change and a number of international aid organizations has applied these concepts to climate change adaptation (International Panel on Climate Change 2012; United Nations Framework for the Convention on Climate Change on Loss and Damage 2012).

Although a number of international groups and hazards researchers have advocated this position, practitioners at the federal, state, and community level, including land-use planners, have failed to incorporate the concepts of hazard mitigation and disaster resilience into their day-to-day activities on a widespread basis. The ability to link these concepts through multiobjective planning can result in mutually reinforcing outcomes and a broader coalition of support across stakeholder groups advocating complimentary positions (Smith and Wenger 2006). Acting on the implications of a changing climate on hazard risk represents both a daunting challenge and an opportunity to link hazard mitigation, disaster resilience, and climate change adaptation scholars and practitioners (Berke et al. 2012; Glavovic and Smith 2014; Schipper 2009; Schipper and Pelling 2006).

Godschalk et al. (1999) provides a good description of a disaster-resilient community and its connectivity to sustainable development principles:

> Resilient communities may bend before the extreme stresses of natural hazards, but they do not break. They are consciously constructed to be strong and flexible rather than brittle and fragile. This means that their lifeline systems of roads, utilities, and other support facilities are designed to continue functioning in the face of rising water, high winds, and shaking ground. It means that their neighborhoods and businesses, their hospitals and public safety centers are located in safe areas rather than in known high-hazard areas. It means that their buildings are constructed or retrofitted to meet building code standards based on the threats of natural hazards faced. It means that their natural environmental protective systems, such as dunes and wetlands, are conserved to protect their hazard mitigation functions as well as their more traditional purposes (p. 526).

As the description suggests, disaster-resilient communities are more sustainable than those that do not develop a comprehensive strategy that incorporates hazard mitigation into their current and ongoing construction, design, and planning activities. Tim Beatley's description of resilience includes the added dimensions of learning from past actions and adapting to changing conditions, both of which are particularly germane to not only reducing current hazards risk but also preparing for and adapting to changes in risk over time, including that due to or influenced by climate change.

Taking appropriate action to ensure greater resilience and sustainability requires gaining a greater appreciation for the hazards prevalent in the area. Hazards analysis as described in Chapter 1 represents the ongoing and systematic process of identifying and defining the physical (magnitude, scope, and intensity) and temporal (timing, speed of onset, duration) characteristics of hazards, assessing their likelihood of occurrence, and estimating their potential impacts or consequences. Understood in the context of achieving disaster-resilient communities, hazards analysis provides a rational basis for individuals, groups, and organizations to make informed decisions based on that knowledge (see Cutter et al. 2007; Deyle et al. 1998). However, an international assessment of hazard mitigation plans has shown that the degree to which vulnerability science was used to inform the formulation of goals and mitigation policies was among the weakest elements in plans (Berke and Godschalk 2009). This important finding highlights the fact that good hazards analysis cannot be performed in isolation and should be part of a systematic management strategy.

Disaster resilience is achieved through hazards risk management as Shaw describes in Chapter 9. Hazards risk management represents the adoption of a comprehensive and integrative series of practices, policies, and behavior that recognizes how routine and planned actions taken by individuals, groups, and communities affect their level of hazard vulnerability. Decision-making processes are shaped by a number of factors, including access to accurate and timely information, the effectiveness of risk communication and outreach strategies, resource availability, political power and influence, leadership, and the practice of planning. The land-use planning profession offers a meaningful integrative role using tools, techniques, and processes needed to link the concepts of hazards mitigation, disaster resilience, and sustainable communities.

Increasingly hazard scholars and a growing number of practitioners have recognized that sustainable communities include those that can bounce back from natural hazard events and disasters (Beatley 2009; Burby 1998; Godschalk et al. 1999; Mileti 1999; National Academies 2012), including those that embrace a new normal, that includes safer, more equitable development practices (Godschalk 2003; Paton and Johnston 2006). The inclusion of resilience in this discourse serves a boundary spanning function, linking social, economic, and environmental themes to pre-event hazard mitigation planning and postevent adjustments to the impacts of disasters (see Figure 9.2). Incorporating hazard mitigation into the routine activities of individuals, governments, businesses, nonprofits, and others also represent an important manifestation of sustainable development principles. The failure to confront hazards through pre-event planning can cause disastrous consequences as was dramatically evident in New Orleans following Hurricane Katrina. Hazard

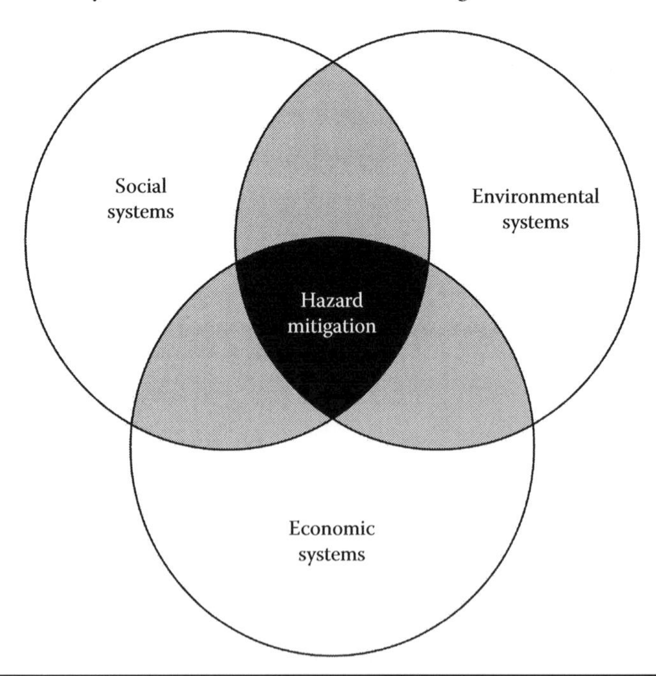

Figure 9.2 Hazard mitigation and sustainable communities.

mitigation planning represents an action-oriented framework used to identify hazards, their expected impact, and measures that can be taken to lessen or eliminate their effects. The practice of multiobjective planning provides a procedural vehicle through which complimentary objectives can be achieved before and after disasters.

Hazards and disasters represent a powerful means to understand the preexisting characteristics of communities as they tend to uncover or highlight social and economic problems that are often tied to issues of race, class, power, equity (see Chapters 4 and 5), and environmental concerns associated with natural system protection and hazard vulnerability (see Chapter 2) (Colten 2005; Costanza et al. 2006). Disasters can serve as a forcing mechanism among communities, causing them to confront problems previously left unaddressed. Examples may include the construction or repair of affordable housing, the incorporation of energy-efficient design principles into new development standards, or changing policies that encourage sprawl into known hazard areas. In the agenda setting literature, this is referred to as a window of opportunity (Kingdon 1984) or focusing event (Birkland 2006). Major disasters can also provide opportunities to encourage more, not less development in areas prone to hazards and disasters. Powerful economic interests may use the event, and the ensuing flow of federal assistance, to further a profit-driven agenda that does not advance a public good, rebuilding communities as quickly as possible, effectively negating the chance to incorporate hazard mitigation or sustainable development principles into the recovery process (Freudenberg et al. 2009).

The assessment of the political landscape and the impact of competing agendas on the adoption of hazard mitigation strategies should be incorporated into decision-making activities (see Chapter 9), including postdisaster recovery. Disaster recovery, which can be defined as the "differential process of restoring, rebuilding, and reshaping the physical, social, economic, and natural environment through pre-event planning and post-event actions" (Smith and Wenger 2006, p. 237), provides an opportunity to incorporate new hazard mitigation initiatives, including those that involve the modification of existing policies and programs to better reflect hazard risk. A number of factors can influence the degree to which this occurs, including past disasters that have fostered a greater awareness of risk among elected officials and individuals living in a community, an awareness of and ability to take advantage of existing postdisaster hazard mitigation grant programs, the existence of good hazard mitigation plans and a robust pre-event planning process, and policies that require adopting higher standards after a disaster occurs.

Critical Thinking: Other than a major disaster, can you think of a focusing event that has occurred in the United States or elsewhere that resulted in significant policy change? Discuss the specific federal, state, and local implications associated with your answer. Is the State of Louisiana's coastal land loss over the past 50 years a good example of a focusing event? Why or why not? How do rapid-onset extreme events like hurricanes or earthquakes differ in terms of their ability to serve as a focusing event versus slow-onset hazards like sea level rise or drought? Based on these differences, what types of policies would you propose to address rapid- and slow-onset hazards and disasters?

Natural hazards are part of the environmental sphere in which we live. Hurricanes, floods, winter storms, and earthquakes play an important role in the regulation of larger natural systems on which we all depend. Attempts to physically modify our environment often have severe consequences, including an increased level of hazard vulnerability and damages following disasters. The modification of our climate and the manner in which these changes are causing or exacerbating extreme events represents perhaps one of the most pressing problems in the field of natural hazards risk management (Glavovic and Smith 2014).

Disasters are a human construct and occur when natural hazards intersect with human settlements or the natural resources on which people depend. The failure to recognize this reality has resulted in development patterns that are inherently unsustainable (Beatley 2009; Burby 1998; May et al. 1996). Many environmentally sensitive areas such as wetlands, barrier islands, steep-sloped or mountainous areas, and wildlands are prone to natural hazards such as floods, coastal storms, landslides, and wildfire. Limiting development in these areas reduces the exposure of individuals and communities to the impacts of hazards. For example, wetlands provide a natural buffer against landfalling hurricanes, a reservoir for excess water following floods, a filtration system for pollutants and excess sediment, a recharge area for ground water, wildlife habitat, and a site for water-based recreational activities such as canoeing and bird watching.

Conducting a land suitability analysis is regularly used by land-use planners to assess and categorize land according to the type of use that is "most appropriate" based on a series of intrinsic characteristics. Historically, this has been done as a means to measure the ecological impact of differing land uses. Planning scholar and practitioner Ian McHarg, in his seminal text, *Design with Nature* (1969), was one of the first to codify this process, incorporating environmental data into the planning process and mapping the results. This method served as the precursor to the development of the geographic information system (GIS) discussed in Chapter 6. The ability to layer and analyze environmental and natural hazards information provides a powerful tool to link land-use decisions to both complimentary geospatial products.* Land suitability analysis may be determined using ecological and natural hazard indicators such as the following:

- Topography/elevation/mean sea level (flood, storm surge, sea level rise)
- Hydric and alluvial (flood, mudslide), volcanic soils (volcano, mudslide)
- Fill material (liquefaction tied to earthquakes, landslide)

---

* In *Design with Nature*, McHarg discusses the connection between design principles that respect coastal ecology and the damages associated with a Nor'easter that struck the New Jersey shore in 1962 (see pp. 15–17). **Critical Thinking**: How are the tools like land suitability analysis and GIS being applied in New Jersey following Hurricane Sandy? Consider reviewing the work being undertaken by the teams involved in the United States Department of Housing and Urban Development-sponsored Rebuild by Design competition.

- Karst (sinkholes)
- Slope (landslide, avalanche, flash flood, and lava flow)
- Vegetative type (wildfire, mudslide, erosion, flash flood)
- Wetland delineation (flood, sea level rise)
- Areas subject to coastal or riverine erosion, subsidence (flood, storm surge, sea level rise)
- Barrier island topography, width, and orientation (flood, storm surge, erosion, sea level rise)

This approach also holds promise as a means to capture and visually display the implications of protecting or failing to protect our local, regional, national, and global ecological capital as discussed in Chapter 2, Hazards Identification.

Critical Thinking: Can you think of other data layers that should be added to the land suitability/natural hazards analysis? What about social and economic data layers? Are there other analytical tools that can be used to help conduct a comprehensive hazards analysis? Consider those used in your profession or area of study. Are they currently being used for this purpose? Why or why not? Can this tool be used to assess those natural hazards that are not geographically defined? How would you map hazards that may not exist today, but are likely to exist at some point in the future (e.g., sea level rise)?

The potential to achieve complimentary benefits must recognize that land prone to natural hazards is often among the most coveted places to live. Examples include the wildland–urban interface, steep-sloped areas, and ocean or riverfront properties. Market demands, reluctance among local officials to limit development in known hazard areas, the inappropriate application of hazard mitigation strategies, and access to large-scale postdisaster assistance programs has encouraged, rather than discouraged risky development. Further hindering the adoption of a sound hazard mitigation strategy is the fact that the true societal costs associated with development in hazardous areas are not effectively measured, nor are they accounted for in a unified hazards policy. The question, who should bear the costs of living in known high-hazard areas, remains one of the most challenging to answer.

Postdisaster assistance policies contribute to this problem. States and local governments tend to view postdisaster assistance as entitlement programs and sought following disasters regardless of pre-event actions taken at the local level that may have increased exposure and vulnerability. The federal government has played an important role in fostering this dependency as the increasing number of federal disaster declarations are partially the result of political patronage (Platt 1999). Natural hazards scholar Ray Burby has coined the terms "safe development paradox" and "local government paradox" to help explain some of the underlying problems associated with good hazard mitigation policy making. The safe development paradox describes how the adoption of some risk reduction measures (such as engineering-based solutions like levees and seawalls) designed to reduce vulnerability can actually encourage development in high-hazard areas leading to greater losses in the future when the protective design parameters are exceeded.

The local government paradox describes a situation in which individuals residing in known hazardous areas bear the suffering associated with disasters, while local governments often pay little attention to policies that limit hazard vulnerability (Berke and Smith 2009; Burby 2006). To address these problems, it is incumbent on states and local governments to pursue greater local self-reliance relative to the potential impacts of hazards through the adoption of proactive rather than reactive strategies. This concept remains an underemphasized characteristic of sustainable, disaster-resilient communities (Smith and Wenger 2006).

Timothy Beatley (1989) argues that the practice of hazard mitigation also involves a moral dimension that should help frame decision making. Undergirding the conceptual discussion of sustainability is the moral imperative that we should take the actions necessary to ensure the well-being of future generations. The application of ethics to the realm of hazard mitigation requires posing the following questions: (1) How do we reconcile federal, state, and local policies that facilitate rather than hinder choices that increase our vulnerability to hazards? (2) To what extent does local, state, and federal government have a moral obligation to adopt and implement more stringent hazard mitigation programs? (3) Does government have a unique obligation to provide additional assistance to the socially vulnerable? and (4) To what extent is it the obligation of individuals to take action to reduce their vulnerability to natural hazards versus relying solely on the government for assistance?

Critical Thinking: Do local, state, and federal governments have a moral obligation to protect life and property from the impacts of natural hazards? If so, how do you reconcile the fact that existing policies have the effect of encouraging, rather than discouraging development in known hazard areas? What would you do to address this dilemma given the existence of the safe growth and local government paradox?

Hazard Mitigation Planning Policy Framework

The Disaster Mitigation Act of 2000 requires that state and local governments must develop hazard mitigation plans to remain eligible for pre- and postdisaster hazard mitigation funding. The Disaster Mitigation Act further codifies the federal rules and requirements associated with the development of state and local hazard mitigation plans. Prior to that time, states were required to develop hazard mitigation plans (referred to as 409 plans) as stipulated by the Robert T. Stafford Disaster Relief and Emergency Assistance Act. The Stafford Act, which was passed by Congress in 1988, created three key disaster recovery programs: the Individual Assistance, Public Assistance, and Hazard Mitigation Grant Programs.* The Stafford Act emphasizes the administration of these programs rather than a broad policy framework advancing the concepts of hazard mitigation and planning for disasters (Godschalk et al. 1999; Mileti 1999).

* The Individual Assistance Program provides grants and loans to assist homeowners and renters in making repairs to damaged homes, whereas the Public Assistance Program provides federal funding to assist states, communities, and nonprofits offset the costs of disaster response efforts, clean up disaster-generated debris, and repair damaged public infrastructure.

Godschalk et al. (1999) found in their study of plans predating the Disaster Mitigation Act that the quality of state hazard mitigation plans were weak and their ability to foster the implementation of a comprehensive hazard mitigation strategy were limited. A study conducted by the Government Accounting Office, cited the failure of the federal government to develop a national policy framework to provide guidance on the use of numerous, but disjointed hazard mitigation policies to more effectively reduce future disaster losses (Government Accountability Office, August 2007). Federal Emergency Management Agency (FEMA) developed a national Hazard Mitigation Framework in 2012, which describes a broad conceptual approach to hazard risk reduction. However, the Hazard Mitigation Framework has yet to result in a significant modification of federal policy including those that have had the net effect of increasing hazard risk or improving the quality of state and local hazard mitigation plans.

In the most comprehensive study addressing the quality of state and local hazard mitigation plans and one that postdates the development of Disaster Mitigation Act plans, a number of troubling findings emerge. Although state plans were found to have improved in general relative to 409 plans, local plans exhibited a number of major deficiencies. For instance, most plans place low priority on land-use tools and processes, emphasizing emergency services, property protection, and information and awareness efforts (Lyles et al. 2014). None of the plans evaluated scored high across all of the plan quality dimensions that are also troubling given the importance of internal consistency across these dimensions.

Hazard mitigation plans should provide a framework for action through a series of interrelated programs, policies, and projects designed to reduce the level of hazard vulnerability in a given area. The Berke, Smith and Lyles study found that plans did not contain a clear vision or set of goals designed to attain higher-order concerns like reducing future damages or achieving greater sustainability and disaster resilience. Most local hazard mitigation plans have been created as a means to an end, namely used to gain access to Hazard Mitigation Grant Program funding (that addresses problems created in the past) rather than a future orientation that seeks to guide development and human settlement patterns in a manner that reflects hazard risk and vulnerability (Berke and Smith 2009). The failure to adopt policies that reflect this future orientation is not sustainable, resilient, or reflective of the more recent issues linking natural hazards risk management and a changing climate. Very few plans assessed by Berke, Smith, and Lyles address the important linkage between hazard mitigation and climate change adaptation. The hazard mitigation planning process, which is described next as a means to address these identified limitations, can be greatly improved if tested planning techniques are effectively used.

Hazard Mitigation Plan

The hazard mitigation plan comprises several parts including: hazard identification and analysis, vulnerability assessment, capability assessment, hazard mitigation strategy, and plan adoption and implementation. The hazard mitigation committee is often tasked with the creation and implementation of the hazard mitigation plan.

In many cases, the development of hazard mitigation plans are led by private sector consultants, not mitigation committees. This can lead to the development of hazard mitigation plans with strong risk assessments but weaker public participation elements (Smith et al. 2013). Furthermore, the lack of local government commitment, including the involvement of land-use planners, can lead to plans that do not place a strong emphasis on land use or the integration of hazard mitigation policies and projects with the jurisdictions comprehensive plan, a document that has legal standing and is used to guide future growth (Lyles et al. 2014; Smith et al. 2013).

A hypothetical community hazard mitigation committee might include the following representatives:

- Public Works Director
- Land-Use Planner
- Local Floodplain Administrator
- Local Emergency Manager
- Building Official/Building Inspector
- City Manager/Assistant City Manager
- Finance Officer or Director
- Economic Development Director
- County Representatives
- Citizen/Neighborhood Representatives
- Nonprofit Representatives
- Local Business Representatives
- Media Representatives
- Utility Company Representative
- Regional Planning Representative
- Emergent Group Representative*

The committee may establish subcommittees based on key issues identified over time. Examples may include functional topics associated with housing; infrastructure; future growth or desired changes in existing land-use regulations, building codes, and ordinances; socially vulnerable populations; environmental protection; small business; and the economy.

It is the responsibility of the committee to solicit public involvement throughout the planning process. The public, for example, should play a role in the identification of hazards and their potential impact. This provides for the inclusion of local experiential knowledge that can be missed by local officials or consultants hired

* Emergent groups, which form after a disaster, are informal in nature, tend to focus on a specific issue or group whose needs are not met by others, and often last for a brief period of time. In some cases, these groups can transform into more formal, long-standing organizations. While pre-event hazard mitigation plans cannot include groups that don't exist, they can be added once identified after a disaster (Smith 2011).

to help develop the plan. Examples may include anecdotal information based on past experiences with hazards and disasters or the accumulation of newspaper clippings and photographs taken by individuals. Public input also provides a vehicle to obtain a collective understanding of local risk perception, which should be used to frame education and outreach strategies (see Chapter 7). The power of participatory planning and the influence of the planning process will be discussed later in this chapter.

Critical Thinking: Can you think of others who should be included in a hazard mitigation committee? Should participation or membership change over time? What role might individuals from your area of study or profession play on the hazard mitigation planning committee? Who should be added to address issues tied to climate change adaptation?

Following the creation of a committee, members should begin the background studies (hazard identification and analysis, vulnerability assessment, and capability assessment) that serve as the factual basis for the hazard mitigation plan. To develop a sound hazard mitigation strategy, it is necessary to identify and analyze the hazards present in the study area. Once the hazards are identified, they are analyzed by documenting their historic occurrence and associated impacts. Newspapers, weather service reports, technical documents, and personal accounts are representative of the types of information sources typically used for this purpose. The historical documentation of past events provides some insight into what an area could expect in the future. The review of past events does not provide a comprehensive means to estimate the probability of future disasters. This requires conducting a vulnerability assessment. The vulnerability assessment involves the evaluation of hazard risk, including the likelihood that events of a varied magnitude will impact the study area, causing a series of estimated damages. As discussed in Chapter 4, some segments of a population are more vulnerable than others due to social and economic characteristics. A sound vulnerability assessment estimates future expected losses based on projected development patterns and demographic changes over time.

As noted in Chapter 6 and this chapter, respectively, the creation of hazards base maps are a useful means to identify and assess geographically defined hazards. The delineation of hazards using GIS allows for the georeferenced analysis of hazards relative to the built and natural environment. A GIS platform, coupled with the tools discussed in Chapter 3, enables the modeling of various hazard scenarios, providing valuable "what if" information that should guide the development of the hazard mitigation strategy that follows. The ability to quantify expected losses as described in Chapter 5 provides an important means to establish a rational basis for action discussed in Chapter 8 and communicate risk (see Chapter 7). The real power of this type of assessment is the ability to estimate how the level of vulnerability can be increased or decreased by taking specific actions. For example, continued development in known hazard areas or the adoption of more stringent building codes can be modeled and their negative and positive impacts on hazard exposure and vulnerability can be measured.

The vulnerability of a community is partly a function of the existing technical, fiscal, administrative, legal, and political capabilities of a jurisdiction to reduce, manage, or eliminate the effects of identified hazards. Elements and indicators of local capacity include the following:

- Technical (GIS, visualization, hazard modeling software, planning)
- Fiscal (local budget, grants-in-aid, loans, capital investments)
- Administrative (technical and administrative staff, contractors)
- Legal (existing rules, laws and regulations, legislation)
- Political will (votes cast, policies and regulations adopted, land-use decisions)

Analyzing this capability requires the review of existing documents, plans, programs, and policies. Local documents may include the comprehensive plan, Local Flood Damage Prevention Ordinance, and Capital Improvements Plan, among others. Local policies that target other aspects of a community (i.e., economic development, environment protection, or social services) may impact the vulnerability of a community to natural hazards. The capabilities of additional stakeholders, such as nonprofits and the private sector, should be included in this assessment. Other examples may include federal and state regulations, regional agreements, or the multijurisdictional sharing of resources. A comprehensive capability assessment revolves around two basic questions: (1) Does the community possess the tools needed to confront their vulnerability to hazards? and (2) Do existing policies, plans, programs, or activities currently in place increase or decrease current or future hazard vulnerability?

Technical capability refers to the access to and use of analytical tools, including GIS, visualization, and hazard modeling software. The fiscal capability of a jurisdiction includes their internal and external access to financial resources and the commitment of these resources to hazard mitigation–related activities. Internal resources include the regular operating budgets of participating stakeholder groups, whereas external resources include funds obtained through the procurement of grants and loans. Administrative capability refers to the staff, personnel, or contractors available to create, monitor, and implement the hazard mitigation plan and associated strategies.* Legal capability refers to the type and strength of government rules and regulations that provide the legal standing to act. Federal, state, and local government agencies are responsible for enforcing regulations and policy. The adoption and

* Following disasters, the ability to implement hazard mitigation strategies can be compromised as staff are often overwhelmed with the management of postdisaster aid programs. The postdisaster environment can also provide an opportunity to hire additional staff through grants triggered by a federal declaration. The ability to retain staff and their highly valued experience-based knowledge is often difficult as these skills are in high demand among federal, state, and local government agencies as well as private sector consultants and contractors.

enforcement of most land-use planning techniques are done at the local level.* Local governments possess the legal authority, or what is referred to as "police power," to enact land-use planning measures necessary to protect the health, safety, and general welfare (Nolon and Salkin 2006). Less well recognized is the application of this concept to protect citizens and their property from the impacts of hazards and disasters.

The adoption of local hazard mitigation strategies are often driven by the level of political will present in a community. Political capability is the willingness of elected officials and parties such as developers, realtors, and business leaders to support the adoption of hazard mitigation strategies. It also refers to the level of acceptance among citizens and community groups, including those who may advocate stronger regulations. Adopting policies restricting land-use activities in identified hazard areas or more stringent building codes and ordinances is representative of actions that require a high level of political will. Convincing skeptics who believe that such measures unnecessarily restrict market choices or land use or that the federal government will pay for the costs of reconstruction following a disaster requires developing a sound counterargument backed up with verifiable data and political support. It also requires moving beyond the realities of short election cycles that officials tend to base many of their policy decisions. Effective strategies engage a broad coalition of elected and nonelected individuals in positions of local power or influence and describe the tangible economic, environmental, and social benefits of hazard mitigation in a way that resonates with their own agenda to achieve and sustain changes in the status quo.

In addition to the assessment of technical, fiscal, administrative, legal, and political capabilities, the hazard mitigation plan should identify gaps in existing policies used to confront hazard vulnerability. Identified gaps can be translated into the adoption of policies addressing these issues. For example, during the assessment of existing policies, it may be found that the community is a participating member of the National Flood Insurance Program, but not the Community Rating System (CRS), which requires the adoption of a more comprehensive approach to floodplain management. Based in part on this review of existing policies, the community may choose to join the CRS as a part of their hazard mitigation strategy.

Critical Thinking: Is one element of the capability assessment more important than another? If so, explain why. Discuss how varied elements of the capability assessment are interconnected.

The hazard mitigation strategy should represent a series of policies, programs, and projects chosen to reflect the actions necessary to address the results of the vulnerability and capability assessments. A vulnerability assessment identifies at-risk structures and populations as well as future vulnerabilities based on projected

* The common refrain among practicing planners, "all planning is local" is not entirely accurate. In reality, federal and state policies can significantly impact land use decisions at the local level, as the safe growth and local government paradox described earlier demonstrate.

growth. This information is used to develop specific strategies designed to reduce identified vulnerabilities (Figure 10.6). An effective hazard mitigation strategy requires the use of multiple hazard mitigation projects and policies identified by a diverse team of stakeholders. Potential hazard mitigation projects may include the relocation or elevation of flood-prone properties, the retrofitting of vulnerable community facilities (i.e., the installation of storm shutters to protect against flying debris generated by a tornado or hurricane), and the hardening of public infrastructure to better withstand ground motion associated with earthquakes. Hazard mitigation policies may include joining the National Flood Insurance Program, limiting the placement of public infrastructure and critical facilities in identified hazard areas or adopting building code standards reflecting the latest findings of the hazards analysis.

As techniques are identified, appropriate funding or administrative support is identified. In some cases, the adoption of mitigation techniques will require amending existing regulations or creating new ones. It is therefore incumbent on the hazard mitigation committee to involve elected officials early and often in the process to gain their support. It is also important to make sure that the mitigation techniques chosen can be accomplished given existing capabilities or government officials commit to procure the resources needed to accomplish them. Otherwise, their implementation will be compromised. The following examples of hazard mitigation techniques were adapted from *Keeping Hazards from Becoming Disasters: A Mitigation Planning Guidebook for Local Government*, prepared by North Carolina Division of Emergency Management in 2001.

- Prevention (planning, zoning, and subdivision regulations; open space preservation, floodplain regulations, stormwater management, drainage system maintenance, capital improvements programming, shoreline/fault zone setbacks)
- Property protection (relocation, acquisition, elevation, critical facilities protection, insurance, retrofitting of hazard-prone structures)
- Natural resource protection (floodplain protection, beach and dune preservation, riparian buffers, conservation easements, erosion and sediment control, wetland preservation, slope stabilization)
- Structural projects (reservoirs, seawalls, levees, channel modifications, beach nourishment)
- Public information (outreach projects, hazard map information, real estate hazard disclosure, warning systems, hazards expo)

Critical Thinking: How would you balance the selection of hazard mitigation projects and policies based on the findings of the risk assessment with the level of local capability to implement them? Which hazard techniques do you believe are the most effective? Why? How might the selection of these techniques differ when addressing climate change-related hazards?

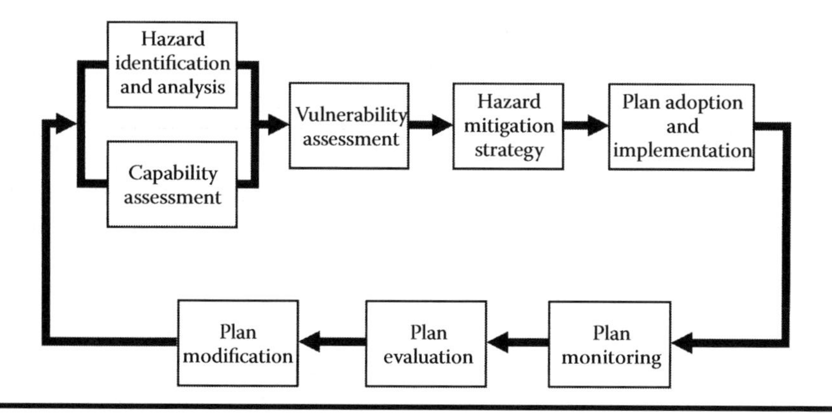

Figure 9.3 The hazard mitigation planning process.

Plan adoption and implementation is not the final step, as the planning process is more accurately characterized as a continually revised series of actions (Figure 9.3). Once written, the adoption of the plan gives it legal standing among those responsible for its implementation and the larger community as a whole. In order for the plan to be an action-oriented document, it should contain a strong implementation strategy. It is critically important that the plan is routinely monitored to make sure that those individuals, agencies, and organizations assigned various tasks are completing them in accordance with the plan schedule.

An important part of any plan requires holding participants accountable. If plan goals and objectives are not being met, the planning committee should take corrective actions. An effective plan monitoring, evaluation, and modification strategy includes the means to integrate the latest understanding of hazard vulnerability and capability, as both change over time. New development, changes in the makeup of the population, or the use of more refined analytical tools influence our understanding of hazard vulnerability. The additional uncertainty associated with climate change-related impacts (e.g., sea level rise, more intense hurricanes, extreme rainfall events, prolonged droughts, more severe wildfires) and the tools used to assess these changes represent the forefront of hazards analysis. Downscaling climate data to subnational, regional, and local scales; developing multiple risk assessment scenarios and an associated set of robust (e.g., "no regrets") and contingent strategies based on differing scenarios; modeling slow-onset hazards; and communicating uncertainty to broad audiences are all important issues to consider in the development of robust hazard mitigation and climate change adaptation plans (International Panel on Climate Change 2012; United Nations Framework for the Convention on Climate Change on Loss and Damage 2012; Berke 2014).

Similarly, the capability to mitigate and adapt to hazards and their effects can change. The formulation of new or modified policies should be incorporated into the document and are subject to change based on new risk assessment findings and changes in capabilities of those tasked with addressing them. Finally, it is important to monitor progress. Attaining mitigation goals should be documented and their benefits measured, when possible (see Chapters 8 and 10). If monetary benefits can be ascertained (i.e., losses avoided due to pre-event actions), this information can garner continued political support, both among local elected officials who may choose to adopt more progressive mitigation policies and federal agencies that may be more willing to provide pre- and postdisaster hazard mitigation funding or other types of assistance.

Critical Thinking: What indicators provide a basis to assess the political will of a community to adopt a comprehensive hazard mitigation strategy? Once identified, how would your mitigation strategy reflect that reality and account for changes associated with election cycles and turnover in personnel? What role do citizens play in shaping the political will of a community to proactively address hazards risk? What about nonprofit organizations (e.g., faith-based groups, foundations, environmental and social justice groups) and members of the private sector (developers, financial investment firms, the insurance industry, consultants, small businesses, corporations)?

Power of Plan Making: Tools and Process

Land-use planning is one of the most powerful tools to advance the aims of hazard mitigation and disaster resilience (Burby 1998, 2006; Burby and Dalton 1994; Berke and Smith 2009). A review of land-use planning can be understood within a process-oriented framework tied to physical and policy-related outcomes associated with the location, type, and density of development (Chapin and Kaiser 1985). Physical planning traces its roots to the development of zoning and the separation of land uses (although this is changing with the advent of smart growth, sustainability, and New Urbanism), whereas policy planning is linked to the traditions of decision sciences and policy analysis (Friedmann 1987). Both rely on the other through the collection, analysis, and display of spatial information used to produce a series of recommended actions.

Land-use planners use a series of land-use planning tools that are directly relevant to hazards risk management (Berke and Beatley 1992; Burby 1998; Godschalk et al. 1989), including those hazards tied to a changing climate (Glavovic and Smith 2014). Most communities have developed a number of planning documents that are intended to guide development. Local plans may include the comprehensive plan, the capital improvements plan, and area plans (central business district plan, and parks and recreational plans). Communities may also participate in the creation of multijurisdictional plans including growth management plans, watershed management plans, and transportation plans. Comprehensive plans provide broad policy guidance regarding future growth.* The ability to guide development in a manner that respects identified hazard areas is an important long-term aspect of hazard mitigation and should be included in a community comprehensive plan (FEMA 2013). The capital improvements plan provides an estimated timeline for future capital development investments such as water, sewer, roads and schools, and the means to finance their construction. The placement and type of public infrastructure choices plays a crucial role in shaping future development and overall vulnerability as roads, water, and sewer projects often guide future settlement patterns, whereas protective infrastructure investments such as levees and seawalls can also influence growth. The degree to which the identification and analysis of natural hazards is used in the "formulation, design or justification of land use and management tools" (Godschalk et al. 1998) varies widely, while conducting a risk analysis as precondition of their use is rarely used as Figure 9.4 suggests.

Planning Process: Building Stakeholder Capacity to Confront Hazards

The ability to effectively convey the implications of the hazards analysis to decision makers and the general public and facilitate the conditions under which appropriate action is taken requires a process-oriented approach. The plan is often superseded in importance by the planning process undertaken to create it (Innes 1995, 2004). The process of collaborative plan-making facilitates locally driven decisions and serves as an important communication venue, educational forum, and means to build a coalition of supportive stakeholders. Public participation, consensus building, and process-oriented learning have a rich history in the field of planning. In Sherry Arnstein's classic Eight Rungs on the Ladder of Citizen Participation (1969), she describes several levels of participation and their implications (Figure 9.5). Manipulation and therapy reflect efforts by those in positions of power to "educate" or "cure" participants rather than allow them to voice their opinion. Informing and consultation are characterized by actions that allow citizens

* Planning naturally involves the act of plan making. Among the most widely recognized community planning documents is the comprehensive plan (see Berke et al. 2006; Kent 1964). The comprehensive plan comprises five defining characteristics: (1) a guide for the physical growth and development of a jurisdiction or area; (2) a long-range orientation based on a broad vision, including the steps necessary to achieve it; (3) comprehensive in nature, addressing city functions (i.e., transportation, housing, public infrastructure and facilities, land use, parks and recreation) and their interconnectivity; (4) a set of coherent policies intended to shape the quality, quantity, type, and density of development, and (5) a tool used by the local governing body to guide decision making (So and Getzels 1988, pp. 60–61). FEMA has developed the document Integrating Hazard Mitigation into Local Planning: Case Studies and Tools of Local Officials that shows hazard mitigation can be incorporated into Local Comprehensive Plans and other local planning initiatives.

Plans and Implementation Tools	Hazard Identification	Vulnerability Assessment	Risk Analysis
Planning			
Comprehensive plan, hazard component of comprehensive plan, and recover/reconstruction plan	CP	S	R
Local emergency management plans	CP	CP	R
Development regulations			
Zoning ordinance	S	R	R
Subdivision ordinance	S	R	R
Hazard setback ordinance	CP	R	R
Building standards			
Building code	CP	R	R
Special hazard-resistance standards	CP	S	R
Retrofit standards for existing buildings	CP	S	R
Property acquisition			
Acquisition of undeveloped lands	CP	R	R
Acquisition of development rights	CP	R	R
Building relocation	CP	CP	R
Acquisition of damaged buildings	CP	CP	R
Critical and public facilities policies			
Capital improvements programs	CP	R	R
Location requirements for critical facilities	CP	S	R
Location of public facilities and infrastructure in less-hazardous areas	CP	R	R
Taxation			
Impact taxes	CP	S	R
Reduced or below-market taxation	CP	R	R
Information dissemination			
Public information program	CP	S	R
Hazard disclosure requirements	CP	S	R

CP = common practice; S = sometimes used; R = rarely, if ever, used

Figure 9.4 Use of hazard assessments in land-use planning and management. (From Table 5.1. Use of Hazard Assessment in Land Use Planning, p. 123, Deyle, R., et al., Hazard assessment: the factual basis for planning and mitigation, In R. Burby (ed.), *Cooperating with Nature: Confronting Land-Use Planning for Sustainable Communities*, Joseph Henry Press, Washington, DC, pp. 119–166, 1998.)

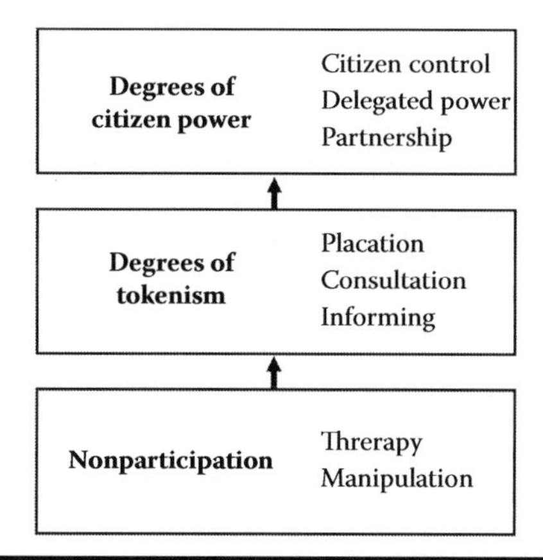

Figure 9.5 Ladder of Citizen Participation.

to provide input, but the information is rarely acted on or implemented. Placation allows for citizens to "advise," but the power to act remains with others in positions of authority. Citizen power includes partnership, delegated power, and citizen control. Partnership allows for negotiated agreements and associated "trade-offs," whereas delegated power and citizen control involves the transfer of decision-making authority. These categories are particularly useful to frame the decision-making process described in Chapters 7 and 8 and the role of dispute resolution techniques that are discussed next.

Critical Thinking: Apply the Eight Rungs on the Ladder of Citizen Participation to the concepts discussed in Chapter 7, Risk Communication. What changes would you recommend to existing risk communication and decision-making strategies based on Arnstein's framework? How can these same concepts be applied to climate change-related decision-making processes and disputes, including the means by which increased uncertainty is conveyed?

Like Arnstein, Paul Davidoff argues that urban planners should move away from their strict belief in the role of the neutral technician, instead embracing the concept of advocacy planning. This approach is intended to provide a voice for the disenfranchised, more socially vulnerable populations described in Chapter 4. Alinsky (1969) notes that planning should be driven by community organizing, a concept often applied postdisaster through emergent groups that form spontaneously based on an identified need that appears to be unmet by existing agencies, organizations, or groups (Drabek 1986). Social Justice and faith-based groups and other nonprofits and professional associations represent organizational types that advocate for the adoption of hazard mitigation practices that address social vulnerability as well as sustainable development principles. This is not always the case, however, as non-profits can unintentionally exacerbate social vulnerability by quickly rebuilding low income housing in known hazard areas before more stringent building codes are adopted or new maps depicting hazard risk are created.

Pre- and postdisaster issues surrounding hazard mitigation are often contentious, particularly when actions are based on governmental mandates rather than collaboratively derived approaches. Examples may include the adoption of building codes, limiting development in identified hazard areas, or other mitigation strategies that place restrictions on the location, type, or density of development. These actions often fall within the purview of local land-use planners, and they should play an important role in the resolution of these conflicts. Following disasters, a tension exists between speeding the reconstruction process and taking a more deliberative approach (Haas et al. 1977; Olshansky 2006; Smith 2011). Development interests and individual citizens place a great deal of pressure on elected officials and their staff to streamline policies and planning processes to regain a sense of normalcy. The postdisaster environment can also provide a unique window of opportunity to embrace the tenets of sustainable redevelopment, including the integration of hazard mitigation concepts into recovery policy and the physical reconstruction of an impacted area (Smith and Wenger 2006). The adoption of hazard mitigation measures in the predisaster environment enables communities to bounce back following disaster more quickly than those that have failed to adopt similar practices all else being equal (Beatley 1998). Achieving a disaster-resilient and a more sustainable community is not accomplished by a single act, or a series of actions, but rather it is characterized as an ongoing process.

Understood in the context of hazards and disasters, the planning process necessarily involves the following: (1) a continual assessment and reassessment of hazards, including their likelihood of occurrence and expected impacts based on regularly updated hazard information; (2) the adoption of new polices, development choices, and settlement patterns that increase or decrease hazard vulnerability; and (3) the creation and maintenance of a participatory decision-making framework based on the best available information regarding hazards, the meaningful involvement of stakeholders, and the use of participatory methods based on local political, social, and institutional conditions.

A number of proven dispute resolution techniques can be used to advance what Arnstein refers to as "citizen power." Traditional ad hoc participatory approaches (e.g., public hearings) often fail to advance egalitarian objectives (Kemp 1985), provide a sound means to communicate hazard risk (Kasperson 1986), or address the conflict-laden disaster recovery process (Smith 2011, p. 292). The power of dispute resolution lies in its ability to seek consensus among competing groups, including those historically excluded from decision making. Care must be taken by those using these techniques to ensure that the approach does not become a tool used by technical experts to advance the status quo advocated by those in power, which Arnstein (1969) refers to as "tokenism." Planners have used these techniques with increasing regularity and represent one of several groups, including diplomats, organized labor, and attorneys, who consider these skills a part of their professional training.

Dispute resolution techniques are representative of tools that are applicable to multistakeholder dilemmas, including those associated with hazard mitigation (Godschalk et al. 1998) and disaster recovery (Smith 2011). Specific methods include facilitation, negotiation, mediation, and policy dialogue. The role of the facilitator is to provide the best possible conditions to foster effective communication between parties. This may include establishing clear rules and conditions, ensuring appropriate stakeholder representation, keeping the group focused on the issues rather than personalities, and asking questions of participants to expose possible areas of agreement. State officials may assume this role as they are often required to bridge policy debates between federal programs and their impact on local communities.

Negotiation is a bargaining process conducted between two or more parties, each seeking a satisfactory agreement. In practice, negotiation can lead to suboptimal outcomes defined by the zero-sum game principle in which the benefit one party derives is necessarily subtracted from the other. Thus, the challenge becomes, under what conditions can negotiations lead to mutually beneficial agreements or a non-zero-sum game (see Raiffa 1982)? Crafting desired changes in federal hazards policy is often the result of local and state officials who possess effective negotiation skills (Smith 2004, 2011). This may include, for example, the development of post-disaster hazard mitigation policies that reflect local needs (Smith 2011).

Disputes among multiple parties can prove difficult given the multitude of interests and their differential access to information and power. Mediation involves the use of a trained, neutral third party who attempts to assist conflicting parties reach a mutually beneficial agreement, recognizing and neutralizing power imbalances (i.e., leveling the playing field), and emphasizing the sharing of information.* Although the use of mediation has not been widely adopted as a means to resolve disputes associated with hazards, it offers promise and merits further study (Rubin 2007; Smith 2011).

The procedural nature of policy making routinely includes the use of policy dialogue. This technique emphasizes ongoing discussions in which participants gain a greater understanding of the other's positions, but more importantly their interests and use this knowledge to seek resolution to policy dilemmas. Iterative communication among stakeholders can elicit cooperation (Axelrod 1984). In those cases where disputes are not resolved, participants often claim to have gained valuable insights into differing points of view while improving communication channels (Bingham 1986). Continued interaction can also elicit the identification of unique and far-reaching decisions that may remain otherwise hidden. As knowledge among stakeholders grows and is shared with others, collectively derived solutions become possible. As noted in the earlier discussion of the capability assessment, a participatory process should lead to the development of realistic policies and

* Mediated agreements are not legally binding. This differs from arbitration, which results in a legally binding decision reached by a third party such as a judge or professional arbitrator.

procedures and identify gaps that merit attention. The Charlotte Mecklenburg case study clearly shows that the involvement of stakeholders in the decision-making process increases the likelihood of adopting hazard mitigation measures based on information derived from a participatory process.

Critical Thinking: Based on personal experience or observation, provide a positive and negative example of the public participation process. If you were involved in the example discussed, did you feel like your participation or the participation of others made a difference? Why or why not? Did the process use dispute resolution techniques? Were they effective?

Meaningful policy shifts can occur given the appropriate circumstances. Examples may include major disasters that trigger federal recovery funding or local demands to address immediate problems. In 1995, the remnants of Hurricane Danny caused significant flooding in the city of Charlotte and Mecklenburg County, NC, but it did not result in a Presidential disaster declaration, which triggers federal assistance. This caused the city and County to investigate local measures to reduce flood losses while maintaining their ongoing efforts to improve water quality and preserve the natural functions of the floodplain without unnecessarily limiting growth. One public official saw the process as a way to avoid what he termed the "hydo-illogic" cycle—a process characterized by the initiation of a postevent stormwater planning study recommending an expensive means to address the problem. Elected officials did not act on the proposed measures, leaving the city vulnerable to future flooding.

Several factors lead to a successful resolution of this policy dilemma. First, the County sought to create a vision that reflected the identified characteristics of the area. Charlotte is a major southern city that prides itself on being a place that welcomes growth. It is also prone to serious flooding. A broad-based collection of interests were brought to the table to craft an acceptable solution. Participants included developers, environmentalists, community organizations, planners, engineers, County Commissioners, city officials, and their staff. Over 2 months were spent in identifying and clearly defining the problem rather than prematurely discussing possible solutions. Initially, many grew impatient. In hindsight, coalition members realized that the work done up front set the stage for success. Those involved in the policy dialogue sought to resolve identified problems (reducing flood losses and improving water quality) rather than dwell on the symptoms, perceptions, or emotions. For example, developers noted that they did not want to build homes that flood, nor did they want to be blamed for the construction of flood-prone homes in the past. This type of honest dialogue provided room for discussion, and the realization that access to information was the most pressing need to drive meaningful policy change. Efforts were made to stay focused rather that straying too far from the agreed on problem, thereby avoiding "deal killers" (Canaan 2001).

The end result was the creation and adoption of a city/County guidance document that identified six primary strategies, including new floodplain development standards, an enhanced flood warning system, a drainage system maintenance plan, a public information campaign, an interagency steering committee, and the development and implementation of watershed-based hazard mitigation plans. Each strategy was tied to specific action-oriented tasks with associated timelines and identified parties responsible for their implementation. An overriding theme of the guidance document was the reliance on verifiable data to guide policy choices.

The basis for policy formulation was the creation of new flood insurance rate maps (FIRMs). The city and County chose to create a map that was not a static depiction of current conditions, like those typically created by FEMA.* The maps were designed to reflect the future flood hazard conditions present once all allowable development had occurred in the floodplain. This was accomplished by reviewing existing zoning maps and estimating future settlement patterns and land uses. This approach reflected a significant departure from traditional FIRMs since new development can cause changes in flood elevations given the increase in impervious surfaces and the placement of fill material in the floodplain.

The Charlotte Mecklenburg Stormwater Services group used a number of scenarios that clearly demonstrated the relationship between development and future flood risk. Not only did the team show how flood elevations change given differing floodplain buildout scenarios but they also quantified expected flood losses under these conditions. By tying proposed actions or inaction to real dollar figures, developers and County Commissioners took notice. Developers asked for repeated iterations of differing scenarios, ultimately recognizing that mapping the flood hazard as it would look in the future and regulating to an agreed upon standard based on this information made sense. The "future conditions" mapping effort not only increased flood elevations by as much as 11 feet but it also expanded the floodway, the area within the floodplain where development is severely restricted.

Prior to Hurricane Danny, the city and County sought to alleviate significant water quality problems. The County Commissioners established a goal of improving water quality to a level that would allow for "prolonged human contact." To achieve this aim, the commission and the county's Department of Environmental Protection established the Surface Water Improvement and Management Panel. The coalition of environmentalists, citizens, developers, and local officials created a stream buffer

* The City of Charlotte and Mecklenburg County negotiated an agreement with FEMA to develop the FIRMs. The formal agreement designated the city and County a "Cooperating Technical Partnership," which enabled them to use the maps to regulate floodplain development and established defined roles for all parties. The State of North Carolina developed a similar agreement with FEMA following Hurricane Floyd, which struck the state in 1999. Floyd, which proved to be the most destructive disaster in the state's history, proved to be an important catalyst, resulting in the development of the North Carolina Floodplain Mapping Program as well as over 20 state recovery programs addressing issues nor covered by post-disaster federal assistance (Smith 2014, pp. 206–207).

plan that identified buffer widths based on the acreage drained by each creek or stream. The larger the drainage area, the larger the buffer required. If the buffer area exceeded the mapped floodplain, new development was not allowed within the larger boundary. By keeping the area along the creek free of development, the existing vegetation filtered pollutants, while the open space provided for additional water storage. By using information that the developers recognized as valid, County and city officials were able to negotiate an agreed upon regulatory framework.

The new maps served as the basis for the city and County's flood hazard analysis component of their watershed-based hazard mitigation plans. The resulting hazard mitigation strategy emphasized the relocation of flood-prone homes and a number of other complimentary objectives. Once specific homes were identified, the County sought funding to remove the structures from the floodplain. Mecklenburg County received approximately $10 million in Hazard Mitigation Grant Program funds to acquire 116 homes. County stormwater fees provided an additional $2.2 million. The acquired properties were demolished and the land reverted to open space. The open space was used to expand an existing greenway and the stream buffer system, thereby reducing flood risk, providing local recreational opportunities, and improving water quality, all key objectives identified in the early stages of the consensus-building process and representative of the nexus between disaster resilience and sustainable development.

The final chapter describes a proposed hazards management policy framework that links the concepts discussed throughout the text and suggests specific modifications to existing federal hazard mitigation policies and programs that will facilitate the creation of disaster resilient and hence, more sustainable communities. The hazards management policy framework emphasizes several important elements, including a greater emphasis on the nexus between land-use planning and hazards, assessing current and future hazard vulnerability, building local capacity and self reliance, balancing policy incentives and penalties, and measuring success.

Discussion Questions

Should the hazard mitigation plan be incorporated into a community's comprehensive plan or serve as a stand-alone planning document? Why or why not? What about climate change adaptation plans?

This chapter suggests that land-use planning enables the integration of hazard mitigation, disaster resilience, and sustainable development principles. Are there other approaches or disciplines that should be considered? If so, what are they and how would you suggest they function?

What lessons can we draw from the Charlotte/Mecklenburg County case study that is applicable to the power of planning and public engagement? What lessons can we draw from that are germane to climate change adaptation?

Applications

You Be the Planner

What planning tools and processes would you use to address the results of a community hazards analysis and why? Are some tools hazard specific? Are your choices based on local capability? How would you facilitate coalition building and the involvement of multiple stakeholders? How might these tools differ when addressing climate change-induced hazards?

Websites

American Planning Association: http://planning.org/

Association of State Floodplain Managers: http://www.floods.org http://www.asfpm@floods.org

Community Rating System: http://www.fema.gov/nfip/crsapp1.pdf

Federal Emergency Management Agency: http://www.fema.gov/

Federal Emergency Management Agency, Higher Education Project: http://www.fema.gov/EMI/web/edu

Institute for Business and Home Safety: http://www.disastersafety.org/

National Response Framework Resource Center: http://www.fema.gov/NRF

National Emergency Management Association: http://www.nemaweb.org/

National Oceanic and Atmospheric Administration: http://www.noaa.gov

National Oceanic and Atmospheric Administration Coastal Services Center: http://www.csc.noaa.gov/bins/resilience.html

Natural Hazards Mitigation Association: http://nhma.info/

References

Alinsky, S. (1969). *Reveille for Radicals*. New York: Vintage Books.

Arnstein, S. R. (1969). A ladder of citizen participation. *Journal of the American Institute of Planners*, *35*, 216–224.

Association of State Floodplain Managers (2003). *No Adverse Impact: A Toolkit for Common Sense Floodplain Management*. Madison, WI: Association of State Floodplain Managers.

Axelrod, R. (1984). *The Evolution of Cooperation*. New York: Basic Books.

Beatley, T. (1984). Applying moral principles to growth management. *Journal of the American Planning Association*, *50*(4), 459–469.

Beatley, T. (1989). Towards a moral philosophy of natural disaster mitigation. *International Journal of Mass Emergencies and Disasters*, *7*(1), 5–32.

Beatley, T. (1995). *Promoting Sustainable Land Use: Mitigating Natural Hazards Through Land Use Planning*. Hazard Reduction and Recovery Center Publication No. 133A. College Station, TX: College of Architecture, Texas A&M University.

Beatley, T. (1998). The vision of sustainable communities. In R. J. Burby (ed.), *Cooperating with Nature: Confronting Natural Hazards with Land-Use Planning for Sustainable Communities*. Washington, DC: John Henry Press.

Beatley, T. 2009. *Planning for Coastal Resilience: Best Practices for Calamitous Times.* Washington, DC: Island Press.

Berke, P. (1995). Natural hazard reduction and sustainable development: a global assessment. *Journal of Planning Literature*, *9*(4), 370–382.

Berke, P. and Beatley, T. (1992). *Planning for Earthquakes: Risk, Policy and Politics*. Baltimore, MD: Johns Hopkins University Press.

Berke, P. and Godschalk, D. (2009). Searching for the good plan: a meta-analysis of plan quality studies. *Journal of Planning Literature, 23*(3), 227–240.

Berke, P., Godschalk, D., Kaiser, E., and Rodriguez, D. (2006). *Urban Land Use Planning*, Fifth Edition. Champaign, IL: University of Illinois Press.

Berke, P. and Smith, G. (2009). Hazard mitigation, planning, and disaster resiliency: challenges and strategic choices for the 21st century. In Urbano Fra Paleo (ed.), *Building Safer Communities. Risk Governance, Spatial Planning and Responses to Natural Hazards*, pp. 1–20. Amsterdam, The Netherlands: IOS Press.

Berke, P., Smith, G., and Lyles, W. (2012). Planning for resiliency: evaluation of state hazard mitigation plans under the Disaster Mitigation Act. *Natural Hazards Review*, *13*(2), 139–150.

Berke, P. (2014). Rising to the Challenge: Planning for Adaptation in the Age of Climate Change. In B. C. Glavovic and G. P. Smith (eds.), *Adapting to Climate Change: Lessons from Natural Hazards Planning*, pp. 171–190. New York: Springer.

Bingham, G. (1986). *Resolving Environmental Disputes*. Washington, DC: Conservation Foundation.

Birkland, T. A. (2006). *Lessons of Disaster: Policy Change after Catastrophic Events*. Washington, DC: Georgetown University Press.

Burby, R. (1998). *Cooperating with Nature: Confronting Natural Hazards with Land-Use Planning for Sustainable Communities*. Washington, DC: Joseph Henry Press.

Burby, R. (2001). Building Disaster Resilient and Sustainable Communities. Federal Emergency Management Agency, Emergency Management Institute, Higher Education Project. The course is available online at http://training.fema.gov/emiweb/edu/completeCourses.asp.

Burby, R. (2006). Hurricane Katrina and the paradoxes of government disaster policy: bringing about wise governmental decisions in hazardous areas. *Annals of the American Academy of Political and Social Science*, *604*, 171–191. (Special issue: *Shelter from the Storm: Repairing the National Emergency Management System after Hurricane Katrina*, W. L. Waugh, ed.).

Burby, R. and Dalton, L. (1994). Plans can matter! The role of land use and state planning mandates on limiting development in hazardous areas. *Public Administration Review*, *54*(3), 229–238.

Burke, E. (1968). Citizen participation strategies. *Journal of the American Institute of Planners*, *34*(5), 287–294.

Canaan, D. (2001). Director, Mecklenburg County, North Carolina, Stormwater Services. Interview.

Chapin, F. S. and Kaiser, E. J. (1985). *Urban Land Use Planning*, Third Edition. Urbana, IL: University of Illinois Press.

Colten, C. (2005). *An Unnatural Metropolis: Wresting New Orleans from Nature*. Baton Rouge, LA: Louisiana State University Press.

Costanza, R., Mitsch, W. J., and Day, J. W. (2006). A new vision for New Orleans and the Mississippi Delta: applying ecological economics and ecological engineering. *Frontiers in Ecology and Environment*, *4*(9), 465–472.

Cutter, S. L., Adams, B. J., Huyck, C. K., and Eguchi, R. T. (2007). New information technologies in emergency management. In W. L. Waugh and K. Tierney (eds.), *Emergency Management: Principles and Practice for Emergency Management*, Second edition, pp. 279–297. Washington, DC: International City/County Management Association.

Davidoff, P. and Reiner, T. (1962). A choice theory of planning. *Journal of the American Institute of Planners, 28*, 103–115.

Deyle, R., French, S., Olshansky, R., and Patterson, R. (1998). Hazard assessment: the factual basis for planning and mitigation. In R. Burby (ed.), *Cooperating with Nature: Confronting Land-Use Planning for Sustainable Communities*, pp. 119–166. Washington, DC: Joseph Henry Press.

Drabek, T. (1986). *Human System Responses to Disaster: An Inventory of Sociological Findings.* New York: Springer-Verlag.

Federal Emergency Management Agency (2013). *Integrating Hazard Mitigation Into Local Planning: Case Studies and Tools for County Officials.* Washington, D.C.: FEMA.

Friedmann, J. (1973). *Retracking America: A Theory of Transactive Planning.* Garden City, NY: Anchor Press.

Friedmann, J. (1987). *Planning in the Public Domain: From Knowledge to Action.* Princeton, NJ: Princeton University Press.

Freudenberg, W., Gramling, R., Laska, S., and Erickson, K. (2009). *Catastrophe in the Making: The Engineering of Katrina and the Disasters of Tomorrow.* Washington, DC: Island Press.

Glavovic, B. and Smith, G. P. Forthcoming. *Adapting to Climate Change: Lessons from Natural Hazards Planning.* Springer Environmental Hazards Series.

Godschalk, D., Beatley, T., Berke, P., Brower, D., and Kaiser, E. (1999). *Natural Hazard Mitigation: Recasting Disaster Policy and Planning.* Washington, DC: Island Press.

Godschalk, D., Brower, D., and Beatley, T. (1989). *Catastrophic Coastal Storms: Hazard Mitigation and Development Management.* Durham, NC: Duke University Press.

Godschalk, D., Kaiser, E., and Berke, P. (1998). Integrating hazard mitigation and land use planning. In R. J. Burby (ed.), *Cooperating with Nature: Confronting Natural Hazards with Land-Use Planning for Sustainable Communities*, pp. 85–118. Washington, DC: Joseph Henry Press.

Godschalk, D. R. (2003). Urban hazard mitigation: creating resilient cities. *Natural Hazards Review, 4*(3), 136–142.

Government Accountability Office (2007). *Natural Hazard Mitigation: Various Mitigation Efforts Exist, but Federal Efforts Do Not Provide a Comprehensive Strategic Framework.* Report to the Ranking Member, Committee on Financial Services, House of Representatives. GAO-07-403. Washington, DC: United States Government Accountability Office.

Haas, J., Kates, R., and Bowden, M. (1977). *Reconstruction Following Disaster.*Cambridge, MA: MIT Press.

Hoch, C. (1994). *What Planners Do: Power, Politics and Persuasion.* Chicago, IL: APA Planners Press.

Innes, J. (1995). Planning theory's emerging paradigm: communicative action and interactive practice. *Journal of Planning Education and Research, 14*, 183–189.

Innes, J. (2004). Consensus building: clarification for the critics. *Planning Theory, 3*(1), 5–20.

International Panel on Climate Change (2012). Managing the risks of extreme events and disasters to advance climate change adaptation. In C. B. Field, C. B., Barros, V., Stocker, T. F., Qin, D., Dokken, D. J., Ebi, K. L., Mastrandrea, M. D., Mach, K.

J., Plattner, G.-K., Allen, S. K., Tignor, M., Midgley, P. M. (eds.), *A Special Report of Working Groups I and II of the Intergovernmental Panel on Climate Change*. Cambridge and New York: Cambridge University Press.

Kasperson, R. (1986). Six propositions on public participation and their relevance for risk communication. *Risk Analysis, 6*(3): 275–281.

Kemp, R. (1985). Planning, public hearings, and the politics of discourse. In J. Forester (ed.), *Critical Theory and Public Life*. Cambridge, MA: MIT Press.

Kent. T. J, Jr. (1964). *The Urban General Plan*. San Francisco, CA: Chandler Publishing Company.

Kingdon, J. (1984). *Agendas, Alternatives, and Public Policies*. Boston, MA: Little, Brown.

Kunreuther, H. (1998). A program for reducing disaster losses through insurance. In H. Kunreuther and R. J. Roth (eds.), *Paying the Price: The Status and Role of Insurance Against Natural Disasters in the United States.*. Washington, DC: Joseph Henry Press.

Lyles, W., Berke, P., and Smith, G. (2014). Do planners matter? Examining factors driving incorporation of land use approaches into hazard mitigation plans. *Journal of Environmental Planning and Management, 57*(5): 792–811.

May, P., Burby, R., Ericksen, N., Handmer, J., Dixon, J., Michaels, S., and Smith, D. (1996). *Environmental Management and Governance: Intergovernmental Approaches to Hazards and Sustainability*. London: Routledge.

McHarg, I. L. (1969). *Design with Nature*. New York: John Wiley and Sons.

Mileti, D. (1999). *Disasters by Design: A Reassessment of Natural Hazards in the United States*. Washington, DC: Joseph Henry Press.

National Academies (2012). *Disaster Resilience: A National Imperative*. Washington, DC: National Academies Press.

Nolon, J. and Salkin, P. (2006). *Land Use in a Nutshell.* Saint Paul, MN: Thompson/West.

Olshansky, R. B. (2006). Planning after Hurricane Katrina. *Journal of the American Planning Association, 72*(2), 147–153.

Paton, D. and Johnston, D. (2006). *Disaster Resilience: An Integrated Approach*. Springfield, IL: Charles C. Thomas Publisher.

Platt, R. (1999). *Disasters and Democracy*. Washington, DC: Island Press.

Raiffa, H. (1982). *The Art and Science of Negotiation*. Cambridge, MA: Belknap Press.

Rubin, M. (2007). Disaster mediation: lessons in conflict coordination and collaboration. *Cardozo Journal of Conflict Resolution, 9*, 351–370.

Schipper, L. (2009). Meeting at the crossroads? Exploring the linkages between climate change adaptation and disaster risk reduction. *Climate Development, 1*(1), 16–30.

Schipper, L. and Pelling, M. (2006). Disaster risk, climate change and international development: scope for, and challenges to, integration. *Disasters, 30*(1), 19–38.

Schwab, J., Topping, K., Eadie, C., Deyle, R., and Smith, R. (1998). *Planning for Post-Disaster Recovery and Reconstruction*. PAS Report 483/484. Chicago, IL: American Planning Association Press.

Smith, G. (2004). *Holistic Disaster Recovery: Creating a Sustainable Future*. Federal Emergency Management Agency. Emergency Management Institute, Higher Education Project.

Smith, G. (2011). *Planning for Post-Disaster Recovery: A Review of the United States Disaster Assistance Framework*. Washington, DC: Island Press.

Smith, G. (2014). Applying Hurricane Recovery Lessons in the United States to Climate Change Adaptation: Hurricane Fran and Floyd in North Carolina, USA. In B. Glavovic, and G. Smith (eds.), *Adapting to Climate Change: Lessons from Natural Hazards Planning*, pp. 193–229. Washington, D.C.: Springer Press.

Smith, G., Lyles, W., and Berke, P. (2013). The role of the state in building local capacity and commitment for hazard mitigation planning. *International Journal of Mass Emergencies and Disasters, 31*(2), 178–203.

Smith, G. and Wenger, D. (2006). Sustainable disaster recovery: operationalizing and existing agenda. In H. Rodriguez, E. Quarantelli, and R. Dynes (eds.), *Handbook of Disaster Research*. New York: Springer.

So, F. and Getzels, J. (1988). *The Practice of Local Government Planning*. Washington, DC: International City Managers Association.

United Nations Framework for the Convention on Climate Change on Loss and Damage (2012). Approaches to address loss and damage associated with climate change impacts. In *Developing Countries That Are Particularly Vulnerable to the Adverse Effects of Climate Change to Enhance Adaptive Capacity. Agenda Item 9*. http://unfccc.int/resource/docs/2012/sbi/eng/l12.pdf. Accessed May 30, 2014.

Walker, B. and Salt, D. (2006). *Resilience Thinking: Sustaining Ecosystems and People in a Changing World.* Washington, DC: Island Press.

World Commission on Environment and Development (1987). *Our Common Future.* Oxford, UK: Oxford University Press, p. 383.

Chapter 10

Creating Disaster-Resilient Communities: A New Natural Hazards Risk Management Framework

Gavin Smith

Objectives

1. Link the concepts discussed throughout the text to hazard mitigation policy making.
2. Understand the strengths and weaknesses of existing hazard mitigation policies and programs, including the need for a Natural Hazards Risk Management Policy Framework to include natural hazards caused or exacerbated by a changing climate.
3. Understand the proposed Natural Hazards Risk Management Policy Framework.

Key Terms

Adaptive capacity
Aggregate risk reduction
Biggert-Waters Act

Climate change adaptation
Community Rating System
Contingent policies
Flood Damage Prevention Ordinance
Flood Insurance Act of 1968
Flood insurance rate map
Four phases of emergency management
Losses avoided study
Multi-Hazard Mitigation Council
National Flood Insurance Program
National Mitigation Framework
Natural Hazards Risk Management Policy Framework
No Adverse Impact
Plan quality principles
Project Impact
Risk governance
Robust policies
Scenario planning
Whole of Community

Introduction

As recently as 2007, the U.S. government Accountability Office stated that a comprehensive Natural Hazards Risk Management (NHRM) Policy Framework did not exist in the United States, nor were programs and policies guided by a widely recognized philosophy or doctrine (Government Accountability Office 2077; see also Mileti 1999). Since that time, a National Mitigation Framework has been developed in an effort to rectify this problem. Yet the means used to reduce the risks associated with natural hazards are still most accurately characterized as a collection of disjointed policies directed by various federal and state government agencies, private sector stakeholder investments, nonprofit assistance strategies, and community-level choices without a coordinative mechanism to guide collective action. Among those federal policies that are most germane to this discussion include the Stafford Act, National Response Framework, National Mitigation Framework, Presidential Policy Directive-8, the Disaster Mitigation Act (DMA) of 2000, and the National Disaster Recovery Framework. The need to include climate change-related effects in this policy discourse and a heretofore emergent national adaptation strategy adds both complexity and opportunities for collaboration through an expanded network of stakeholders.*

* President Barack Obama released a Presidential Order "Preparing the United States for the Effects of Climate Change," November 1, 2013 (see http://www.whitehouse.gov/the-press-office/2013/11/01/executive-order-preparing-united-states-impacts-climate-change. Accessed November 6, 2013).

These policies will be discussed next including their potential to assist local communities become more disaster resilient in an era of climate change. As noted in Chapter 9, resilience provides a powerful means to expand on the definition of hazard risk management to include learning from past events and building an enhanced adaptive capacity. The chapter will conclude with a proposed NHRM policy framework and a series of recommendations to improve the nation's commitment to taking action in a more comprehensive and systemic manner based on a sound understanding of hazard vulnerability including that exacerbated or caused by a changing climate.

Hazard Mitigation Policies

The Disaster Mitigation Act of 2000 and the more recent National Mitigation Framework represent a movement toward a more proactive set of activities guided by state and local hazard mitigation planning efforts rather than the reactive post-disaster grant and aid programs spelled out in the Stafford Act. The DMA stipulates that state and local governments must develop hazard mitigation plans to remain eligible for certain types of pre- and postdisaster mitigation funding provided by the Federal Emergency Management Agency (FEMA). In concept, the connection between the development of a hazard mitigation strategy and federal assistance represents a step in the right direction. In practice, communities that have failed to develop hazard mitigation plans prior to a disaster have been allowed access to post-disaster assistance, whereas many plans created pre-event have failed to adequately address unsound land-use practices in known hazard areas or implement specific projects that target vulnerable structures identified in their hazard vulnerability analysis (Lyles et al. 2014).

These problems can be traced, in part, to the limited emphasis placed on local capacity building by state officials, the preponderance of federal hazard mitigation assistance provided to local governments after a disaster in the form of grants versus pre-event investments in training and education, and the underemphasized role of land-use planners in the development of hazard mitigation plans (Lyles et al. 2014 Smith et. al 2013; Smith 2011). The creation of the Pre-Disaster Mitigation Program, which has funded the development of hazard mitigation plans, is not sufficiently capitalized to implement local capacity building efforts nationwide in addition to the demand for funds to address at-risk structures, which routinely exceeds that which is available annually (Lyles et al. 2014).

The federal program, Project Impact, which sought to develop predisaster public–private partnerships and proactively build local capacity to address hazard vulnerability was discontinued and replaced with the Pre-Disaster Mitigation Program (Birkland 2006; Smith 2011, pp. 184–185). Positive lessons from the short-lived program include the value of providing flexible resources in the form of seed money, workshops, and training to communities to achieve higher-order federal goals (e.g., disaster resistance) through locally championed initiatives; engaging in sustained capacity-building efforts (e.g., education, outreach, and training); and branding the initiative to gain public support. The development of strong working relationships with some constituents (Project Impact–designated communities—over 200 across the United States by 1999—and members of Congress who had Project Impact communities in their districts) and not others (e.g., state Emergency Management Directors who felt that the program effectively bypassed states and sent funding directly to local governments, FEMA staff who had funds shifted from their programs to fund Project Impact, and the Office of Management and Budget who found it difficult to identify clear metrics of program effectiveness) ultimately influenced the staying power of the program during a change in Presidential administrations (Smith 2011, pp. 184–185).

The creation of local hazard mitigation plans offers promise as a process (codified in law through the DMA) through which a sustained capacity can be built by actively engaging multistakeholder partnerships. Specific examples of how the planning process can help to achieve this aim includes gaining a greater understanding of hazard risk, conveying the findings to diverse audiences, and collectively identifying locally grounded risk reduction strategies that reflect community goals and values. However, for this to prove successful across communities of varying capabilities, a greater commitment is required among federal and state agencies to institute targeted training programs while educating stakeholders about the merits of embracing a more holistic approach that draws on the collective strength of diverse coalitions.

FEMA's (2011) Whole of Community concept is similar in some ways to Project Impact in that it strives to foster a broader approach to NHRM that is inclusive of public, private, and nonprofit stakeholder groups. However, unlike Project Impact, FEMA does not provide resources to local governments to help achieve this aim. Rather communities are encouraged to pursue the strategy as part of their day-to-day activities with limited guidance as to how this will be achieved. One way to strengthen the Whole of Community is to draw lessons from those programs that once existed, like Project Impact, as well as those that remain, like the National Flood Insurance Program (NFIP).

The NFIP, established by the Flood Insurance Act of 1968, represents the nation's most comprehensive program to reduce hazard-related losses in the United States.* The NFIP requires participating communities to adopt a Local Flood Damage Prevention Ordinance that stipulates the type and location of development that can occur in a regulated floodplain. The flood insurance rate map designates estimated floodplain boundaries, approximates expected flood depths associated with differing flood return periods, and is used to calculate flood insurance rates. By regulating development that occurs in the floodplain, members of the community are able to purchase flood insurance. Some communities choose to enact additional measures

* Flooding represents the most significant natural hazard in the United States. Water-related damages account for over 75% of federal disaster declarations and over $6 billion in annual losses (Association of State Floodplain Managers 2003).

Credit Points	CRS Class	Premium Reduction (SFHA)
4500+	1	45%
4000–4499	2	40%
3500–3999	3	35%
3000–3499	4	30%
2500–2999	5	25%
2000–2499	6	20%
1500–1999	7	15%
1000–1499	8	10%
500–999	9	5%
0–499	10	0

Figure 10.1 Community Rating System categories and affiliated flood insurance rate reductions.

to reduce flood hazard-related losses. For instance, the Community Rating System (CRS) is an incentive-based program that results in reduced flood insurance rates for those policyholders living in a CRS-designated community. Points are given for various hazard mitigation activities and as they reach cumulative thresholds, flood insurance rates are reduced in accordance with assigned percentages (Figure 10.1).

Even though the NFIP represents the most advanced program available to confront natural hazards, it has had the effect of encouraging development in flood hazard areas and increasing community exposure (Burby and French 1981; Burby et al. 1985; Freitag et al. 2009). As noted in Chapter 9, Ray Burby refers to this phenomenon as the safe development paradox, wherein protective measures like levees or seawalls or higher codes and standards in inherently risky places are adopted by communities, which provide a false sense of security, thereby encouraging additional development in these areas. This approach has contributed to flood-related losses, particularly when flood events exceed the NFIP regulatory standards adopted by local governments.*

In an attempt to disincentivize development in flood hazard areas and reduce the escalating trend in flood-related damages, the Association of State Floodplain Managers (2003) has developed the program No Adverse Impact (NAI). This initiative is designed to encourage the local adoption of a more comprehensive flood

* The adoption of codes and standards that reflect our current understanding of risk (e.g., 100-year flood return periods or hurricane-induced storm surge inundation areas associated with differing storm intensities, size, and orientation at landfall) is likely to underestimate flood hazard risk in many locations in the future due to more intense rainfall events and the increased intensification of hurricanes associated with climate change. Recent changes to the CRS includes providing more points for proactive flood risk reduction planning. Changes were made due to the findings of a study of Flood Mitigation Plans which found that land use remained an underutilized risk reduction tool (Berke et al. 2011).

loss reduction strategy that includes higher NFIP standards, joining the CRS program, the practice of multiobjective management, adopting the principles of sustainability and resilience, and the implementation of broad floodplain regulations (i.e., land-use and building codes, storm water management, protection of natural systems, and planning) (Association of State Floodplain Managers 2003). Although the NAI suggests an approach that includes key themes described in the previous chapter, they are not linked to a national, codified strategy associated with specific incentives and penalties that increase the likelihood of compliance. This requires improving the connectivity between federal, state, and local policies and capacity-building initiatives that are designed to foster forward-looking, multiobjective planning initiatives reflective of local conditions and capabilities while exploring the expansion of incentive-based programs like CRS to include all natural hazards, including those affected by climate change.

The next section proposes a new NHRM policy framework that requires communities to implement a more rigorous pre- and postdisaster hazard mitigation strategy while developing the local capacity necessary to implement and sustain the strategy over time accounting for risk, uncertainty, and the dynamism of natural hazards.

Emergent National Climate Change Adaptation Policy

On November 1, 2013, President Barack Obama issued an Executive Order titled "Preparing the United States for the Impacts of Climate Change." Natural hazards are prominently discussed as primary threats, and resilience is described as an organizing principle guiding proposed actions. Specific measures include modifying federal programs to increase the Nation's resilience, including the "reform [of] policies and Federal funding programs that may, perhaps unintentionally, increase the vulnerability of natural or built systems, economic sectors, natural resources, or communities to climate change related risks." The Executive Order also encourages "more climate-resilient investments" among states and local governments through the provision of capacity-building efforts such as grants, technical assistance, and the provision of "easily accessible, usable, and timely data, information, and decision-support tools on climate preparedness and resilience" for use in assessing climate change-related impacts (President Barack Obama, November 1, 2013).

Activities will be coordinated through the newly established Council on Climate Preparedness and Resilience. More specifically, the council will coordinate interagency efforts and track progress; support state, regional, and local actions to assess vulnerability; "cost-effectively increase climate preparedness and resilience of communities, critical economic sectors, natural and built infrastructure, and natural resources" and integrate climate science in policies and planning initiatives (President Barack Obama, November 1, 2013).

Many of the initiatives proposed in the President's Executive Order are similar to the issues discussed throughout this chapter. However, the Executive Order proposes techniques and relies on concepts that have been described by many in the natural hazards community, with mixed results in terms of long-standing policy change and interagency coordination. The current status of the national hazard mitigation policy milieu provides a powerful example of underachievement. This too provides important lessons for the expanding NHRM community to include those who are now looking at risk through the lens of climate change.

New Natural Hazards Risk Management Policy Framework

The analysis of natural hazards provides the factual basis for action manifested in plans, policies, programs, and actions taken by individuals, groups, and organizations. The effective transfer of knowledge to action through hazards analysis and planning must recognize the existing social, political, and organizational conditions that shape outcomes. The characterization of hazards, the likelihood of their occurrence, the differential exposure of people and the built environment, the tools used to model their impacts, the methods chosen to communicate risk, and actions taken to reduce the impact of hazards are based on human constructs and decision-making processes (see Chapter 2). The proposed NHRM described next integrates the concepts discussed throughout the text and provides an improved natural hazards policy that captures the lessons learned across NHRM strategies used in the United States, including, in particular the NFIP and CRS, DMA, and the National Mitigation Framework and the National Disaster Recovery Framework.

We suggest a framework that includes 10 important elements: (1) expand natural hazards analysis more explicitly to take into account sustainability and resilience themes; (2) develop improved risk assessment tools that address current and expected future exposure, vulnerabilities, and losses; (3) use the findings of the assessment to develop plans, programs, and policies that emphasize land-use and human settlement pattern adjustments; (4) improve federal, state, and local level capacity building, community self-reliance, and collaborative risk governance; (5) modify policies that incentivize unsound pre- and postevent development relative to hazards; (6) limit federal assistance to state and local governments that fail to comply with the proposed framework; (7) hold communities increasingly accountable over time for their actions once new policies are created and the capacity necessary to effectively implement them is developed; (8) develop a natural hazards insurance program; (9) document the merits of the framework at the community and regional level through the quantitative and qualitative measures of hazard loss avoidance and incorporate the findings into local hazard mitigation planning standards and mitigation strategy reward structure (e.g., premium discounts, additional pre- and postdisaster assistance); and (10) build a broad coalition of support for the creation and maintenance of the framework.

Expand Natural Hazards Analysis to Include Sustainable Development and Disaster-Resilient Themes

The process of natural hazard analysis should incorporate factors that define sustainable and disaster-resilient communities, including economic, environmental, and social conditions as well as measures of local capacity and self-reliance. Expanding the definition of a hazards analysis requires applying the methods discussed throughout the text, including the incorporation of a community's fiscal, technical, administrative, legal, and political capability. A high level of local capability increases a community's ability to effectively confront hazards and their related impacts. It also reduces the need for outside assistance before and after disasters, making them more self-reliant, an important but often underemphasized part of sustainable and disaster-resilient communities (Smith and Wenger 2006). Substantial, yet measured and ongoing organizational and institutional changes reflected, for example, in human settlement pattern adjustments should be made over time based on the results of this broader assessment.

Understood in the context of resilience and more recently applied to climate change-related challenges, adaptive capacity provides a bridge between hazard mitigation and the broader concept of resilience. Resilience implies learning from past events and modifying policies accordingly to include those actions that represent a more flexible approach capable of change over time. Scenario planning is gaining acceptance among planners as a sound way to plan for the uncertainties associated with climate change, while drawing lessons from hazard mitigation planning policies and plans (Berke 2014; Quay 2010).

Use Risk Assessment Findings to Guide Land Use and Scenario-Based Planning That Assesses Current and Future Vulnerability

The use of risk assessment findings to shape future land-use and human settlement patterns benefits from a procedural and spatially oriented approach common to planning practice. Gaining acceptance from those impacted by proposed changes in land-use policy is enhanced when they are involved early and often in decision-making processes as the Charlotte/Mecklenburg case study in Chapter 9 aptly demonstrates. Land suitability analysis provides a sound way to visually and analytically link sustainable development themes to geographically defined hazards and comparatively evaluate the appropriateness of various development scenarios. The more recent advent of scenario planning provides another vehicle to address multiple policy options based on varied hypothetical outcomes.

The ability to adopt these approaches requires a greater commitment from state and federal agencies to assist communities with the data collection and analysis needed to effectively use these tools and planning processes. It also requires building the local capapacity needed to sustain an ongoing hazards analysis process as part of a larger hazard mitigation and disaster-resilient community program. An enhanced level of local capacity should include enabling communities to conduct a regular reanalysis of hazard vulnerability and a recalibration of policies that are reflective of the implementation of a community's hazard mitigation plan, changes in hazard vulnerability associated with new development, and observed changes as identified by recent disasters and slower onset threats.

Scenario planning assumes that the traditional planning paradigm is inadequate due to heightened levels of uncertainty and longer time horizons required to assess emerging risks tied to a changing climate. Important lessons can be drawn from hazard mitigation planning to better address these shortcomings. Lessons include the following: (1) anticipating disaster-related impacts, (2) recognizing the future orientation of good hazard mitigation plans, and (3) addressing rapid- and slow-onset events (Berke 2014). Berke argues that new climate change adaptation (CCA) plans should include the following: (1) multiple futures based on collaboratively developed scenarios, (2) clear monitoring and implementation strategies, and (3) a flexible strategy composed of robust and contingent policies (2014). Robust policies refer to those actions that are applicable to multiple futures or scenarios, whereas contingent policies are linked to an individual scenario. Good adaptation plans contain a mix of robust and contingent policies, thereby fostering a flexible strategy that can be implemented over time based on evolving conditions (Chakraborty et al. 2011). A sound portfolio of policies also requires the evaluation of existing policies, programs, and plans, including the degree to which they support or conflict with one another and collectively advance higher-order goals like sustainability and resilience.

Assess Hazard Risk Management Policies, Programs, Plans, and Projects as Part of a Larger Effort to Build Local Capacity and Self-Reliance through a Risk Governance Strategy

Hazards analysis requires assessing the ability and willingness of organizations to adopt policies, programs, and plans addressing hazards and their potential effects. The process is also intended to identify weaknesses and develop a strategy to address them. Increasing local capacity and self-reliance should be the long-term aim of this approach. Conducting an inventory of existing tools help to frame how members of a community perceive risk, including its political salience relative to other policy agendas. Actions taken by groups can decrease or increase exposure to hazards. For example, economic development policies may create incentives for growth while increasing hazard vulnerability by placing new development in known hazard areas. Environmental preservation programs may limit development in environmentally sensitive areas that are also prone to hazards. To capture the collective impact of policies, programs, and plans, they should be assessed across federal, state, and local government agencies; businesses; non-profits; financial and lending institutions; the insurance industry; and regional planning organizations. All play a role in shaping a community's vulnerability to natural hazards. A significant weakness in the current United States system is that it fails to adequately incorporate these interests into the hazard mitigation planning process.

The concept of risk governance provides a way to frame this issue and guide possible solutions. Risk governance implies a collective effort spanning public, private, and nonprofit actors who strive to confront the root causes of risk (e.g., poverty, institutional incentives that encourage investments in high-risk locations, environmental degradation, lack of public awareness, etc.) and develop integrated strategies to confront these drivers (Glavovic and Smith 2014). Actions may include the formation of committees, the development and implementation of plans, and the crafting of policies that address common issues or themes. Among the most challenging issues to confront, and one certainly worthy of a multistakeholder approach, is how to strike a balance between incentives and mandates focused on the location, type, and density of human settlements relative to hazardous areas.

Balance Incentives and Penalties Affecting Human Settlement Pattern Adjustments: A Critical Look at Existing Hazard Mitigation Programs

A number of policies, programs, and plans exists that address NHRM. Yet their effectiveness has been significantly limited due to three primary reasons: (1) Policies have been developed in isolation from one another without a broader framework or plan to guide collective action focused on reducing our nation's vulnerability to natural hazards and disasters. (2) The formulation of hazard mitigation policy has not been coupled with a sincere federal commitment to build capacity at the state and community level to implement hazard mitigation strategies. (3) Existing NHRM policy has not effectively linked reducing future hazards losses to existing and future land-use activities. As described in Chapter 9, the practice of planning is uniquely situated to play a key role in addressing these critical weaknesses. Planning is process oriented, participatory, and coordinative in nature. Building local capacity through participatory planning activities is a widely practiced skill among professional planners, whereas land-use planning is the fundamental domain of the planner.

Existing hazard mitigation policies are described next, including how they can be more effectively integrated into a comprehensive policy framework. The DMA and National Mitigation Framework, the NFIP and CRS, and the National Disaster Recovery Framework include elements from which a substantially improved approach can emerge. In the first edition of this book, we proposed that the framework could form the basis of a resurrected National Mitigation Strategy. The original National Mitigation Strategy outlined steps taken by the federal government to support community-based hazard mitigation (Federal Emergency Management Agency 1995). Since that time, FEMA has created the National Mitigation Framework. To address the challenges cited here, the National Mitigation Framework needs to be further refined to include tangible objectives that address the weaknesses identified in this chapter and monitored over time to ensure their completion and modification as needed. The strategy should also recognize the uncertainty associated with our changing climate and include a set of integrated robust and contingent strategies. In its current form, the DMA does not address the means to coordinate hazard mitigation policies across federal programs, including those associated with a changing climate.

The DMA emphasizes planning, narrowly defined, rather than codifying the process necessary to integrate programs and policies or tackle fundamentally important questions of land use, hazard vulnerability, and adaptation. FEMA is the lead federal agency for the DMA and yet has been reluctant to address land use and its role in hazard mitigation policy. This is not surprising as the agency does not have a codified federal mandate to actively pursue the often contentious issues associated with federal involvement in local land-use planning. Nor does the agency have a clear climate change adaptation strategy, although some progress has been made given the creation of FEMA's adaptation plan and the recent Executive Order requiring the formation of a council to improve federal planning in the face of a changing climate.

To confront these obstacles, the DMA, as well as the NFIP and the pre- and postdisaster recovery plans to be promulgated under the National Disaster Recovery Framework should be amended to require states and communities to adjust human settlement patterns based on the findings of a more rigorous hazards analysis. This could be accomplished through a combination of pre- and postdisaster projects and the formulation of a land-use strategy that discourages development in known high-hazard areas. The failure to shift land-use patterns based on this information would result in limited or no federal disaster relief following a presidentially declared disaster.* The DMA, which mandates the development of hazard mitigation plans, underemphasizes land use and focuses instead on the identification of hazard mitigation projects. These projects tend to address structures that are at risk due to poor land-use decisions made in the past. Thus, plans become retrospective rather than forward looking. The potential power of plan making is grounded in

* The DMA stipulates that communities are not eligible for postdisaster hazard mitigation funds unless they have developed a federally approved hazard mitigation plan. This requirement is sometimes ignored following disasters. Instead, communities are allowed to develop the plan postdisaster to gain access to Hazard Mitigation Grant Program funds. Developing a plan in the aftermath of a disaster is often problematic considering the number of competing agendas, including the desire to return to normal as soon as possible, rather than taking the time required to develop a plan based on a deliberative, participatory process.

its future orientation and should be used for this purpose. Either a project-based or land-use-based strategy done in isolation is not representative of a comprehensive mitigation strategy that is both retrospective and prospective in nature.

The DMA does not clearly articulate, nor require action based on the results of the hazards analysis. Instead, proposed policies and projects may or may not be closely aligned with identified community vulnerabilities or adjacent jurisdictions that share a common regional threat. Improving the link between hazards analysis and mitigation strategies should become more clearly codified and assessed relative to local and regional hazard conditions. First, communities should be required to describe the nexus between the results of the hazards analysis and identified actions and be held accountable for taking the steps necessary to reduce future expected losses. Second, communities should be required to measure the losses avoided over time as recommended projects and policies are completed or adopted.

The ability to quantitatively measure the impact of hazard mitigation activities provides benchmarks for success and evidence of tangible (i.e., monetary) benefits to those who may question the necessity or efficacy of these activities. Like one of the central problems with mitigation plans (which focus on structures at risk rather than limiting future losses through good land-use planning) currently used efforts to assess losses avoided are too narrow in focus. Losses avoided studies take a project-based approach rather than assessing exposure and risk in a more comprehensive way that includes the effects of current and projected land-use and development patterns, changes in the hazards themselves over time, changes in policies, and the implementation of risk reduction projects.

Losses avoided studies should be enhanced to include assessing the reduction in aggregate vulnerability across a geographic area or region. Aggregate vulnerability can be defined as the level of exposure and risk that accounts for the adoption of hazard mitigation policies, the implementation of hazard mitigation projects, and the effects of ongoing development in known hazard areas. This approach would provide a more meaningful measure of the efficacy of a comprehensive hazards risk management program and could be used to measure multijurisdictional efforts to address regional hazards (Smith et al. 2013).

The National Disaster Recovery Framework, through its encouragement of pre- and postdisaster recovery planning and capacity-building efforts, provides another opportunity to inject hazard mitigation and closely related adaptation measures into the recovery and reconstruction process (Smith 2014). The ability to do this benefits from good pre-event planning, including an honest discussion of risk among community members and the adoption of proactive risk reduction policies and the implementation of projects (Smith 2011). Good recovery plans are anticipatory too, identifying possible hazard mitigation projects that may be implemented after a disaster when sufficient federal assistance may be available and communities may be more willing to adopt new policies, including those advancing land-use-based tools and techniques.

We propose a new NHRM framework that emphasizes the adoption of an all-hazards insurance program as a way to operationalize many of the key themes discussed in this chapter and throughout the text. A key part of the implementation strategy draws on lessons from the CRS and expands this incentive-based program to natural hazards beyond floods. At the time the first edition of this textbook was being written, FEMA was in the process of finalizing programmatic rules linking CRS and DMA planning requirements,* the consolidation of mitigation grant programs, and developing a strategy to update their floodplain mapping program to include the application of map products to risk assessments conducted for other natural hazards (i.e., RISKMAP), new loss estimation tools, mitigation planning, and risk communication (Federal Emergency Management Agency 2007).

We propose taking this approach a step further, linking hazard mitigation, disaster recovery, and climate change adaptation planning activities to reduce all-hazards insurance rates for those residing in communities that have developed plans that meet new requirements. Specific improvements would include the following: (1) confronting human settlement pattern adjustments in hazard-prone areas, (2) establishing a more direct correlation between the findings of the hazards analysis and the mitigation strategies adopted and implemented before and after disasters, (3) developing the means to assist local communities' build capacity and self-reliance, (4) measuring losses avoided, and (5) using this information to revise and update local hazard mitigation plans and recalibrate insurance rates based on changes in the exposure to hazards.†

The development of an all-hazards insurance program offers promise and has become an increasing topic of conversation among policy makers following Hurricane Katrina. In the United States, natural hazards insurance programs are uncoordinated, they tend to be disaster specific, premiums do not always reflect actual risk, nor are adequate economic incentives provided to stimulate the adoption of hazard mitigation measures. Kunreuther (2006) notes that an all-hazards insurance approach offers two key benefits: a reduction in coverage uncertainties associated with hazard-related impacts (e.g., the wind versus water debate following hurricanes), and an opportunity to spread risk, thereby reducing the likelihood that insurance payouts following a major disaster will exceed insurance premiums collected from across the country. The

* In a study conducted by Berke et al. (2011), the authors suggest that a greater emphasis should be placed on the adoption of land-use measures in CRS evaluative criteria. These suggestions have since been added to the federal program and offer one example of how FEMA can begin to more actively encourage the adoption of sound land-use strategies as a way to reduce future natural hazards risk.

† The CRS program has been criticized as too labor intensive and costly to implement. This is a real concern for communities with limited technical, fiscal, and administrative capabilities. A key part of an all-hazards CRS program would require a sincere commitment of federal and state level technical assistance in the form of ongoing training and capacity building. Financial assistance, in the form of Pre-Disaster Mitigation funding and postdisaster Hazard Mitigation Grant Program funds should be available for associated start-up costs. Consideration should be given to the financial commitment of the insurance industry as sound pre-event planning should reduce future losses and associated insurance payouts following disasters.

sustained success of a program of this type requires developing an insurance strategy, where rates accurately reflect the risk in known hazard areas. Significant challenges to this idea should be recognized and addressed. This will require gaining the buy-in of insurance companies, elected officials, and others* and monetizing risk and benefits across an integrated portfolio of multiple perils, including how a changing climate alters return periods, intensity, and uncertainty.

Insurance can be a powerful hazard mitigation tool, particularly when it helps to foster the adoption of risk reduction measures while encouraging those directly impacted by hazards and disasters to pay for the costs associated with recovery when a disaster strikes (Kunreuther 1998, p. 214). Encouraging "good behavior," including individual disaster resilience and self-reliance, require ensuring that policyholders have access to information about the actions they can take to reduce risk and the monetary benefits of doing so. Underlying this dialogue is gaining a greater understanding of risk and reflecting that knowledge in sound public policy. An additional factor worthy of note and one that further complicates matters is the incorporation or monetization of climate change-related risk.

Assess Losses Avoided and Build Disaster-Resilient Communities in the Age of Climate Change

The efficacy of any federal program should be evaluated over time to ascertain whether it is meeting established national goals and measureable objectives. Losses avoided studies measure the benefits of reduced expected losses associated with the implementation of hazard mitigation projects and policies. The evaluation and assessment of hazard mitigation is in need of a more systematic means to measure its effectiveness and apply the results to the larger hazards risk management policy framework as discussed in the context of aggregate risk. The Multi-Hazard Mitigation Council, a Congressionally-appointed collection of hazard scholars tasked with assessing the benefits of hazard mitigation activities, found that hazard mitigation projects and programs like Project Impact provided a 4 to 1

* This approach may prove problematic as shown by the recent efforts to weaken the Biggert-Waters Act following Hurricane Sandy. The act, which was signed into law in July 2012, reauthorizes the NFIP through September 2017 and introduces substantive changes to the NFIP. For example, premium discounts will no longer be given to properties that are below the base flood elevation, even if those properties were up to code when they were originally built. The Biggert-Waters Act also removes the NFIP's subsidy for: (1) newly purchased property; (2) property where NFIP coverage was deliberately allowed to lapse; (3) properties receiving an offer of mitigation assistance following a major disaster or in connection with a repetitive loss property; (4) repetitive loss or severe repetitive loss properties; (5) businesses; (6) nonprimary residences; (7) substantially damaged property; and (8) property (at least) 30% improved. Higher flood insurance rates coupled with significant reconstruction costs associated with higher standards have been cited by members of Congress whose constituents live in Sandy-affected areas as stifling recovery.

benefit–cost ratio when comparing losses avoided versus program costs (Multi-Hazard Mitigation Council 2005). Although promising, the results of the Multi-Hazard Mitigation Council study have not led to the widespread integration of hazard mitigation into the larger disaster preparedness or homeland security directives of the federal government. Crafting policy that will be accepted by multiple stakeholders who maintain differing perspectives requires taking the time necessary to collect and analyze relevant information and build coalitions of support. These coalitions should include technical experts, practicing professionals, and politicians to develop the institutional conditions needed to sustain the effort over time and across political election cycles at the federal, state, and local level.

Although losses avoided studies tend to focus on reduced damages to physical structures, a more robust analysis should also measure how existing (and proposed) policies and projects impact economic, environmental, and social factors associated with the concept of sustainability, resilience, and adaptation.* The use of loss estimation models should also account for future development and measure changes in aggregate vulnerability to natural hazards (Smith et al. 2013). This requires assessing how the collective application of hazard mitigation and adaptation strategies (i.e., land use, building codes, relocation of at-risk structures, retreat from the nation's shorelines, etc.) and projected development patterns affect community and regional vulnerability. For example, a community may adopt a more stringent building code (which requires new construction to be built to a higher standard) and relocate at-risk structures while they place new development in harm's way. Expanding the example to include the incorporation of climate change adaptation measures means expanding on the future-oriented nature of scenario planning to include design standards, land-use patterns, and infrastructure investments that account for increased uncertainty.

Critical Thinking: What additional challenges may limit the likelihood of adoption and implementation of the proposed framework beyond those mentioned in this chapter? Can you think of other elements that may be missing in the proposed framework?

Recommendations for Action

The recommendations described next are intended to frame the continued debate calling for a coordinated NHRM strategy. We have proposed a new framework, achieved through the expansion of existing policies and programs including a still emerging national climate change policy. In order for the framework to function effectively, several important procedural items should be addressed. They include the following: (1) the need to assess the efficacy of local hazard mitigation, disaster recovery, and climate

* President Barack Obama's Executive Order "Preparing the United States for the Impacts of Climate Change" suggests developing "cost-effective" climate preparedness and resilience strategies. The ability to develop and measure these strategies will require developing new or modified methods, tools, and data.

change planning in an integrative manner, particularly their role in shaping land-use decisions and settlement patterns before and after disasters, (2) the need to develop an enhanced training program targeting local capacity-building strategies, (3) the provision of tangible benefits to those communities that adhere to enhanced standards and withholding access to postdisaster monetary benefits if they do not, and (4) increasing the involvement of professional land-use planners in the natural hazards management framework, including their role in facilitating social change and policy learning.

Draw Lessons from a National Assessment of Local Hazard Mitigation Plans

The passage and implementation of the DMA has resulted in the development and approval of local and all state plans across the country. Have they made communities safer? Who was involved in their creation and what role has this had on plan quality? What are the key characteristics of successful hazard mitigation planning processes? Which techniques have proven most successful in reducing hazard vulnerability? Godschalk et al. (1999) assessed the quality of state hazard mitigation plans prior to the adoption of the DMA. They found that the quality of the plans varied widely and their implementation strategies were weak. This study predated the development of more rigorous planning requirements. Have the new rules promulgated in the DMA proven effective?

The 6-year study of the quality of state and local hazard mitigation plans conducted by Berke, Smith and Lyles have uncovered several important findings. Key findings include the following: (1) plans have improved relative to those evaluated by Godschalk et al. in 1999, but they remain generally weak, (2) no plans scored well across all dimensions of plan quality, (3) policies and projects identified in plans did not systematically address the findings of the risk assessment, (4) few local hazard mitigation plans addressed climate change, (5) land-use planners were often uninvolved in the development of plans (those who did involve land-use planners produced higher quality plans), (6) state agencies struggled to assist local governments to build local capacity and commitment to hazard mitigation, and (7) plans did not effectively include land-use techniques as part of their risk reduction strategy.

The primary means used to assess state and local plans involved the application of plan quality principles. Plan quality principles represent a set of widely recognized components of plans that can be broken into internal and external principles. Internal principles include the following: (1) issue identification and vision; (2) a set of coherent goals; (3) a fact base, which undergirds goals and associated policies; (4) a means to implement identified actions; (5) a process used to monitor and evaluate progress over time; and (6) a means to ensure internal consistency across the plans vision, goals, and policies. External principles include a plan's: (1) organizational structure and the means by which information is presented, (2) interorganizational coordination across horizontal (i.e., local stakeholders) and vertical (federal, state, and local) parties, and (3) the degree to which the plan complies with existing mandates (Figure 10.2).

Internal Principles
Issue identification and vision: Description of community needs, assets, trends, and future vision of resiliency. 1.1. Assessment of major issues, trends, and disaster impacts associated with forecasted change. 1.2. Description of major opportunities for and threats to resilient land-use and development patterns. 1.3. A vision that identifies what the community wants to be vis-a-vis disaster resiliency.
Goals: Reflections of public values that express desired future land-use and development patterns. 2.1. Statements of desired conditions that reflect the breadth of community values (equity, economy, and environment).
Fact base: Analysis of current and future conditions, and explanation of reasoning. *Vulnerability assessment:* 3.1. Delineates type, magnitude, duration, speed of onset, and frequency of hazard occurrence. 3.2. Includes current and projected future population and employment exposed to hazards. 3.3. Includes current and projected capacity and demands for facilities and services that support vulnerable populations (shelters, transportation, and medical). *Techniques that clarify, explain, and illustrate facts:* 3.4. Includes maps that visually portray location of different population groups, housing, and facilities. 3.5. Includes tables that aggregate data by vulnerable population groups, land-use activities, and infrastructure. 3.6. Uses facts to support reasoning and explanation of issues and action strategies. 3.7. Identifies data sources.
Policies: Specification of principles to guide public and private land-use decisions to achieve goals. 4.1. Sufficiently specific (not vague) to be tied to definite mitigation actions. 4.2. Spatial designs that specify future land use, infrastructure, and transportation infrastructure that avoid or at least limit development in hazard areas.

Figure 10.2 Principles of plan quality for hazard mitigation. (From Berke, P., and G. Smith, Hazard mitigation, planning and disaster resiliency: challenges and strategic choices for the 21st century, *In Building Safer Communities: Risk, Governance, Spatial Planning and Responses to Natural Hazards,* NATO Science for Peace and Security Series, E, Human and Societal Dynamics, Volume 58, ISO Press, Amsterdam, the Netherlands, pp. 14–15, 2009.)

Implementation: Commitments to carry out policy-driven actions. 5.1. Timelines for actions. 5.2. Organizations identified that are responsible for actions. 5.3. Sources of funding are identified to supporting actions.
Monitoring and evaluation: Provisions for tracking change in community conditions. 6.1. Goals are based on measurable objectives (e.g., desired percentage or number of housing units exposed to hazards). 6.2. Indicators of objectives to assess progress (e.g., annual change percentage or number of housing units exposed to hazards). 6.3. Organizations identified responsible for monitoring. 6.4. Timetable for updating plan based on monitoring of changing conditions.
Internal consistency: Issues, vision, goals, policies, and implementation are mutually reinforcing. 7.1. Goals must be comprehensive to accommodate issues and vision. 7.2. Policies must be clearly linked back to goals and forward to implementation actions. 7.3. Monitoring should include indicators to gauge goal achievement and effectiveness of policies.
External Principles
Organization and presentation: Provisions to enhance understandability for a wide range of readers. 8.1. Table of contents, glossary of terms, executive summary. 8.2. Cross referencing of issues, vision, goals, and policies. 8.3. Clear visuals, for example, maps, charts, pictures, and diagrams. 8.4. Supporting documents, for example, video, CD, web page.
Inter-organizational coordination: Integration with other plans/policies of public and private parties. 9.1. Vertical coordination with plans/policies of federal, state, and regional parties. 9.2. Horizontal coordination with plans/policies of other local parties within/outside local jurisdiction.
Compliance: Consistent with the plan mandates. 10.1. Required elements are included in plan. 10.2. Required elements fit together.

Figure 10.2 (*Continued*) Principles of plan quality for hazard mitigation. (From Berke, P., and G. Smith, Hazard mitigation, planning and disaster resiliency: challenges and strategic choices for the 21st century, *In Building Safer Communities: Risk, Governance, Spatial Planning and Responses to Natural Hazards,* NATO Science for Peace and Security Series, E, Human and Societal Dynamics, Volume 58, ISO Press, Amsterdam, the Netherlands, pp. 14–15, 2009.)

Place a Greater Emphasis on Land-Use Decisions and Human Settlement Patterns

Among the greatest weaknesses of existing hazard mitigation plans is their failure to confront development in known hazard areas. This is due in large part to perverted hazards policies that encourage, rather than discourage communities from making choices that reduce exposure and risk. Addressing this reality will not be easy and requires a combination of integrated mandates and incentives that modifies or eliminates counterproductive policies while building federal, state, and local capacity to systematically address problems identified in this chapter.

Establish a Robust Training and Capacity-Building Approach

The adoption and routine update of hazard mitigation plans, policies, and programs require an improved, more systematic means to train local, state, and federal officials as well as other stakeholders who are often excluded from the process of hazards risk management. An adequate training regimen must emphasize the need for the direct involvement of local land-use planners throughout the hazard mitigation planning process. Kartez and Faupel identified the lack of coordination among land-use planners and emergency managers in a national assessment they conducted in 1994. It remains an ongoing problem and is particularly troubling when discussed in the context of hazard mitigation planning, considering the Disaster Mitigation Act of 2000 was passed 15 years ago. (Lyles et al. 2014).

It is incumbent on planners and emergency managers to lead efforts to develop an improved training approach that emphasizes the technical aspects of plan making and capacity building. Hazards analysis techniques should emphasize a locally derived approach reflecting local capabilities, vulnerabilities, and solutions. At the same time, the hazard mitigation strategy must meet more stringent standards than those currently in place today. To achieve these aims, local officials must be provided improved support from states and FEMA (Smith et al. 2013; Lyles et al. 2014) to assist in the planning process, which can help to build greater local capacity.

Enact Sanctions for Low-Performing States and Communities that Underperform and Provide Benefits to Communities and States that Develop Enhanced Hazard Mitigation Plans

The national framework should couple local capacity-building efforts with strong incentives, including clear benefits for those who actively participate. The DMA stipulates that states and local governments must develop and adopt a hazard mitigation plan to remain eligible for pre- and postdisaster hazard mitigation funding.

As shown in the 6-year study of the quality of state and local hazard mitigation plans, plans were weak and did not provide a sound platform in their current state to systematically reduce natural hazard risk.

Federal assistance should be reduced for communities that fail to adopt a hazard mitigation plan that meets higher standards. The bulk of federal assistance following most federally declared disasters are associated with debris clean up and the repair of damaged infrastructure. These funds should be withheld postdisaster if a community fails to adopt an approved plan. On the other hand, communities that adopt higher standards, including those that address land use relative to hazards should be eligible for additional types of assistance, including reduced all-hazards insurance premiums described in the proposed hazards risk management framework.

States are eligible to receive an increased level of Hazard Mitigation Grant Program funds postdisaster if they develop an "enhanced plan." However, enhanced plan status has been disproportionately tied to an evaluation of a state's ability to administer pre- and postdisaster mitigation funds, not necessarily their ability to create a comprehensive state-level mitigation strategy, nor does the enhanced plan status measure the degree to which states are able to assist local governments to build local capacity. Rather, Smith et al. (2013) found that states struggle to build local capacity due to wide variations in staffing, training, and the availability of state resources. One way to address these issues is to amend the definition of enhanced state plan status to include an evaluative criterion that addresses capacity building. Layering additional requirements on states cannot succeed without a commitment to enhance the federal delivery of pre-event resources to states.

Proactive local governments should be similarly rewarded through the creation of an enhanced planning status, which does not currently exist. A percentage of pre- and postdisaster assistance should emphasize training and capacity building, thereby avoiding the rich get richer syndrome where larger, wealthier communities are able to meet higher standards due to their preexisting financial, technical, and administrative capabilities. The federal government should also use community officials who have developed plans that meet higher standards to train others as part of a comprehensive peer-to-peer outreach effort.

Engage Professional Land-Use Planners in the Implementation of the Proposed Natural Hazards Management Framework

Professional land-use planners rely upon an array of tools, methods, and processes that have been refined over a century of codified practice (Hall 1988). These tools, while directly applicable to the challenges facing communities prone to natural hazards, have been used sporadically as part of a comprehensive hazards management strategy (Burby 1998; Mileti 1999; Lyles et al. 2014). Chapter 9 described both plan making and the procedural nature of planning. These activities are the professional domain of land-use planners. This realization is not intended to imply that planners are the only one capable of leading planning activities. Rather, their involvement as technical expert and facilitator should not be underused.

Many land-use planners mistakenly believe it is the responsibility of emergency managers to develop hazard mitigation plans, or they fail to recognize the connection between hazard resilience and sustainable development, of which the latter is widely acknowledged as a key aim of planning practice (Berke et al. 2006). To capitalize on this underused and fundamentally important resource base, an extensive outreach campaign should be implemented by the American Planning Association, FEMA, the International Association of Emergency Managers, the National Emergency Management Association, and others to publicize this partnership and actively engage planners, working in concert with emergency managers in this process.

Facilitate the Use of Planners as Agents of Social Change and Policy Learning

The recommendations noted in this chapter reflect principal themes of planning practice, including plan making, data collection and analysis, participatory planning and dispute resolution, and capacity building. Yet when discussed in the context of hazards, those individuals and organizations trained in these skills, namely land-use planners, are often excluded. Perhaps this provides some understanding as to why we have not done a better job of systematically addressing natural hazards in the United States. It also points out that planners have not proven as effective when confronting basic land-use challenges that are often driven by those in positions of power, namely what Logan and Molotch refer to as the "land-based rentier elite" (1976, 1987). Indeed, the Marxist view of planning defines the planning process as a means to maintain the capitalist city through the organization of private interests while reducing social conflict (Castells 1977). Planners and others supporting the concepts of sustainable development and disaster resilience must become advocates as Arnstein and Davidoff suggest, forming an effective coalition capable of advancing complimentary agendas (Arnstein 1969; Davidoff 1968).

Social learning theorists assume that lessons are translated into policy adaptations based on experience and the formulation of coalitions advocating change (Sabatier and Jenkins-Smith 1993). These concepts have been applied to hazard mitigation (Olson 2000) disaster recovery (Birkland 1997 2006; Smith and Wenger 2006). The increasingly ubiquitous concept of resilience has been adopted by planners and emergency managers as a call to action. Although problems remain, including the use of the term as a slogan rather than a clearly defined concept with indicators of success, the articulation of resilience in the academic arena has added important features tied to learning from past events and adapting to changing conditions (Walker and Salt 2006).

Birkland (2006) has emphasized the potential for disasters to serve as "focusing events" that can elicit change or highlight the failings of the policy system as Hurricane Katrina so dramatically demonstrated. In order for planners to be effective change agents, it is important to understand both the impact of the preexisting conditions found in an area subject to natural hazards and disasters as well as the triggers that may be elicited when an event occurs. Armed with this information, planners can serve several key roles—collecting and synthesizing information, displaying and analyzing spatial data, facilitating dialogue among stakeholders, and creating action-oriented recommendations as part of the plan-making process.

Include Hazards Analysis in Planning for Climate Change

The global implications of hazards have become increasingly salient as scientific consensus has been achieved on several key dimensions of climate change. Similarly, citizens in the United States, including areas that have traditionally remained skeptical are acknowledging that the climate is changing. The realization that we are altering the planet in a way that could drastically affect our way of life, our economies, and future human settlement patterns brings us back to the basic tenets of sustainable development described in Chapter 9. The findings of the Intergovernmental Panel on Climate Change describe a world in which we can expect significant changes in ecosystems and biodiversity, the spread of infectious disease, increases in extreme heat and drought, and abnormally heavy precipitation in other locations (Intergovernmental Panel on Climate Change Report 2007, 2012, 2014). In coastal areas, where over 50% of the United States population currently resides, we can expect a rise in sea levels and an increase in tropical cyclone activity, including those of greater intensity. This will result in greater damages due to coastal erosion, flooding, and high winds in areas experiencing rapid development over the last several decades.

Globally scaled models display a wide range of possible impacts. The analysis of these hazards merits continued research to gain a greater understanding of the influence of multiple variables alone and in combination with others as well as the development and refinement of techniques that allow for the downscaling of data to the community and parcel level.

The current U.S. NHRM approach, which tends to focus on events that are well understood in terms of their physical characteristics and their impacts on individuals, groups, organizations, and institutions, does not use an integrated policy framework. How then do we expect to engage in the type of long-term scenario-based planning necessary to address climate change-related threats? Anecdotal evidence suggests that the climate change dilemma provides an opportunity to expand the hazards risk management network to include those tackling the broad challenge of climate change and sustainable development. A new group of scientists, scholars, government officials, international organizations, activists, and citizens are placing greater attention on natural hazards and disasters through research initiatives, plans, proposed policy changes, and tangible adaptive actions.

Slow-onset catastrophic hazards provide both the substantial threat and the time necessary to develop appropriately scaled plans and develop the coalitions of support needed to implement recommendations that will require an unprecedented level of commitment. On the other hand, slow-onset hazards may offer even greater challenges as changes in exposure and vulnerability occur over time periods that exceed local election cycles and typical planning horizons. As a result, federal, state, and local officials may be resistant to adopt forward-looking policies that may prove politically controversial and whose benefits may not be accrued immediately. Plans are typically created to address decadal timescales, whereas climate change-induced hazards will require longer time horizons.

Advancing an agenda capable of addressing these complex issues and uncertainties requires building and sustaining a broad-based coalition that includes elected officials, local advocates, and organizations that derive their influence from grassroots support and not just the results of election returns. Figure 10.3 describes how those engaged in NHRM and climate change adaptation can work together to address common issues that are undergirded by the themes of governance, sustainability, and resilience.

Figure 10.3 shows that climate-based risk represents the intersection of weather and climactic events, exposure, and vulnerability.* Climate risk, which is viewed as one component of disaster risk, is also shaped by the natural variability of hazards and their alteration due to anthropogenic changes in climate as well as a vulnerability context shaped by episodic shocks, trends, and seasonally influenced events. Further, the intersection of climate, local development, and risk management is influenced by existing institutions (i.e., culture, laws, and policies) as well as a number of barriers and opportunities to foster resilience and sustainability, including those taken by multiple actors that are part of a larger risk governance network. The options available to these actors are, in turn, affected by their access to a range of assets, including financial (F), physical (P), human (H), social (S), and natural (N) capital. The access to and effective coordinated use of these assets is mediated by existing and emergent institutions and established processes.

Local communities possess a range of authorities, including planning and development, infrastructure and asset management, emergency management activities, and others. The strategies used to address NHRM and CCA at the local level is

* Natural hazards include those that are not affected by a changing climate, such as earthquakes and tsunamis. Lessons can be derived from the response to, mitigation against, and recovery from these types of hazards and applied to climate change adaptation, (Glavovic and Smith 2014).

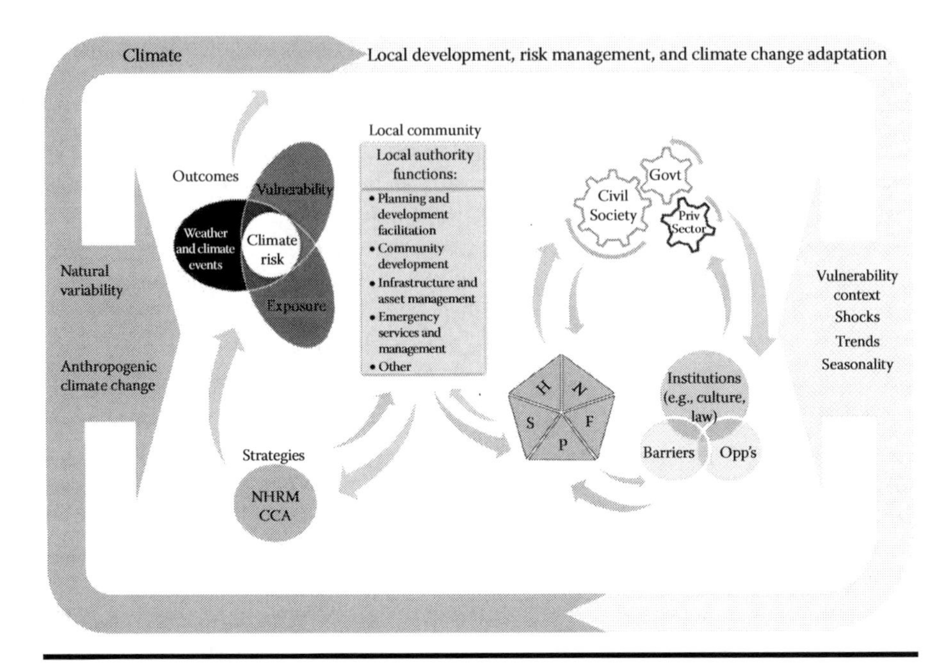

Figure 10.3 Local development, risk management and adapting to climate change. (From Glavovic, B., and G. Smith (eds.), *Adapting to Climate Change: Lessons from Natural Hazards Planning*, Springer Environmental Hazards Series, New York, 2014.)

often shaped by the degree to which local actors are able to work with others in the risk governance milieu.

Critical Thinking: Which argument do you agree with, that questioning the ability to plan for slow-onset catastrophic events or the idea that events of this magnitude will foster the level of collaboration needed to tackle them? What role does scenario-based planning play in this discussion?

Summary and Conclusions

This chapter linked the concepts described throughout the text to existing and proposed hazard mitigation policies, including the description of key elements comprising a proposed NHRM policy framework that includes natural hazards caused or exacerbated by a changing climate. The ability to create such a framework requires drawing lessons from existing policies and modifying them to meet higher-order goals tied to disaster resilience and sustainability. This approach also requires investing more resources in the predisaster time period, focusing on the development of an enhanced collective capacity among risk governance networks while holding states and communities accountable to higher standards over time.

Discussion Questions

Do you think the creation of an all-hazards insurance program is a good idea? Based on what you know about hazards, their geographic distribution, and their associated impacts, who do you think will support or oppose this concept? How would you propose to incorporate climate change-induced or exacerbated hazards into an all-hazards insurance program?

The primary means used to assess the efficacy of hazard mitigation is through the use of benefit-cost analysis and associated losses avoided studies, which monetize the value of averting future damages associated with natural hazards. What other social, environmental, and economic benefits can you think of that hazard mitigation measures provide?

What approach would you suggest to encourage changes in existing land-use patterns relative to hazards?

Do you think a federal mandate is the most effective way to facilitate change at the local community level? Can you think of other alternatives?

Can you think of any recommendations in addition to those posed at the end of this chapter that merit attention? If so, what are they and why do you think they are important?

Applications

You Be the Policy Analyst

Based on your reading of the text and other sources of information, what other policies offer promise when addressing the challenge of reducing our vulnerability to natural hazards? Critique the proposed policy framework, identifying strengths and weaknesses while offering specific improvements. How do the implications of climate change affect your perspective?

Following Hurricane Sandy, the U.S. Department of Housing and Urban Development created the Rebuild by Design Competition in which architects, planners, and engineers developed various redevelopment scenarios for cities impacted by the event. Select a completed project design and critique the result based on the recommendations found in this chapter and other information found throughout this text. See http://www.rebuildbydesign.org/.

You Be the Planner

Use the plan quality principles described in Figure 10.2 to evaluate the quality of select local hazard mitigation plans. Note: The table represents a summary of plan quality principles. A more detailed local hazard mitigation plan quality evaluation protocol can be found at http://coastalhazardscenter.org/ and at http://www.ie.unc.edu/cscd/projects/dma.cfm.

References

Arnstein, S. R. (1969). A ladder of citizen participation. *Journal of the American Institute of Planners, 35*, 216–224.

Association of State Floodplain Managers (2003). *No Adverse Impact: A Toolkit for Common Sense Floodplain Management*. Madison, WI: Association of State Floodplain Managers.

Berke, P. (2014). Rising to the challenge: planning for adaptation in the age of climate change. In B. Glavovic and G. Smith (eds.), *Adapting to Climate Change: Lessons from Natural Hazards Planning*, pp. 171–190. New York: Springer Environmental Hazards Series.

Berke, P., Godschalk, D., Kaiser, E., and Rodriguez, D. (2006). *Urban Land Use Planning*, Fifth Edition. Champaign, IL: University of Illinois Press.

Berke, P., Smith, G., Salvesen, D., and Lyles, W. (2011). *An Evaluation of Floodplain Management Planning Under the Community Rating System*. Final Report for the Federal Emergency Management Agency, Department of Homeland Security Coastal Hazard Center of Excellence, Chapel Hill, NC.

Birkland, T. A. (1997). *After Disaster: Agenda Setting, Public Policy and Focusing Events*. Washington, DC: Georgetown University Press.

Birkland, T. A. (2006). *Lessons of Disaster: Policy Change after Catastrophic Events*. Washington, DC: Georgetown University Press.

Burby, R. and French, S. (1981). Coping with floods: the land use management paradox. *Journal of the American Planning Association, 47*, 289–300.

Burby, R., French, S., Cigler, B., Kaiser, E., Moreau, D., and Stiftel, B. (1985). *Flood Plain Land Use Management: A National Assessment*. Boulder, CO: Westview Press.

Burby, R. (1998). *Cooperating with Nature: Confronting Natural Hazards with Land-Use Planning for Sustainable Communities*. Washington, DC: Joseph Henry Press.

Castells, M. (1977). *The Urban Question: A Marxist Approach*. London: Edward Arnold.

Chakraborty, A., Kaza, N., Knaap, G., and Deal, B. (2011). Robust plans and contingent plans scenario planning for an uncertain world. *Journal of the American Planning Association, 77*(3), 251–266.

Davidoff, P. (1965). Advocacy and pluralism in planning. *Journal of the American Institute of Planners, 31*(4), 48–63.

Federal Emergency Management Agency (1995). *National Mitigation Strategy*. Washington, DC: FEMA.

Federal Emergency Management Agency (2007). *Higher Education Project*. Washington, DC: FEMA.

Federal Emergency Management Agency (2011). *A Whole Community Approach to Emergency Management: Principles, Themes and Pathways for Action.*, FDOC-104-008-1. Washington, DC: FEMA.

Freitag, B., Bolton, S., Westerlund, F., and Clark, J. L. S. (2009). *Floodplain Management: A New Approch for a New Era*. Washington, DC: Island Press.

Glavovic B. and Smith, G. (forthcoming). *Adapting to Climate Change: Lessons from Natural Hazards Planning*. New York: Springer Environmental Hazards Series.

Godschalk, D., Beatley, T., Berke, P., Brower, D., and Kaiser, E. (1999). *Natural Hazard Mitigation: Recasting Disaster Policy and Planning*. Washington, DC: Island Press.

Government Accountability Office (2007). *Natural Hazard Mitigation: Various Mitigation Efforts Exist, but Federal Efforts Do Not Provide a Comprehensive Strategic Framework*. Report to the Ranking Member, Committee on Financial Services, House of Representatives. GAO-07-403. Washington, DC: United States Government Accountability Office.

Hall, P. (1988). *Cities of Tomorrow*. Oxford, England: Basil Blackwell, Ltd.
Intergovernmental Panel on Climate Change (2007). Summary for policymakers. In: M. L. Parry, O. F. Canziani, J. P. Palutikof, P. J. van der Linden and C. E. Hanson (eds.), *Climate Change 2007: Impacts, Adaptation and Vulnerability*. Contribution of Working Group II to the Fourth Assessment Report of the Intergovernmental Panel on Climate Change. Cambridge, England: Cambridge University Press.
International Panel on Climate Change (2012). Managing the risks of extreme events and disasters to advance climate change adaptation. In C. B. Field, V. Barros, T. F. Stocker, D. Qin, D. J. Dokken, K. L. Ebi, M. D. Mastrandrea, K. J. Mach, G.-K. Plattner, S. K. Allen, M. Tignor, P. M. Midgley (eds.), A Special Report of Working Groups I and II of the Intergovernmental Panel on Climate Change. Cambridge and New York: Cambridge University Press.
Kartez, J. and Faupel, C. (1994). *Comprehensive Hazard Management and the Role of Cooperation Between Local Planning Departments and Emergency Management Offices*. Unpublished data.
Kunreuther, H. (1998). A program for reducing disaster losses through insurance. In: H. Kunreuther and R. J. Roths (eds.), *Paying the Price: The Status and Role of Insurance Against Natural Disasters in the United States*. Washington, DC: Joseph Henry Press.
Kunreuther, H. (2006). Has the time come for comprehensive natural disaster insurance? In R. J. Daniels, D. F. Kettl, and H. Kunreuther (eds.), *On Risk and Disaster: Lessons from Hurricane Katrina*. Philadelphia, PA: University of Pennsylvania Press.
Logan, J. R. (1976). Notes on the growth machine: Toward a comparative political economy of growth. *American Journal of Sociology. 82*(2), 349–352.
Logan, J. R. and Molotch, H. L. (1987). *Urban Fortunes: The Political Economy of Place*. Berkeley, CA: University of California Press.
Lyles, W., Berke, P., and Smith, G. (2014). Do planners matter? Examining factors driving incorporation of land use approaches into hazard mitigation plans. *Journal of Environmental Planning and Management, 57*(5), No. 2:792–811.
Mileti, D. (1999). *Disasters by Design: A Reassessment of Natural Hazards in the United States*. Washington, DC: Joseph Henry Press.
Multi-hazard Mitigation Council (2005). *Natural Hazard Mitigation Saves: An Independent Study to Assess the Future Savings from Mitigation Activities. Vol. 1, Findings, Conclusions, and Recommendations*. Washington, DC: National Institute of Building Sciences.
Olson, R. S. (2000). Toward a politics of disaster: Losses, values, agendas, and blame. *International Journal of Mass Emergencies and Disasters, 18*, 265–287.
President Barack Obama. Executive Order. *Preparing the United States for the Impacts of Climate Change*. November 1, 2013. http://www.whitehouse.gov/the-press-office/2013/11/01/executive-order-preparing-united-states-impacts-climate-change. Accessed November 6, 2013.
Quay, R. (2010). Anticipatory governance: a tool for climate change adaptation, *Journal of the American Planning Association, 76*(4), 496–511.
Sabatier, P. A. and Jenkins-Smith, H. C.(1993). *Policy Change and Learning: An Advocacy Coalition Approach*. Boulder: Westview Press.
Smith, G. (2011). *Planning for Post-Disaster Recovery: A Review of the United States Disaster Assistance Framework*. Washington, D.C.: Island Press.
Smith, G., Ward, L., and Berke, P. (2013). The role of the state in building local capacity and commitment for hazard mitigation planning. *International Journal of Mass Emergencies and Disasters, 31*(2), 178–203.

Smith, G., Whitehead, J., Kaza, N., Park, J., Pine, J., Kolar, R., Sandler, D., and Thomas, E. (2013). *Aggregate Flood Hazard Risk Reduction Scoping Project.* Chapel Hill, NC: Department of Homeland Security-Coastal Hazards Center of Excellence.

Smith, G. and Wenger, D. (2006). Sustainable disaster recovery: operationalizing and existing agenda. In H. Rodriguez, E. Quarantelli, and R. Dynes (eds.), *Handbook of Disaster Research.* New York: Springer.

Walker, B. and Salt, D. (2006). *Resilience Thinking: Sustaining Ecosystems and People in a Changing World.* Washington, DC: Island Press.

Index

D

E

F

G

H

I